MICRODIFFUSION ANALYSIS AND VOLUMETRIC ERROR

BY

EDWARD J. CONWAY

M.D., D.Sc., Sc.D., F.R.I.C., F.R.C.P.I., F.R.S.A,, F.R.S.

PROFESSOR OF BIOCHEMISTRY, UNIVERSITY COLLEGE, DUBLIN;
MEMBER OF THE ROYAL IRISH ACADEMY;
CHAIRMAN OF THE DUBLIN INSTITUTE OF ADVANCED STUDIES;
MEMBER OF THE ACADÉMIE SEPTENTRIONALE, PARIS;
MEMBER OF PONTIFICAL ACADEMY OF SCIENCES.

1963

CHEMICAL PUBLISHING CO., INC.
212 Fifth Avenue New York, N. Y.

Microdiffusion Analysis and Volumetric Error

ISBN: 978-0-8206-0151-9

Chemical Publishing Company:
www.chemical-publishing.com
www.chemicalpublishing.net

First American Edition: Chemical Publishing New York 1963

Printed in the United States of America

CONTENTS

CONTENTS

CONTENTS

LIST OF ILLUSTRATIONS

CHAPTER I

INTRODUCTORY

In the development of modern biological research large numbers of observations become increasingly necessary in determining the questions at issue. It becomes essential to have analytical methods of a micro kind in which, without sacrifice of accuracy with respect to macro procedure, the labour and time expended with the single analysis are reduced to a minimum.

In the search for such a method, primarily for ammonia and urea, the microdiffusion principle of analysis was elaborated here and a special diffusion apparatus and burette designed to give it effect. It was subsequently found applicable to a variety of micro determinations by the author and other workers.

The method in general would appear to be the simplest possible consistent with the maximum attainable accuracy in the handling of micro volumes. Distillation and aeration are eliminated, the passage of ammonia or other substance taking place by diffusion from one chamber, in which it exerts a certain tension, into an absorbing fluid in another chamber in which its tension is reduced to zero. With a knowledge of the diffusion conditions controlling this passage and used in the design of the standard microdiffusion apparatus or unit, full absorption times with many substances run from about half an hour to two hours.

With serial determinations the time expended on each determination need be only a few minutes ; one of the special advantages of the method being the ease with which large numbers of accurate data can be assembled.

With the standard unit, the accuracy of ammonia and consequently, of urea and other determinations, is limited only by the accuracy of delivering and titrating fluid volumes of the order of 1 ml. With the development of the titration principles described in the text, this accuracy can be brought to any desirable level in practice, so that with comparative ease the percentage error need not exceed that for ordinary macro titrations on the 25 ml. level.

The following is a list of the determinations developed so far in which the microdiffusion procedure has been found of value :

(1) Ammonia, including special application to normal blood ammonia determinations.
(2) Total nitrogen.
(3) Urea, with applications to renal function tests.
(4) Adenosinetriphosphoric acid.
(5) Adenylic acid.
(6) Adenosine.
(7) Amines.
(8) Amides.
(9) Glutamine.
(10) Histaminase.
(11) D-amino oxidase.
(12) Nitrate.
(13) Acetaldehyde.
(14) Threonine
(15) Lactic acid (revised).
(16) Formaldehyde.
(17) Formaldehydogenic steroids.
(18) Glycine.
(19) Ethanol.
(20) Urethane.
(21) Methanol.
(22) Isopropanol.
(23) Phenol.
(24) Cyanide.
(25) Azide.
(26) Sulphide.
(27) Chloride.
(28) Bromide.
(29) Iodide.
(30) Halogen mixtures.
(31) Halogenated hydrocarbons.
(32) Chloroform.
(33) Acetic acid and volatile fatty acids.
(34) Acetylcholine esterase.
(35) Carbon dioxide—with applications to determination of (*a*) blood bicarbonate or ' alkali reserve ', (*b*) oxidation rates of organic substance.
(36) Glucose.
(37) Carbon monoxide—with application to determination of oxygen capacity of blood.
(38) Separation of crystals and gums.
(39) Total molecular concentration in 3–4 milligrams of fluid.

Many of these methods are now in considerable use in other laboratories, and as mentioned in the preface several have been developed elsewhere.

With the standard ' unit ' a special burette was developed, the use of which may be extended to general micro titrations (e.g. its use for arsenic determinations in biological material by Levvy, 1943, and by Ramsay, 1944, in the determination of iron).

For the microdiffusion technique evolved it may now be claimed that it represents a new and special section of micro technique with a wide range of application and the purpose of the present volume is to give an account of the methods which have been tested in practice.

The general error of volumetric procedure. During the development of this technique the more general question of the error involved in volumetric procedure was studied firstly in order to compare the error of the new technique with those already established. Certain principles, however, were in this way discovered of practical help in assigning standards of attainable accuracies, and likely to prove useful in micro work.

To anyone with the laboratory experience it will seem that there are few more practical questions of micro technique than the comparative accuracies in the use of glass ware in volumetric procedure and titration. The constant error of general titrimetric procedure has no doubt been expounded by Bjerrum (1914) and others, and the question of exact calibration long considered. The variable error, however, has not hitherto been equally appreciated, and for the biological worker with his comparative series it is the more important ; not only this, but a correct understanding of the variable error affects in turn the question of calibration, and forms much of the rationale of micro-volumetric procedure. This latter turns on an equation representing the addition of the variable glass and the chemical errors, which serves also as a means whereby, for given quantities, we can choose ideal volumes.

In the study of the variable error the exact influence of certain factors on the constant error was also discovered, which may be of interest to laboratory workers in general, applying a volumetric technique.

Note on the description of micro-concentrations. As is well known the description of the concentrations of standard solutions is generally made in terms of normality or molarity (with the symbols N or M associated with the descriptive number). The immediate results of analysis, however, are more usually stated in terms of the absolute units of mass and volume, such as grams per ml., per 100 ml. or per litre.

Macro-concentrations may in this way be considered as those most conveniently expressed as grams per 100 ml. or per cent., and such description may be retained no matter how low the quantity analysed provided the relation of quantity to volume remains on the same level, though for the actual designation of the minute quantities or volumes themselves we have the milligram (mg.), and the microgram (μg) also termed the 'gamma' (γ or 0·001 mg.), and the lambda (λ or 0·001 ml.—introduced by Kirk).

Micro-concentrations may be expressed as milli-equivalents or as millimols per litre, but analytical results are more immediately given as milligrams per litre (or per cent.), parts per million, or μg (or γ) per ml.

These three units of concentration are alike and the term 'gammil' as the equivalent of the three was suggested by the author in the previous edition as a convenience in the laboratory. In relation to more accepted conventions this might be better termed 'microgrammil' shortened conveniently to 'micril'.

The distinction between the micro-quantity (or volume) and the micro-concentration is in any case one worth drawing, because of its significance in relation to the theory of titration and the variable chemical error (as expounded in Part III of the present volume). From a consideration of the latter the analytical practice with small quantities should be to avoid or raise the micro-concentration by scaling down on the micro-volume. There is a 'micril' level, however, above which there is little or no practical value in increasing the concentration (taking certain standard precautions of acidimetric micro-titration), and this for ammonia is approximately 10 'micrils', though at 1 'micril' the error need not exceed 2%.

For colorimetry or spectrophotometry of the ordinary kind 0·3 'micril' is the order of limit for accurate working (within 2%). This concentration of ammonia nitrogen after Nesslerisation and viewed through a stratum depth of 50 mm. gives an extinction value (E) of approximately 0·3, with filter *S* 43 of the Pulfrich photometer. An extinction of 0·3 is towards the lower end of the extinction scale for accurate observation.

In comparing such 'micril' limits, the relation of the molecular weights must of course be borne in mind when these diverge greatly.

Scale and accuracy of the microdiffusion methods described. Using the standard unit or absorption apparatus, accuracies of the order of 0·5% are easily obtainable for ammonia, urea, chloride, bromide and carbon dioxide at the 20, 50, 250, 500 and

500 'micril' level respectively. Accuracies of the order of 1–2% can be obtained with one-tenth these amounts, and of 5% with as low as 0·4 'micril' of ammonia and a few 'micrils' of chloride and bromide.

With 1 ml. volumes used in the standard unit, the 'micril' represents the 'microgram' as the quantity analysed. By scaling down the dimensions of the apparatus and the volumes used much lower quantities could be dealt with having the same concentrations and with the same accuracies. It is in short the 'micril' level (i.e. the concentration) that sets the possible accuracy and not the quantity or volume considered separately.

While such suggestions as to terminology are made to draw attention to the significance of *concentrations* as distinct from *quantities* in micro or ultra micro work, what is here termed the 'micril' will be represented in the usual symbols of μg/ml.

The present volume is divided into three parts. Part I deals with the apparatus and principles used in microdiffusion analysis. Part II gives an account of the methods for laboratory usage. In Part III is expounded a new treatment of volumetric error with particular reference to micro-quantities and the rationale of micro-volumetric technique.

In an Appendix to the volume an account is given of the best as well as the easiest method of applying blood and urine urea determinations to obtain a measure of renal function. This is expressed as a urea *normality ratio* which is very similar to but not quite the same as the 'clearance test' of Van Slyke.

The student desiring a rapid account of the essential microdiffusion technique will find this in Chapters II, V and the first part of Chapter VI, describing the horizontal burette; he will find in Part II a detailed practical account of the various methods.

PART I

APPARATUS AND PRINCIPLES USED IN MICRODIFFUSION ANALYSIS

CHAPTER II

A STANDARD MICRODIFFUSION APPARATUS OR UNIT

THE standard microdiffusion apparatus or 'unit' consists of a pyrex glass container resembling a small Petri dish with thick glass walls, within which a second chamber is formed by a circular wall of glass arising from the floor. The wall of the inner chamber is about half the height of the outer.

The apparatus is covered during determinations by a square glass lid on which is smeared a small amount of fixative, such as vaselin or a mixture of vaselin and paraffin wax, and in this way makes an air-tight compartment.

Figs. 1 and 1*a* illustrate the standard pyrex unit (No. 1) at present available from Messrs. A. Gallenkamp, Finsbury Square, London.

Principle of use. The general principle of the microdiffusion methods consists in the absorption by simple gaseous diffusion of the volatile substance from the outer chamber where it exerts a certain tension into the inner chamber where its tension is zero on the surface of the absorbing fluid. Though this on a first consideration would seem a very slow process, yet from the manner of construction of the 'unit' the rate is comparatively fast, and determinations may be carried out in about 10 minutes to two hours for a number of different substances. Where large numbers are being carried out the procedure can be so arranged that the average working time for the single determination is only a few minutes. Ammonia analyses by simple gaseous diffusion were carried out by Schlösing and by Shaffer (as given in Neuberg's *Der Harn*,

1911) some fifty years ago. Separate vessels were used, both being covered by a large bell jar, and upwards of 48 hours with the Schlösing method and 3 to 4 days with the Shaffer modification were required for the absorption, which even then gave somewhat

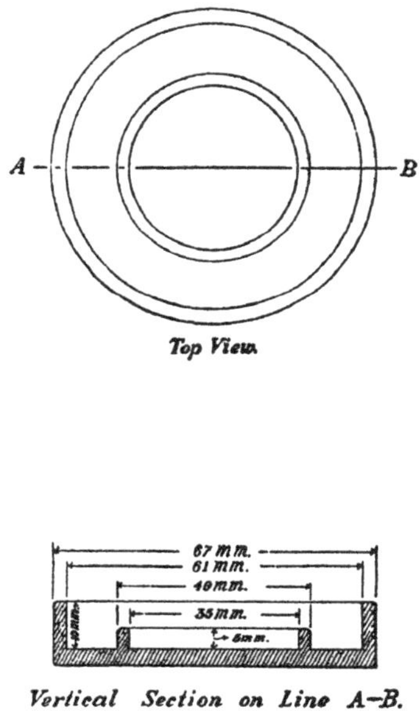

FIG. 1. Plan of standard unit, Conway and Byrne.
(From *Biochemical Journal*, **27**, 420, 1933.)

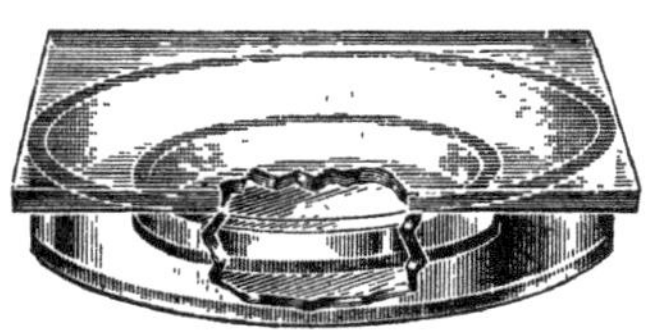

FIG. 1*a*. Standard unit (as supplied by Messrs. A. Gallenkamp & Co., Finsbury Square, London).

low results. Diffusion in this form was of little analytical value, the controlling factors were not understood and the method was abandoned. The recognition that the process could rapidly reach completion as a microdiffusion led to the introduction of this as a new principle of analysis the general application to volatile substances being rationally elaborated. Full absorptions could be obtained in as short a time as 10 minutes with the standard 'unit' using 0·2 ml. ammonium sulphate solution and shorter times may be obtained with small drops on enclosed surfaces—of paraffin or other material.

Procedure as exemplified by ammonia determinations. 1 ml. of standard acid is introduced into the central chamber. This may be introduced either from an Ostwald or from a straight tube pipette, with a delivery time not less than about five seconds, or what is often convenient for routine work, from a 2 c.c. standard burette of the Bang type (1920) (*vide* Fig. 10) with comparatively large reservoir. Into the outer chamber is run a measured volume of the fluid to be investigated—usually ranging from about 0·1 to 2·0 c.c. For this also an Ostwald pipette or straight tube pipette may be used. The lid is suitably smeared with the fixative (vaselin, vaselin plus paraffin or a mixture of paraffins) and placed in position on the 'unit'. The apparatus is then slightly tilted—most conveniently by resting one edge on a spare lid. The tilting collects the fluid in the outer chamber away from where the alkali is introduced. The lid is horizontally displaced so that a small opening appears into the outer chamber sufficient to allow the introduction of the tip of a pipette containing a measured quantity of reagent liberating the volatile substance, which for ammonia determinations is usually one ml. of saturated potassium carbonate.

This is quickly run in *without blowing* and—as its delivery does not require to be very accurate—it is most conveniently run in from a simple pipette made from a piece of glass tubing somewhat drawn out and with a mark at the required volume. The lid is at once replaced and the mixture in the outer chamber rotated about 10–20 times. This is easily done by gripping the unit between thumb and fingers and tilting slightly from side to side. It will be found that—at least with very little experience—there is no danger of spilling the contents of the chambers, when the volumes described in the determinations are used.

After a suitable time for the absorption the lid is detached and the contents of the inner chamber titrated. For other determinations it may be necessary to remove it for colorimetric investigation.

Variations in the above procedure occur in the different determinations described in text, particularly in the manner of introducing the reagents into the outer chamber. Instead of one such, several may be required, the last reagent liberating the volatile substance.

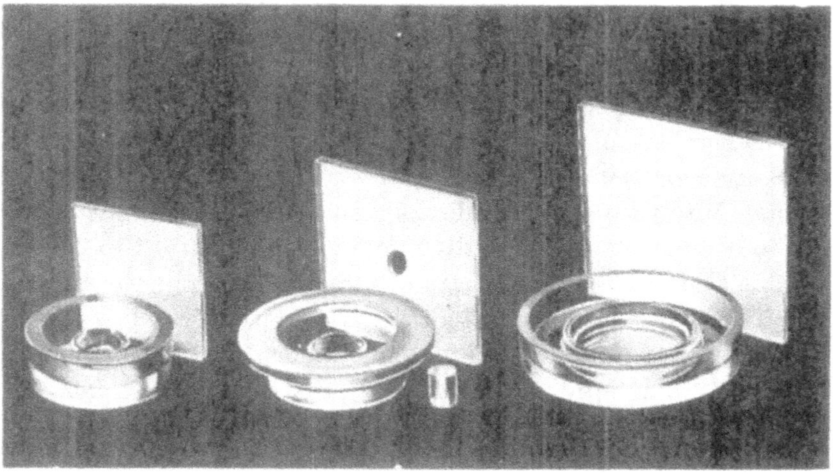

FIG. 2. Standard 'unit' with 'units' No. 2A and 2 (as supplied by Messrs. A. Gallenkamp, Finsbury Square, London).

In developing the method for blood glucose as later described, a smaller 'unit' was designed than the No. 1. This 'unit'—No. 2—could be more easily secured from any leakage inwards of atmospheric CO_2, and much lower ranges of CO_2 could likewise be dealt with. A modification of this No. 2 'unit' was devised later—No. 2A—which had a much wider rim and a drilled hole through the lid, made air-tight by a small cylinder of glass treated with fixative. This 'unit' may be preferred for fine carbon dioxide work, and is very efficient, when determining, say, the carbon dioxide content of 0·1 ml. blood, which may be done with a high degree of accuracy. These modifications of the original 'unit' are shown in Fig. 2 and 2*a*. The No. 2 and 2A types have the same dimensions, except for the wide rim in 2A.

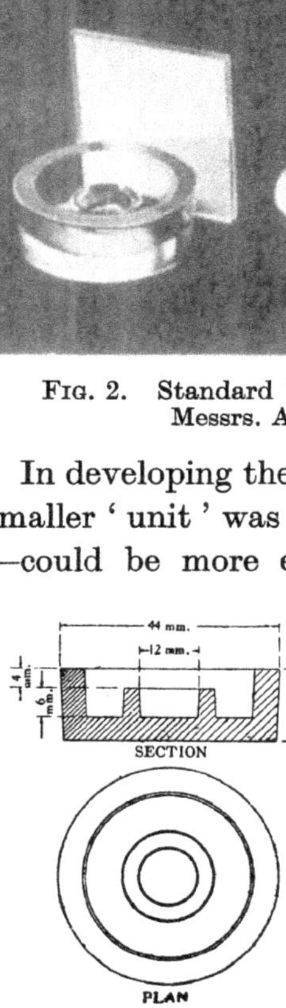

FIG. 2*a*. Plan of 'unit' No. 2 (from the *Biochemical Journal*, **36**, 655).

These smaller 'units' may be used, of course, for any of the other microdiffusion determinations as well as No. 1, the volumes being suitably scaled down, and their usage allows of determinations at lower micro levels. For the less soluble gases such as carbon dioxide, the halogens, etc., the times with the small 'units' are scarcely altered. For the highly soluble gases or vapours such as ammonia, amines, etc., the times will be very dependent on the fluid volumes in the outer chamber and much influenced also by the reduced emission and absorbing surfaces. The diagram for the times required for ammonia absorptions is given later. Throughout the present volume when the microdiffusion 'unit' is mentioned, the No. 1 type is to be understood unless the No. 2 and No. 2A 'units' are definitely stated.

Cleaning the 'units'. The success of the methods—particularly for the minute ammonia determinations—very largely depends on a satisfactory cleaning. This cleaning is quite easy, but requires the carrying out of a definite routine.

The following routine of cleaning is found suitable for all determinations.

1. The 'units' after use are well washed out under the cold-water and then under the hot-water tap.

2. They are then cleaned of any grease with a test-tube brush or pledget of cotton wool, using a little detergent, and subsequently well rinsed under the tap.

3. After this they are filled with dilute sulphuric acid ($N/200$ roughly) containing indicator. This acid may be made up in bulk in a large distilled water bottle and filled into each 'unit' from the outlet of the bottle. They are allowed to soak in the acid for fifteen minutes.

4. They are now well rinsed under the cold-water tap and finally with distilled water. The drops are shaken out (the interior of the apparatus intended for immediate use should never be dried with a cloth) and the 'units' are left to dry. For quick drying it will often be found very convenient to range them along the top of a radiator, resting upside down along the ridges. They also dry comparatively quickly after shaking out the drops well by mere exposure to the air.

Important points for correct usage in the general procedure. (*a*) The 'units' when used for determinations where the final measurement is acidimetric should be perfectly clean in an acid-alkali sense and a routine similar to that given must be followed in cleaning.

(*b*) The surface of the inner chamber must be fully covered. With the use of a detergent in the cleaning process, as mentioned above, 1 ml. fluid will very easily cover the surface. The

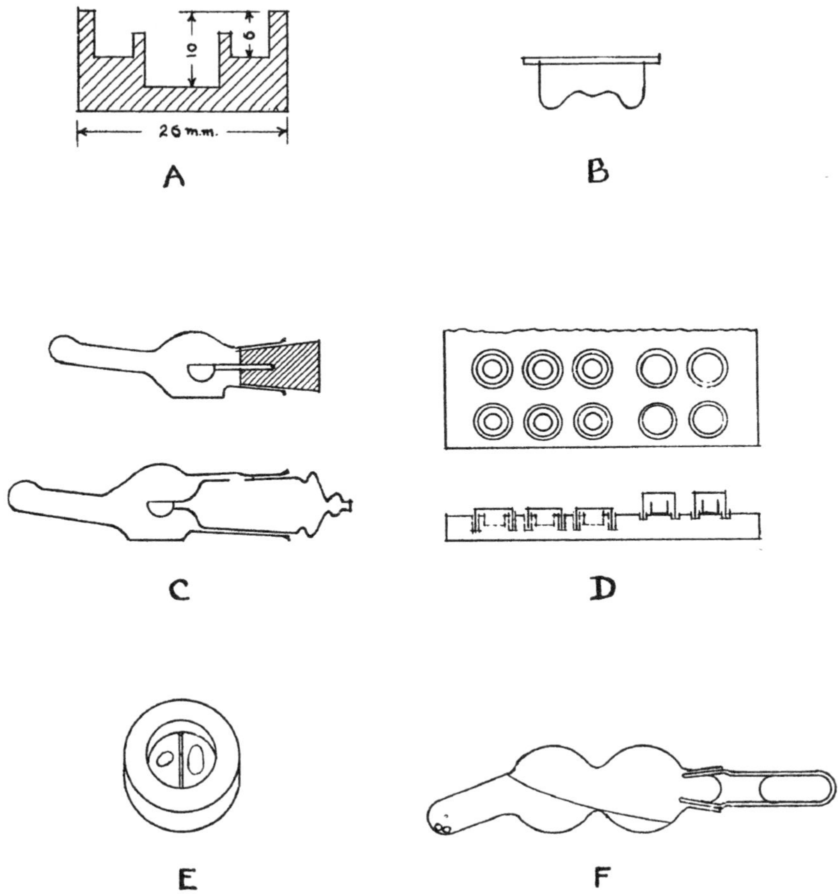

FIG. 3. Other forms of microdiffusion apparatus, for minute amounts of ammonia determination.

A—from Borsook and Dubnoff (1939), see text, page 140.
B—from Gibbs and Kirk (1934) ; 32 mm. outside diam.
C—from Tompkins and Kirk (1942), see text, page 144.
D—from Kinsey and Robison (1946), upper, top view ; lower, side view.
E—from Conway and O'Malley (1943) ; see text, page 68.
F—from Needham and Boell (1939), see text, page 143.

introduction of soap means at the same time the introduction of alkali, and hence the need for soaking in acid for some time.

(*c*) The fluid volume in the outer chamber should not exceed that described in the various determinations. (The fluid volume in this chamber, as shown in the next chapter, has in general a very marked effect on the absorption rate, the time for a 99·5 clearance being proportional to this volume.)

Other forms of microdiffusion apparatus. Since the description of the standard unit (Fig. 1) was published, various forms of microdiffusion apparatus, applying the new principle have been elaborated and are shown in Figs. 3 and 3a. Recently Öbrink has

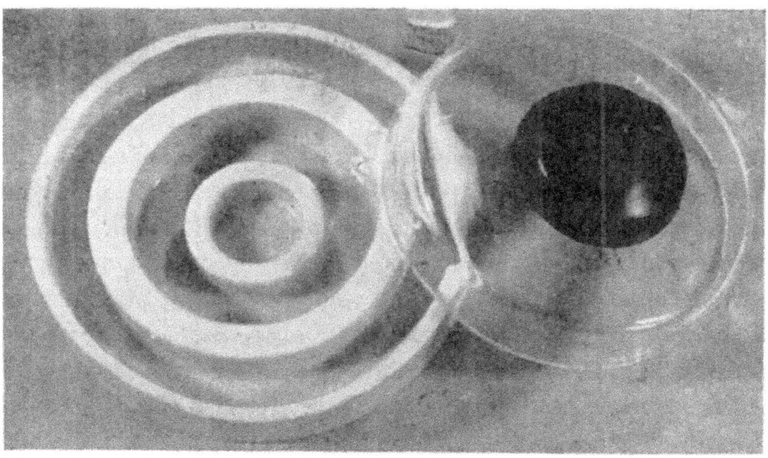

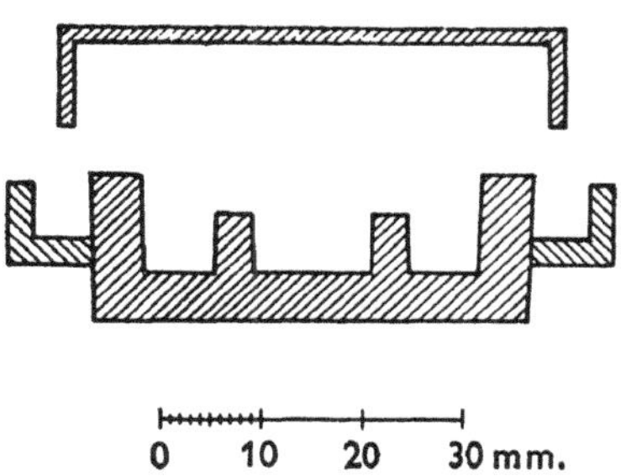

FIG. 3*a*G. Modified unit with Petri dish and rubber suction cup and cross-section (Öbrink, *Biochemical Journal*, **59**, 134, 1955).

introduced a modification of the No. 2 unit, as represented in Fig. 3*a*G. In this no fixative is required as the lid fits into a circular groove containing some of the fluid used to liberate the volatile substance.

Another type of unit, termed a ' distillation diffusion unit ' (Kirk, 1950) is illustrated in Fig. 3*a*H.

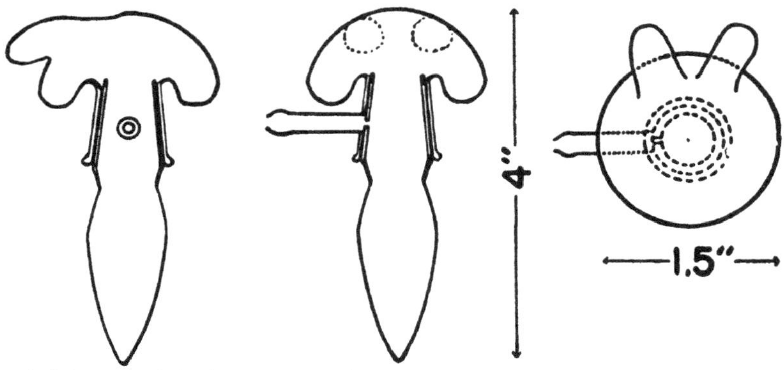

Kirk's Distillation-Diffusion Unit **Cross-section and top views**

FIG. 3*a*H (from *Anal. Chem.* **22**, 611, 1950).

This is described as follows : ' It consists of a top portion in which a liquid phase is allowed to spread in a thin layer around the entire periphery without spilling into the extended tube below. Bulbous protuberances blown on the side allow the sample and any reagent which is to be mixed with the sample to be kept separated until the cell is sealed and evacuated. The receiver compartment is attached to the top compartment by means of a ground-glass joint which is arranged to allow evacuation of the interior. This is necessary to prevent pressure development by heating during the distillation, which can blow the seal open. The unit is held on a sheet of wood or metal having holes drilled in it, so that the receiver can extend downward into an ice-water bath. The top portion containing the sample and added reagents is exposed to heat from an infra-red heat lamp, which maintains the temperature of the liquids and volatilization of any dissolved gases.'

Kirk remarks that ' the shape and configuration of the unit are such that titration is possible but not convenient without transfer of the receiver material. It is therefore adapted better to those determinations which are finally colorimetric in nature than to those which are volumetric.

Testing with standard ammonia solutions showed a complete transfer in 10 to 15 minutes.

Here it may be noted that complete transfers from 0·2 ml. fluid in such terms can also be obtained at room temperature with the standard unit using the ‘ vibrating table ’ or the older rocking apparatus as described below.

Fig. 3a also includes a unit (Fig. 3*a*I) used by Black (1949) for the microdiffusion determination of volatile fatty acids, using temperatures of 100–105°C.

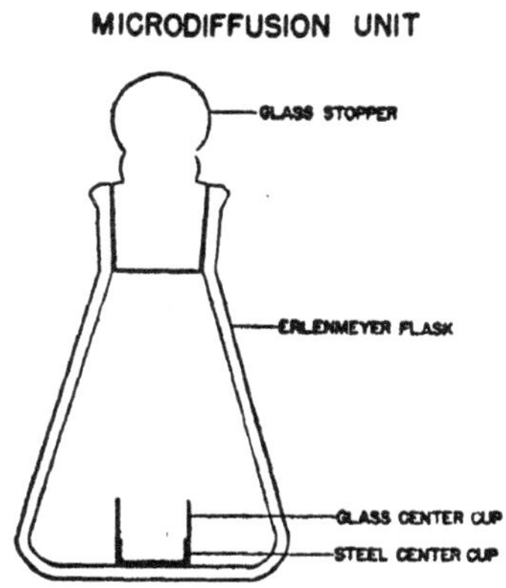

Fig. 3*a*I. Black's microdiffusion vessel: a 50 ml. glass-stoppered Erlenmeyerflask (from *Archives of Biochemistry*, **23**, 3, 1949).

Rather recently Berka (1959) introduced a microdiffusion unit with a special modification for higher temperatures and a removable inner container (Fig. 3*a*J).

These units were made out of polyvinyl chloride (novodur) and polymethakrylate (plexi-glass).

For room temperatures the inner container constituted the only essential difference from the standard unit. For high temperatures the unit was provided with two or three small metal spikes which fitted into grooves on the lid. (Fig. 3*a*K). A rubber or polyethylene gasket is placed between the lid and the container and by turning the lid air-tightness is ensured.

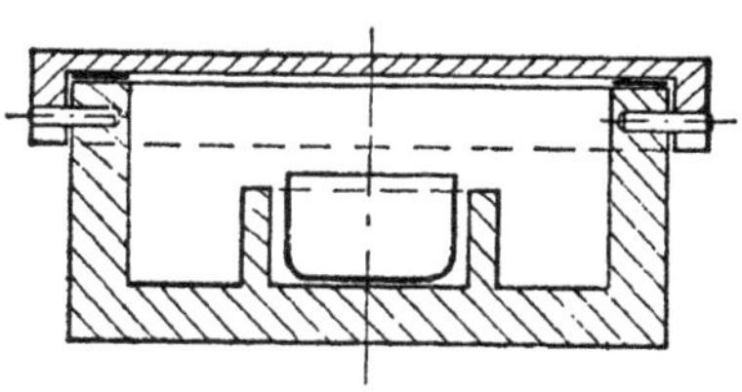

Fig. 3*a* J.

Fig. 3*a* K.

Berka's microdiffusion unit with special modification for higher temperatures.

CHAPTER III

FACTORS INFLUENCING THE ABSORPTION RATE FROM OUTER TO INNER CHAMBER, WITH SPECIAL REFERENCE TO AMMONIA

IN the practical descriptions of the various methods in their standard form, these factors need not be further considered, but if the conditions are in any way altered, or if new methods are tried out, it is essential to note their general effects.

The effect of time on the absorption rate. The rate at which the volatile substance comes across is directly proportional to the tension it exerts in the outer chamber. For ammonia with a comparatively slow absorption rate the tension in the fluid in outer chamber after time 't' will be proportional to $(a-x)$, where a is the original amount in the outer chamber, and x the quantity that has diffused therefrom after time 't'.

Writing this relation in the form

$$\frac{dx}{dt}=2{\cdot}3A_1(a-x), \qquad (1)$$

where $2{\cdot}3A_1$ is the absorption rate of ammonia per unit quantity in the outer chamber, we obtain on integrating

$$\frac{1}{t}\ln\frac{a}{a-x}=2{\cdot}3A_1 \qquad (2)$$

$$\text{or}\quad \frac{1}{t}\log\frac{a}{a-x}=A_1. \qquad (3)$$

Provided the absorption is not too rapid under the conditions this equation describes approximately the time-absorption curve of ammonia or other volatile substance when this is liberated all at once in the outer chamber. It will be seen from equation 3 that *the proportion of the ammonia absorbed after time 't' is independent of the amount originally present* (under the same volume, temperature and other such conditions). A more general treatment is given in the subsequent chapter.

Fig. 4 shows the time-absorption curve of ammonia at room temperature when to 1 ml. of ammonium chloride solution in the outer chamber is added 1 ml. of *N*/10 NaOH (Curve *A*, 23° C.) or 1 ml. of saturated potassium carbonate (Curve *B*, 23° C.).

Curve *A* is described fairly closely by equation 3. Curve *C* shows the rate at which ammonia is absorbed from 0·5 ml. ammonium chloride solution plus 0·5 ml. saturated potassium

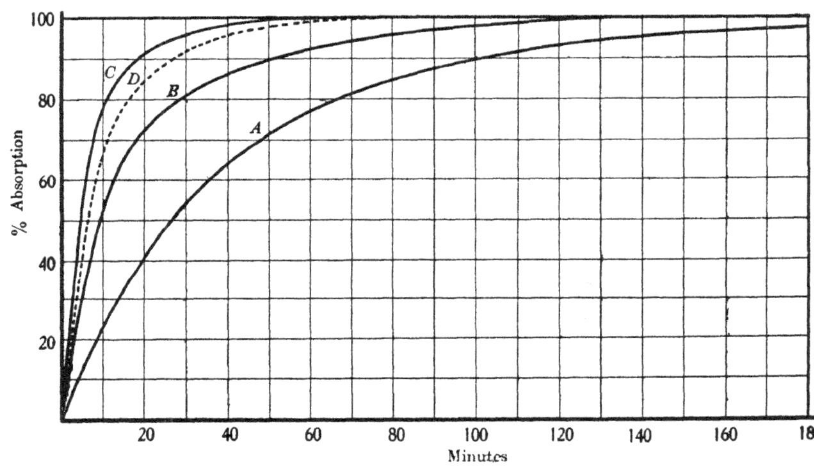

FIG. 4. Ammonia absorption curves in standard unit, No. 1.

Inner chamber containing 1 ml. *N*/200 sulphuric acid, outer chamber containing for

A.	1 ml. *N*/150 am. chlor. plus 1 ml. *N*/10 NaOH.	Temp. 23.
B.	,, ,, ,, ,, ,, ,, sat. K_2CO_3.	,, ,,
C.	0·5 ml. ,, ,, ,, ,, 0·5 ml. ,,	,, ,,
D.	1 ml. *N*/150 am. chlor. ,, 1 ml. ,,	,, 38.

carbonate, and Curve *D* the absorption rate at 38° C. from 1 ml. solution plus 1 ml. carbonate.

Fig. 5 shows some more rapid absorption curves. In curve *A* the absorption, with rocking, is complete in 10 minutes at room temperature. For this the conditions were 0·2 ml. standard ammonium sulphate and 1 ml. saturated potassium metaborate. For *B* the same conditions hold but without rocking, absorption being complete in about 25 minutes.

In the chloride determinations the total amount of chlorine finally absorbed is not liberated at once in the outer chamber. The oxidation reaction proceeds rather slowly, and the resulting curve of absorption is a straight line up to an absorption of about 70%. (See Fig. 60.)

Effect of the fluid volumes on the absorption rate. The fluid volume in the inner chamber has practically no influence on the absorption rate provided the surface is fully covered. This is due to the fact that it operates by reducing the tension of the absorbed substance to zero, and therefore provided it is in excess, a change

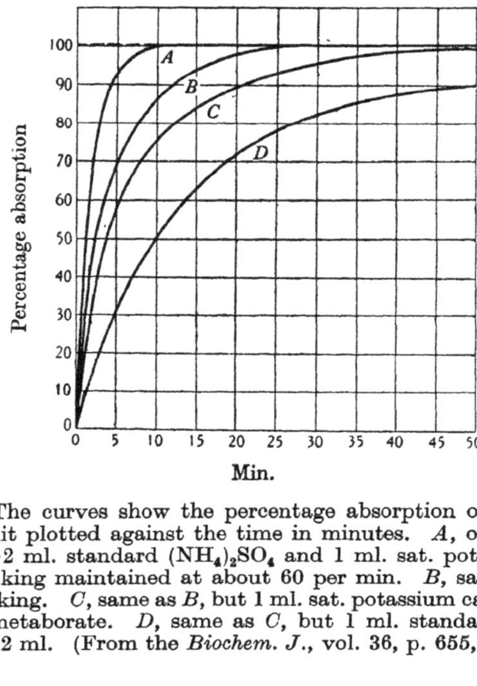

FIG. 5. The curves show the percentage absorption of NH_3 in the standard unit plotted against the time in minutes. *A*, outer chamber contained 0·2 ml. standard $(NH_4)_2SO_4$ and 1 ml. sat. potassium metaborate; rocking maintained at about 60 per min. *B*, same as *A*, but without rocking. *C*, same as *B*, but 1 ml. sat. potassium carbonate used instead of metaborate. *D*, same as *C*, but 1 ml. standard $(NH_4)_2SO_4$ instead of 0·2 ml. (From the *Biochem. J.*, vol. 36, p. 655, 1942.)

of volume has no significant effect. It is essential in the use of the 'unit' that *the surface of the inner chamber should be fully covered.* The 'unit' is so designed that 1 ml. of fluid readily covers the surface and is in general the most suitable volume to use, though for the most delicate determinations 0·7 ml. will be found sufficient.

The fluid volume (V) *in the outer chamber has by contrast a very important effect on the absorption rate, which—other things being equal—is inversely proportional to this volume.* This arises from the absorption rate being directly proportional to the tension, and hence to the concentration of the volatile substance in this chamber. If a certain quantity of substance is dissolved in 1 ml.

fluid it will exert a tension twice as great as when dissolved in 2 ml., the fluid having otherwise the same composition. By comparison the absorption will require twice the length of time in the latter case to clear the substance to the extent of 99·5% from the outer chamber. Mathematically we may express the relation (similar to equation 1) as

$$\frac{dx}{dt}=A_2\frac{a-x}{V}, \tag{4}$$

where A_2 is the absorption rate per *unit concentration* ; $2{\cdot}3A_1$ as in equation 1 being the rate per *unit quantity* under the conditions.

From (4) $$\frac{V}{t}\ln\frac{a}{a-x}=A_2 \tag{5}$$

or from (3) $$A_2=2{\cdot}3A_1V. \tag{6}$$

The relation of absorption rate of fixed quantity to changing volume is shown in Table I.

TABLE I

Quantity of ammonia in outer chamber (millimols).	*V = volume of fluid in outer chamber ml.*	*Time allowed for absorption mins.*	A_1 *value* = $\left(\frac{1}{t}\log\frac{a}{a-x}\right)$.	A_2 *value* = $2{\cdot}30A_1V$.
1·0	1·0	16	0·0153	0·0352
1·0	2·0	26	0·0080	0·0368
1·0	3·0	24	0·0056	0·0386
1·0	4·0	22	0·0041	0·0377

Temperature = 18·5°.

Two estimations were carried out for each volume. The A_2 value represents the calculated absorption rate in millimols per minute per unit concentration of the ammonia from the outer chamber, unit concentration being here 1 millimol per ml.

There are certain qualifying conditions for the simple inverse relation of fluid volume to absorption rate, but it should be borne in mind as an underlying relation. It is necessary, for example, that the surface of the outer chamber be fully covered ; secondly, in practice the fluid in this chamber is generally altered in composition as the total volume alters, and this can affect the relationship ; finally, if the rate of formation of the volatile substance is markedly slower than its absorption rate (as with chlorine from chloride under the standard conditions) the simple inverse relation does not hold.

Effect of the saline content of mixture in the outer chamber. The effects of the saline content of the outer chamber are given in Chap. XI for ammonia absorptions and the results would appear applicable also to the aliphatic amines, which are briefly considered subsequent to the ammonia determinations.

The effect of temperature. All the determinations described may be carried out at room temperature. At the same time the absorption rate can be hastened by placing the ' unit ' in an incubator, and this is suitable for ammonia determinations in general, and doubtless for other absorptions, though with the fixative in present use for the halogens and the most delicate ammonia determinations it is inadvisable, owing to the fluidity of this fixative at the raised temperature.

Incubating at 38 degrees reduces the necessary absorption time for ammonia determinations to about 60–70% of that at room temperature. This will be again considered when dealing with ammonia determinations.

The effect on the absorption rate of altering the dimensions of the apparatus. The standard apparatus at present procurable from the makers is suitable for the various determinations described in Part II. For special purposes, however, alterations in the dimensions may be required, in which case the effect of such alterations will be useful to know. Where the clearance distance between the inner and outer chamber or the distance of the inner chamber from the lid roof is not greatly altered, the following principles apply approximately.

(*a*) The absorption rate is proportional to the geometric mean of the surfaces of the inner and outer chambers.

(*b*) It is inversely proportional to the mean distance of passage of the molecules.

For the application of principle (*a*) the two surfaces must be fully covered.

From the operation of the first of the two principles it will be seen that the absorption rate will be directly proportional to any change in the general linear dimensions of the standard apparatus. An apparatus designed with all dimensions doubled will give twice the absorption rate, and if halved will give half the absorption rate. The utilisation of this to increase the rate of absorption is limited by the fact that as the linear dimensions increase the volume required to cover the surface increases as the square of these dimensions.

As shown above, the time of 99·5% clearance of the volatile substance is proportional to the volume in the outer chamber. It will be seen that if we keep the volume in the outer chamber in all cases *just what covers this surface*, we will get the fastest relative absorptions or the shortest times for clearance of the substance as we *decrease* the linear dimensions.

TABLE II

Area of floor of outer chamber cm.²	*Area of floor of inner chamber cm.²*	*Height of inner chamber cm.*	*Height of outer chamber cm.*	*Distance of molecular passage * cm.*	*Absorption rate found (A_2 value = $2{\cdot}30A_1 \times V$)*	*A_2 divided by the geom. mean of the areas multiplied by the distance of passage of mols.*
66·8	43·5	1·0	2·0	13·9	0·091	0·023
22·7	16·6	0·6	1·6	8·6	0·064	0·028
43·1	14·8	0·8	1·3	9·6	0·062	0·024
7·9	17·8	0·5	1·0	7·0	0·047	0·027
16·7	9·6	0·5	1·0	6·9	0·044	0·024
25·6	8·6	0·85	1·65	8·3	0·037	0·021
8·5	1·76	0·80	1·40	4·4	0·025	0·028

Temperature = 23°.

* The mean distance of the molecular passage from the outer to inner chamber would be very difficult to state, but it has been taken as proportional to the total diameter *plus* the height of the outer chamber *minus* the fluid depths in the two chambers (the total diameter of the outer chamber is taken at the top and thus includes that of the inner chamber).

The absorption rates have been determined with pure ammonia solutions (N) in the outer chambers and N sulphuric acid in the inner. The figures in the last column show a fair constancy and are indicative of the general principles underlying the effect on the absorption rate of changes in the dimensions of the apparatus. Divergences from the mean value of these figures are largely attributable to irregular differences in the depth of the air passage between the inner and outer chambers.

It may be pointed out that the linear dimensions of the first and fifth type of 'unit' investigated above are very nearly in the ratio of 2 to 1, and it will be observed that the absorption rates in the last column but one are also nearly in the same ratio, for these 'units'.

The absorption rates given in the last column but one are the A_2 values representing the rate of absorption per minute per unit concentration in the outer chamber, or $2{\cdot}30A_1 \times V$.

Halving these dimensions will then halve approximately the time required for ammonia absorptions under such conditions.

The best utilisation of these principles will depend on the special analytical conditions. The standard apparatus is designed with the most suitable general dimensions, in which 1 ml. covers the

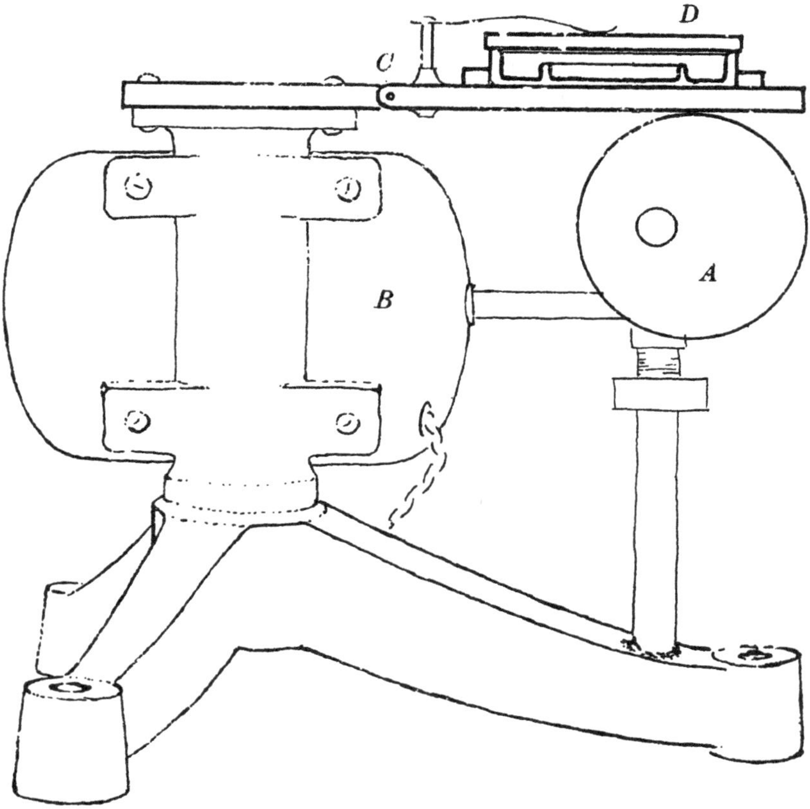

FIG. 6. Rocking device for units if required for special experiments. With the general use of the unit it is not required.

central chamber without any trouble and in which the outer chamber surface is also conveniently covered by 1 ml. of fluid plus 1 ml. reagent. Table II summarises the results obtained with a number of units of different dimensions to the standard apparatus.

Effect of rocking on the absorption rate. An increase in rate of absorption may be obtained by a simple rocking device with motor—Fig. 6—the ' units ' (*D*) resting upon a hinged platform

(*C*) supported by an excentric wheel (*A*) making contact underneath the platform with glass (such as a glass slide fixed underneath).

A shaking device of the following kind (Fig. 7) has been recently devised by Dale (1949). It consists of two glass plates ($\frac{1}{4}$ in. thick) each mounted with three hollow ground slides which, when coinciding, (as indicated in the diagram) can form three shallow cups containing three steel ball-bearings of $\frac{1}{8}$ in. diameter. An

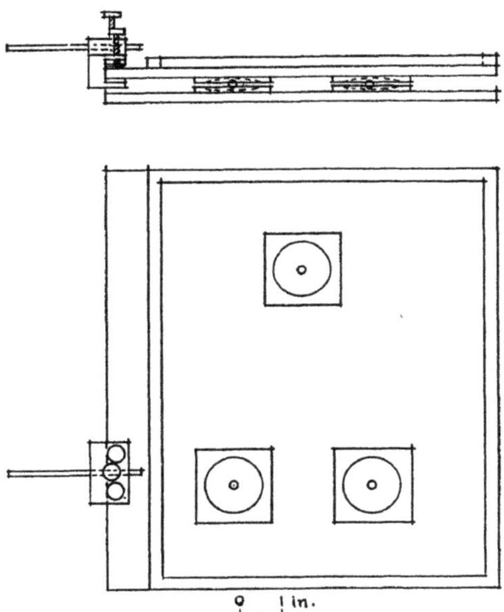

FIG. 7. Dale's shaking device for Conway units. (From Dale, *Biochemical Jour.*, **44**, *xxvii*).
'Oscillating Table' made by Messrs Gallenkamp and Co., Finsbury Square, London.

impulse imparted to the upper plate sets up three-dimensional wobbling movements for a considerable time and imparts a suitable mixing movement to the microdiffusion units placed on the upper plate within a wooden frame. A geared-down motor providing an occasional impulse to the protruding rod suffices for continuous shaking.

It is a curious fact—which is explained by the considerations in the following chapter—that where the conditions already exist for a fast absorption (use of saturated carbonate in outer chamber with ammonia or the absorption of substances like carbon dioxide, etc.), the rocking increases this markedly, but when the absorption is already slow there is little effect.

The rocking device (Fig. 6) converts an absorption rate of ammonia of 0·030 (A_1) to 0·045 at room temperature, or will change the full absorption time of 1½–2 hours to about 1 hour (1 ml. ammonia solution in outer chamber plus 1 ml. saturated carbonate); a much lesser proportionate effect is obtained when $N/10$ sodium hydroxide is used to liberate the ammonia. Increasing the period of rocking beyond a certain point (about 60 per minute) has no further action on the absorption rate.

The Vibrating Table. Recently we had occasion to investigate the most rapid absorptions of ammonia and CO_2 in the standard No. 1 units and designed a vibrating table which produced a striking shortening of the absorption time when 0·1 to 0·5 ml. of standard ammonium sulphate solution was used with 1·0 or 2·0 ml. of saturated K_2CO_3 (Conway and Dolan, 1953).

An improved model than that first used is shown in Fig. 8.

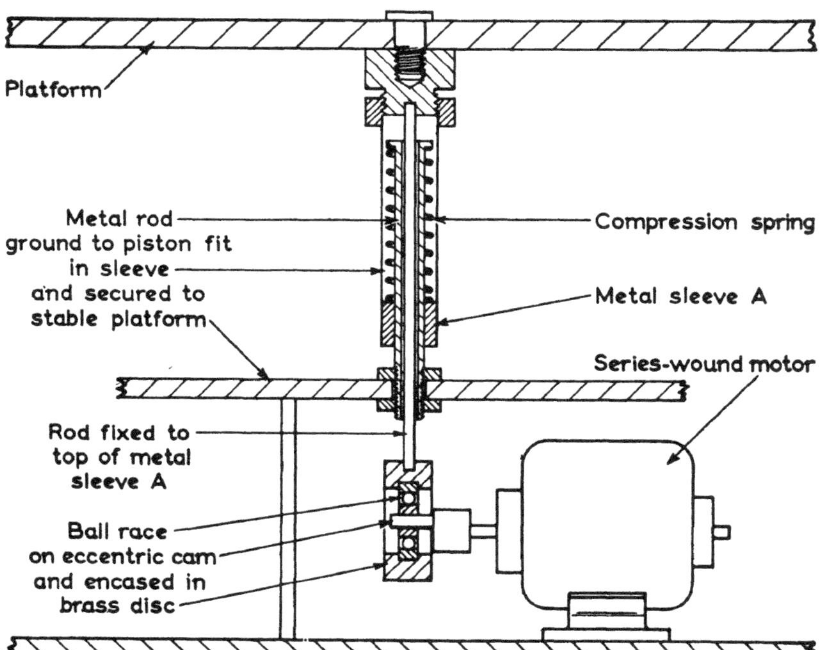

FIG. 8

This consists of a circular wooden platform centrally supported by a metal sleeve, housing a compression spring and piston. A rod fixed to this sleeve is coupled through a ball-race to a cam eccentrically mounted on a shaft of a series-wound motor. (Further improvements using a special application of the interrupted magnet principle are being tested and it is obvious that many different arrangements can be devised.)

TABLE III

Effect on diffusion time using the Vibrating Table

		Without any vibration or shaking		*With the vibrating table*	
Vol. of $(NH_4)_2SO_4$ *solution* *ml.*	*Vol. of satd.* K_2CO_3 *used* *ml.*	*For 95% absorption* *mins.*	*For full absorption* *mins.*	*For 95% absorption* *mins.*	*For full absorption* *mins.*
0·1	2·0	110	190	12	23
0·2	1·0	38	57	8	17
0·2	2·0	61	170	11	19
0·5	1·0	48	122	18	24
0·5	2·0	68	142	13	32
Dale Shaker					
0·1	2·0	110	190	80	120

The effect of the vibrating table in reducing the absorption times is shown in Table III.

It will be seen that for those relatively fast ammonia absorption rates due to low volumes and high K_2CO_3 concentrations in the total fluid vibration has a considerable effect, much greater than that obtained by using the Dale shaker.

In general, rocking is not advised except for special purposes, since it will obviously tend to diminish the working ease and simplicity of the method; where, however, suitable shaking devices are at hand their use may quite often prove helpful.

Other factors. The time absorption relations are more fully treated in the subsequent chapter, and special factors will be referred to when dealing with particular determinations.

CHAPTER IV

GENERAL PRINCIPLES GOVERNING THE ABSORPTION TIME IN MICRODIFFUSION ANALYSIS

IN the previous chapter an account was given of the practical effects on the ammonia absorption of varying the more important factors. This also applies very largely to the absorption of other substances, particularly those like ammonia, which are not only volatile but very soluble.

In the present chapter the general principles governing the absorption rate of any substance are theoretically derived and are summarised in simple formulae.

The full solution of the general problem of the successive and continuous diffusion of a substance through three media, the conditions in each of which are different, is complicated and scarcely lends itself to useful application in analytical procedure. It will be seen, however, that the practical requirements are sufficiently met by a limitation to special cases.

I. Time required for 99·5% clearance of a volatile substance from a solution stratum of 'a' cm. depth when the tension at the upper surface is maintained at zero.

The solution of this problem is contained in the equation,

$$\frac{100-y}{100}=\frac{8}{\pi^2}\left[e^{\frac{-k\pi^2t}{4a^2}}+\tfrac{1}{9}e^{\frac{-9k\pi^2t}{4a^2}}+\tfrac{1}{25}e^{\frac{-25k\pi^2t}{4a^2}}+\ldots\right], \tag{7}$$

where 'y' is the total percentage diffused, 'k' the diffusion coefficient and 't' the time. When y is 99·5% and k of the order of 1×10^{-3} (cm.^{2}/min.) for ammonia and other molecules of a similar magnitude through water, this series is very rapidly convergent and only the first term need be considered. From this,

$$0{\cdot}005=0{\cdot}81\times e^{\frac{-k\pi^2t}{4a^2}}. \tag{8}$$

If we consider ammonia with a diffusion coefficient of $1{\cdot}09\times10^{-3}$ at 20° and $1{\cdot}66\times10^{-3}$ at 38° (from Landolt Bornstein's Tables and Hufner's data in same tables reckoning for 20° and 38°) we obtain

$$t=2{\cdot}05\frac{a^2}{k}$$

$$=1880a^2 \text{ for ammonia at } 20^\circ$$

$$\text{or} \quad 1230a^2 \quad ,, \quad ,, \quad ,, \; 38^\circ \tag{9}$$

Where 'a' is 0·1 cm. (corresponding to 1·7 ml. in outer chamber of 'unit' No. 1), $t=19$ minutes at 20° C. and 12 minutes at 38° C.

For a depth of 0·12 cm. (2 ml. in outer chamber) $t=27$ minutes or about half an hour at room temperature and about 18 minutes at 38° C.

When the layer depth is 5 mm. it would require 307 minutes or five hours for a 99·5% clearance at 38° C., and this, quite independently of the volume of fluid used. *For microdiffusion analysis the stratum depth below the emission surface is therefore of the greatest importance.*

II. Time required to reach a steady state when a gas diffuses from a fixed tension at one end of a cylinder to zero tension at the other.

In a similar way it may be shown that the time for a steady state is given by the relation

$$t=\frac{2l^2}{\pi k}, \tag{10}$$

where 'l' is the cylinder length and 'k' the gaseous diffusion coefficient. The diffusion coefficient of oxygen into air is 10·7 (cm.²/min.) as found by Obermayer (1880) at zero centigrade. On the basis that the diffusion coefficient is inversely proportional to the square root of the molecular weight and that the temperature effect is given by $k=k_0\left(\frac{T}{273}\right)^n$ where n is 2 (Loschmidt 1870; Stefan 1872) we may deduce the gaseous diffusion coefficient of ammonia as 16·9 at 20° and 19·1 at 38°.

With the mean distance of passage in a 'unit' of about 3 cm., the time required is only 0·34 minute at 20° and 0·30 at 38°. It is obvious therefore that only a negligible time is required for reaching a steady state or for the attainment of gradients which are simply proportional to $\frac{p}{l}$, where 'p' is the tension in outer chamber and 'l' the mean distance of molecular passage.

III. The time required for passage of a volatile substance into an absorbing medium.

This we may regard as a variant of I, and consider that a minimal value will be given if we consider the time required for a certain percentage clearance from the absorptive material itself, when one side is maintained at zero concentration.

When acid is being neutralised, then the time required for about 80% neutralisation will be only the fraction of a minute, a layer of 1 mm. depth being considered, it being noted also that the diffusion coefficient of hydrochloric acid through water is over twice that of ammonia.

From the foregoing, it is clear that the only appreciable delay in the total absorption in apparatus of the standard type, occurs from the diffusion through the liquid layer in the outer chamber, and the time required to unload the quantity of volatile substance across a gradient expressed by p/l. The times for the attainment of uniform gradients across the gas layer and the time of absorption in the central chamber are comparatively negligible.

It will also be seen that no matter how high the tension of the volatile substance is, and no matter what arrangement we may make with regard to the disposition of the surfaces, the total absorption time from layers of 1·2 mm. depths (2 ml. in outer chamber of 'unit') and without agitation of the apparatus, will take about half an hour at room temperature. We could, however, work with lesser volumes in the outer chamber and still cover the surface, though this becomes difficult with 1 ml. We could also, if necessary, much decrease the absorption time by rocking.

Factors entering into long absorption times. Where a large amount of the volatile substance produces a relatively small tension the absorption time will be comparatively long. Ammonia will serve as an example. At 38° with 1 ml. of $N/20$ ammonium chloride plus 1 ml. $N/10$ NaOH in outer chamber of 'unit', the total absorption time (99·5%) is about two to two and a half hours. Here the diffusion in the liquid layer of about

1 mm. depth and the time for steady gradients in the gaseous layer are comparatively unimportant.

The following factors mostly determine the absorption. The rate of the absorption at any time from the beginning of the absorption will be determined by the gradient expressed as p/l, or, if diffused quantities are to be calculated, then more conveniently as $\frac{c_a}{l}$, where c_a is the concentration of the gas as g./ml. in the air layer in immediate contact with the outer chamber or diffusing surface, 'l' being the mean distance of passage. The absorption rate will be also proportional to the area of diffusing surface, or when the diffusing and absorbing surfaces are different in extent to $\sqrt{s_l s_a}$, where s_l and s_a are the area of the diffusing and absorbing surfaces respectively (shown experimentally as described in preceding chapter). We have then

$$-\frac{dq}{dt}=\frac{c_a}{l}\times\sqrt{s_l s_a}\times k, \tag{11}$$

where k is the diffusion coefficient of the gas or volatile substance in air. We can express c_a in terms of the solubility as

$$q=c_a\times S\times v, \tag{12}$$

where S is the solubility as g./ml. produced by 1 g./ml. in the gas layer, v is the volume in the outer chamber and 'q' the quantity of volatile substance dissolved therein.

Then $$-\frac{dq}{dt}=\frac{q}{S\times v\times l}\times\sqrt{s_l s_a}k. \tag{13}$$

Integrating, we get

$$2{\cdot}3\log\frac{q_0}{q_0-q_t}=\frac{t}{Svl}\times\sqrt{s_l s_a}\times k, \tag{14}$$

and where 99·5% has been absorbed from the outer chamber

$$2{\cdot}3\times 2{\cdot}3=\frac{t}{Svl}\times\sqrt{s_l s_a}\times k$$

or $$t=\frac{5{\cdot}3Svl}{\sqrt{s_l s_a}\times k}. \tag{15}$$

This equation shows the theoretical proportionality of time to volume in outer chamber and to length of mean distance between the surface as shown experimentally in the previous chapter.

From this equation we may make a calculation of the absorption time for ammonia in the 'unit' at 38 degrees, when 1 ml. of ammonium solution is alkalinised with 1 ml. of $N/10$ NaOH. We have

$v=2 \cdot 0$ ml.

$l=3 \cdot 8$ cm. (as an approximate figure for the mean distance of molecular passage, being the total distance proceeding vertically from the middle of outer chamber to half the distance from top of inner chamber to roof, across and down at right angles touching a diameter of the floor of inner chamber at one-half the distance from edge).

$S=$ the solubility (g./ml. in fluid per g./ml. in gaseous layer). The solubility of ammonia is very much greater at low tensions than at high which latter are usually listed in solubility tables, the gas not obeying Henry's Law at high pressures. For $N/1$ ammonia Abegg and Riesenfeld (1902) found a tension of 13·45 mm. at 25 degrees. The figure would be slightly increased by the presence of $N/20$ NaOH—which increase may be neglected. At 38 degrees, assuming the temperature effect to be similar to that at high pressures, we would get a tension of 21·8 mm. In $N/1$ ammonia we have then 0·017 g./ml. produced by or producing a tension of 21·8 mm. or $\frac{M}{22000} \times \frac{21 \cdot 8}{760}$ in g./ml. The solubility (S value) is therefore $\frac{0 \cdot 017}{2 \cdot 2 \times 10^{-5}}$ or $0 \cdot 77 \times 10^3$.

$k=19 \cdot 1$ (cm.²/min.) at 38°, as given above.

Inserting these values in the equation (15), we get

$$t=\frac{5 \cdot 3 \times 0 \cdot 77 \times 10^3 \times 2 \times 3 \cdot 8}{12 \cdot 7 \times 19 \cdot 1}$$

$$=\frac{31 \cdot 0 \times 10^3}{242}$$

$$=128 \text{ minutes.}$$

This corresponds well with the time found experimentally, which, as already stated, is somewhat over 120 minutes.

If now, instead of ammonia, we consider a gas such as chlorine, the solubility at 38° is 0·48 g./100 ml. for 760 mm. tension. We can convert this to solubility in the above units by multiplying by $\frac{2 \cdot 2}{M} \times 10^2$, the S value being then 1·49 as contrasted with 770 for ammonia.

The k value for chlorine at 38° may be reckoned, from that of ammonia above, as 9·4.

Inserting these chlorine values in equation 15, we get an absorption time of about 0·5 mins. For chlorine and a number of such substances of comparatively low solubility when compared with ammonia, therefore, practically the whole time of the absorption will be taken for the diffusion clearance from the fluid layer in the outer chamber.

For ammonia with 2 ml. in the outer chamber and a coefficient of $1{\cdot}09 \times 10^{-3}$ at room temperature, this is 27 minutes or just half an hour. Keeping the same volume relations, this will be the order of figure for a number of substances since the diffusion coefficient will go as the square root of the molecular weight.

Thus chlorine and bromine from chlorine and bromine water, and carbon dioxide from acidified bicarbonate are found experimentally to be fully or almost fully absorbed after 30 minutes, 40 minutes and about 30 minutes respectively, the diffusion proceeding from 2 ml. volumes. Iodine goes in a similar relation from iodine water or oxidised iodide, being somewhat but not much slower, as may be expected.

Summary of the factors influencing the absorption times in microdiffusion analysis at constant temperature, and in particular for the standard 'unit' (No. 1). (1) In microdiffusion vessels with distance between emission and absorption surfaces of the order of 3–5 centimetres or less, and where the solubility of the substance at 760 mm. pressure is about 0·5 g./100 ml. or less (this being multiplied by $\frac{2{\cdot}2}{M} \times 10^2$ for use in the present treatment), the time for a 99·5% absorption is given by

$$t = 2{\cdot}05\,\frac{a^2}{k},$$

where 'a' is the depth in centimetres and 'k' the diffusion coefficient through water (cm.²/min.).

For 2 ml. fluid in the outer chamber of the 'unit', 'a' is 0·12 cm. and the absorption time for a number of substances will then lie between 30 minutes and 1 hour.

The time absorption curve for such substances will follow a

diffusion curve and not the logarithmic relation given in the previous chapter (equation 3).

(2) Where the solubility is high (upwards of 50 g./100 ml. or more as expressed above), and where the gaseous tension of the volatile substance is low for a standard concentration, the absorption times will be comparatively slow, and the liquid diffusion (with 1 mm. layer) comparatively unimportant. This is so for ammonia when the fluid in the outer chamber is 2 ml. and the tension has not been raised by potassium carbonate or other substance. The time in minutes for a 99·5% absorption is then given by the general relation,

$$t = \frac{5 \cdot 3Svl}{\sqrt{s_l s_a} \times k},$$

where S is the solubility in g./ml. for a gaseous concentration of 1 g./ml. (This is $\frac{2 \cdot 2}{M} \times 10^2$ times the usual figure given as g./100 ml./760 mm. pressure.) s_l and s_a are the surfaces of emission and absorption (cm.^{2}); v is the volume in the outer chamber (ml.); 'l' is the mean distance of passage of molecules, and 'k' the diffusion coefficient (cm.^{2}/min.) of the gas or volatile substance through air. With 2 ml. in the outer chamber this equation for the standard 'unit' reduces to

$$t = 3 \cdot 2 \times \frac{S}{k}. \tag{15a}$$

(Solubilities should here be reckoned for low tensions, since Henry's Law may not be obeyed at high pressures.)

The time absorption curve in this case follows a logarithmic relation of the kind given in Chapter III.

(3) Rocking if used will have a much greater effect on the rapid absorptions than on the slow type, since in the former the time of absorption is almost altogether due to the diffusion gradients through the liquid which are easily broken up by rocking. With the slow type of absorption due to high solubility, rocking has very little effect, for the uniform diffusion gradient through the gas layer will be scarcely disturbed, and the liquid diffusion gradients are comparatively unimportant.

CHAPTER V

PIPETTES (SUITABLE FOR USE WITH THE STANDARD UNITS) AND THEIR DELIVERY ERRORS

HERE it is not proposed to give a general account of pipettes, but chiefly of such as have been used in practice with the many determinations subsequently described, and which may be held also to illustrate some principle, in the application of which various forms at one time or other arise. The use of No. 1 microdiffusion unit is to be understood throughout the chapter, a note on the No. 2 unit being given at the end.

For accounts of micro-volumetric apparatus, the reader is referred to such volumes as those of Emich (1932). More recent reviews are given by Wyatt (1944) and by Wilson (1949), the latter in *Quantitative Micro-Analysis* (collected and edited by Milton and Waters, 1949); also by Glick in *Techniques of Histo and Cyto-Chemistry* (1949).

Deliveries into the central chamber. The volumes required for the central chamber deliveries range from 0·7 to 2·0 ml. Where accuracies of the order of 0·1 to 0·2% (standard deviation) only are required, any of the above may be used, the pyrex tube pipette delivering in not less than 4 seconds. Where accuracies of a higher order are needed (down to 0·02%) the best is the simple tube pipette, and the worst is probably the pipette delivering between marks, since the rate of this delivery is usually irregular.

It may be pointed out that the error involved in the use of any kind of glassware cannot be assumed apart from a statistical determination of the variation occurring in a relatively large series of deliveries, titrations or the like under the stated conditions of usage. For particular types of glassware this need be done only once and considered applicable to the exact conditions occurring. Prior conclusions with respect to accuracy based solely on delicacy of reading or adjustment are likely to be misleading.

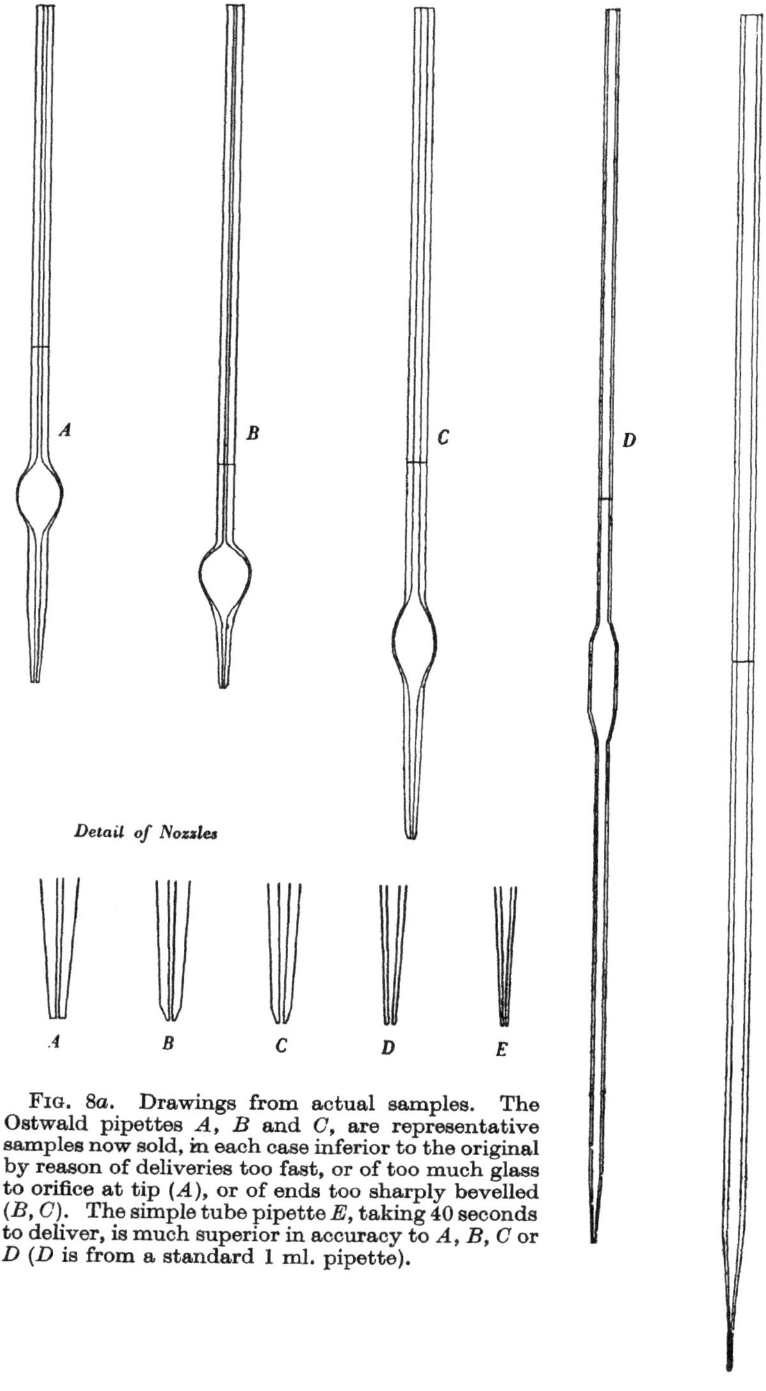

FIG. 8a. Drawings from actual samples. The Ostwald pipettes *A*, *B* and *C*, are representative samples now sold, in each case inferior to the original by reason of deliveries too fast, or of too much glass to orifice at tip (*A*), or of ends too sharply bevelled (*B*, *C*). The simple tube pipette *E*, taking 40 seconds to deliver, is much superior in accuracy to *A*, *B*, *C* or *D* (*D* is from a standard 1 ml. pipette).

The simple tube pipette. (To deliver volumes ranging from 0·7 to 2·0 ml., Fig. 8*a*, *E*.) These are easily made from pyrex tubing with internal diameter of about 3·0 mm. for the 1 ml. delivery. A section of tubing is drawn out in the flame and cut to a length of about 30 cm. The tip should be fine as in Fig. 8*a*, *E*, and the time of delivery down 20 cm. length with tube held upright being noted. If the accuracies of 0·1 to 0·2% are required it should not take less than 3 seconds to deliver, the delivery time being regulated by constriction in the flame. 1 ml. is now sucked up (this 1 ml. being delivered by an Ostwald or 2 ml. burette into a small tube) and a level a little above the meniscus marked with a

TABLE IV

No. of pipette.	*Kind of pipette.*	*Volume ml.*	*Delivery time. Secs.*	*Variable error as coeff. of variation.*	*Manner of using.*
1.	Straight tube, fine tip	0·73	40	0·029	Drop blown out gently after delivery, rotating tip against side.
1.	Same straight tube, some tip removed	0·73	4·8	0·096	As in previous.
2.	Ostwald	1·0	2·2	0·12	As in previous.
3.	Straight tube	0·96	2·5	0·13	As in previous.
4.	Straight tube	1·0	1–2	0·16	15 seconds drainage with tip touching side.
5.	Straight tube, wide tip	1·0	0–1	0·40	Drop blown out sharply after delivery with tip against side.
5.	Straight tube, wide tip	1·0	0–1	0·41	Tip touching glass floor after delivery, drop blown out sharply.
1.	Straight tube, more of tip removed	0·73	0–1	0·48	Drop blown out sharply after delivery, tip against side.
5.	Straight tube, wide tip	1·0	0–1	0·76	Tip touching fluid for 2 seconds after delivery.

The various straight tube pipettes had lengths from 13–18 cm. from mark to tip.

little gummed paper. The pipette is then standardised for delivery. Whatever volume is delivered is divided by the length of the pipette, and in this way any small difference in level of mark to bring to 1 ml. exactly is noted, and a mark made at this point with a small sharp file. The pipette is then finally standardised. From Table IV it is obvious that by far the most important factor in accuracy is time of delivery. The coefficient of variation with the first pipette listed—0·029—is practically the same as that found with a 25 ml. standard pipette. The pipette was a simple tube pipette with fine tip made in the laboratory in a few minutes. The total time taken for its delivery was not more than 41 seconds, that of the 25 ml. pipette—45 seconds, including drainage. From this it would not be surprising if a 1 ml. technique came to supersede entirely the 25 ml. macro technique.

For reading the meniscus a carefully made file mark will suffice, but an etched ring may also be made by first coating with melted paraffin, rotating as this cools, ringing around with a fine metal point or knife edge and then painting with hydrofluoric acid.

A background formed by a small piece of white label gummed behind the mark will be found suitable.

Elaborate attempts to secure increased accuracy with the meniscus reading are largely waste of time, particularly if the main factors influencing the pipette error are not appreciated.

With regard to the accuracy of delivery, we may consider briefly what exact meaning is to be attached to accuracy in such matters. In pipetting 1 ml. by any such means we deliver in the process a definite amount of fluid, which can be weighed and its exact volume ascertained. When, however, we come to make a second delivery under the same conditions, it is a fact that this may deviate from the first, and if we were to assemble a large number of such observations, it would be found that plotting as ordinates the number of times a given deviation from the mean occurred, against the plus or minus deviations, we would obtain a more or less regular curve according to the number obtained. This curve (Fig. 71, Part III) would be symmetrical around the mean and tail down rapidly on each side. It may also be constructed if we know the standard deviation from the mean, and the practical limits of the curve are included in plus or minus three times the standard deviation. Consequently the standard deviation, or alternatively the probable error (0·67 time the standard deviation), gives us the required measure of accuracy.

The standard deviation as a percentage of the mean (coefficient of variation) is given for a simple tube pipette—as shown in Part III--by the relation

$$v_p = \frac{0{\cdot}022}{\sqrt{dt}},$$

where 'd' is the internal diameter of the tube in cm. and 't' the mean time in seconds for the meniscus to fall 1 cm.

For reaching high degrees of accuracy with any such pipette it is essential that the tip be fine, with a small proportion of glass surface to orifice at tip, and a fine taper with no sharply shelving shoulders. Fig. 8*a*, *E* illustrates the kind of tip desirable, which is a little narrower than that in use with the standard 1 ml. pipette and much finer than that for many Ostwald pipettes sold commercially.

Immediately after the delivery, the drop in the tip, which should touch the glass, is gently blown out, the pipette being rotated.

Results from pipettes of such a kind and also of a coarser type of pipette are given in Table III. It appears from this table that our error will decrease from 0·48% to 0·029% (standard deviation) merely by altering the time of delivery ; and that the error found with a 0·73 ml. simple tube pipette delivering in 40 seconds is practically as low as the error with the standard 25 ml. pipette. This delivers with a coefficient of variation of 0·025%, taking altogether about 45 seconds when drainage is included.

It is obvious however that time and care will often be economised by considering the required accuracy with reference to the maximum unavoidable error in the whole analysis.

The Ostwald pipette. The original Ostwald pipette, introduced over forty years ago, had the form shown in Fig. 9, which is taken from the 2nd edition of the *Physiko-Chemische Messungen*, by Ostwald and Luther (1912) (1st edition in 1893). This original Ostwald pipette is still the best of its kind, subsequent variations or attempted improvements merely diminishing its accuracy.

The following is a translation of its description: "The delivery end is made from the same glass tube as the body of the pipette, the neck which is about 1 mm. wide, being joined on. It is essential that the delivery end should be narrow, passing down in a slightly conical manner with walls not too thin; the tip is narrowed in the flame—considerably increasing the durability—and so regulated that the free delivery of the contents takes $\frac{3}{4}$ minute. With such pipettes the procedure previously described with blow out and with the tip touching the wall is essential for exact measurements."

FIG. 9. Original form of Ostwald pipette, from Ostwald-Luther's *Physiko-Chemische Messungen*. Leipzig, 1902.

In many subsequent representations of the Ostwald pipette the original delivery time is considerably diminished (as in pipettes from which *A*, *B*, and *C* of Fig. 8*a* are drawn), thereby removing its essential feature; one also meets with inconvenient relations of bulb to delivery tube (Fig. 8*a*, *B*), and of too great a proportion of glass to orifice at the tip (Fig. 8*a*, *A*). One may find also that the bulb is provided with rather sharp angles tending to collect air bubbles.

The accuracy of the original Ostwald pipette depends very largely on its slow delivery and fine delivery tip tapering without any sharp bevelling at the end; but the fact remains that a simple tube pipette delivering in the same length of time gives an equal accuracy. The lesser internal surface of the Ostwald pipette would seem at first to be an advantage, but the great variation in rate of fall of meniscus acts against this advantage, and in any case a slight extra increase of delivery in the single tube time would compensate.

It may be held that the 1 mm. bore of the neck gives a lesser reading error than the 3 mm. bore of the simple tube. This question of the reading error is here of negligible importance. With tubes of such a bore and slow delivery, it is very easy to arrange the meniscus sharply at the mark. The semicircular meniscus is clearly demarcated and a deviation from the mark of 0·1 mm. is as great as need ever be present, and with experience it should be less. A possible deviation of the meniscus from the mark of 0·1 mm. in a 1 ml. tube of 20 cm. length means a standard deviation of 0·014%. This in adding to 0·03% will increase it only to 0·034% (since they add as the square root of the sum of the squares).

A further advantage of the straight tube pipette is that the residual fluid can be always calculated if necessary, and in accordance with the formula (as shown in Part III):

$$F = 3{\cdot}45dl\sqrt{\frac{d}{t}},$$

where F is the wall fluid in c.mm., l the length in cm. and d and t have the same significance as in the previous formula.

The 2 ml. standard Bang burette and the graduated pipette. Both of these are alike in that a delivery of the order of 1 ml. is made from one mark to another, but the burette is more convenient for routine fillings, since the meniscus is controlled by a tap and not by the finger, and filling from the reservoir makes it also agreeable to work with. Neither is superior to the simple 1 ml. tube.

Fig. 10 illustrates a convenient form of the 2 ml. Bang burette. (See next chapter for its description.)

The Trevan Pipette or Burette (Syringe driven by a screw and used for pipetting). In Trevan's paper (1925) which describes a slightly modified form of his earlier model (1922) it is noted that a syringe driven by a screw for the measurement of small volumes of fluid had been used by Lister and the same principle had also been used by Kelvin for measuring relatively large volumes of gas in the constant pressure thermometer. Trevan independently discovered the advantage of this principle in measuring small volumes and introduced it into microchemical work both as a burette and a pipette.

Modifications of the Trevan design (1925) relate to the type of syringe and screw used and the manner of holding these in position. The use of such an arrangement with the piston of the syringe working between two stops was described by Trevan (page 1114 of his 1925 paper), the upper stop being a fixed position of the micrometer head (which was adjusted to give the required volume)

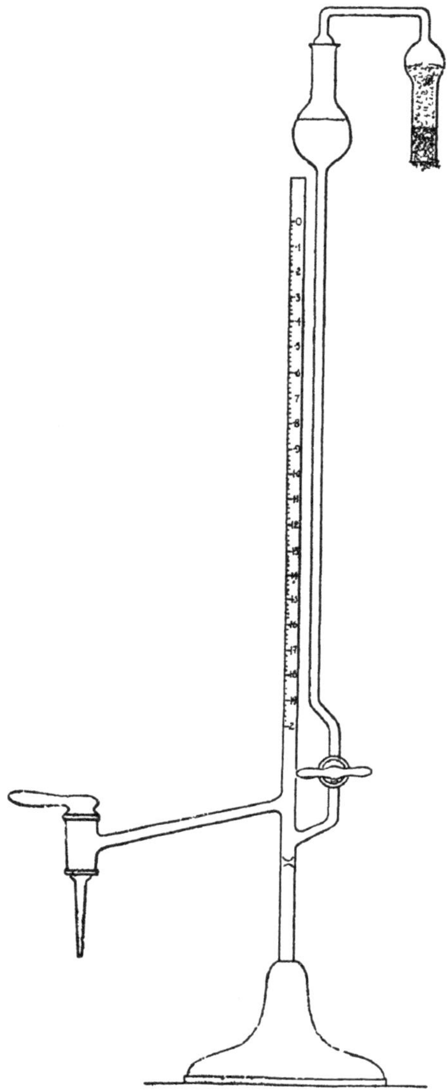

FIG. 10. 2 ml. Bang burette, with soda-lime tube attachment.

and the lower the upper surface of the barrel of the syringe. The Trevan pipette or burette is shown with simplified or diagrammatic metal holder arrangement in Fig. 12.

In using the pipette fluid is drawn into the syringe until the head of the piston meets the micrometer plunger, and the fluid is then ejected by allowing the piston of the syringe to fall until the syringe is emptied.

Trevan has given eight successive weighings of approximately 1 c.c. of $N/10$ sulphuric acid. From these the coefficient of variation may be calculated to be 0·05 per cent. This may be compared with the deliveries from the straight tube pipette taking 45 secs. to deliver in total or 0·029 per cent. At the same time from the syringe pipette samples ranging from 0·1 ml. upwards can be pipetted with a high degree of accuracy by suitable arrangement of the upper stop or micrometer head.

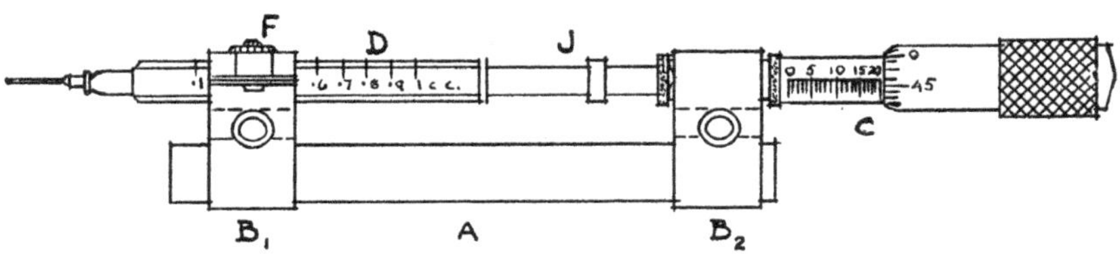

Fig. 11. The Trevan syringe pipette, (slightly modified after Trevan, *The Lancet*, 1922 ; the holding arrangement was further improved by Trevan, *Biochem. J.*, 1925, and the improved model is sold by Messrs. Borroughs Wellcome & Co., as the "Agla" micrometer syringe).

Trevan notes with regard to his own practice that when titrating with acid the steel needle is replaced by a glass tube drawn out to a point, and ground to fit the nozzle of the syringe, the point being covered on the outside with paraffin wax to prevent the fluid running back. Calibration was carried out by measuring the diameter of the piston of the syringe with a micrometer gauge and working out from the arc of the piston the length this must traverse to displace 1·0 c.c. Calibration by weighing mercury ejected and titration of fluids was also used.

The Krogh syringe pipette.

The nature of Krogh's modification (Krogh and Keys, 1931) will appear from Fig. 12 showing an all glass syringe of 2 c.c. capacity. The micrometer head is replaced by a screw of ordinary type. At ' c ' there is an arrangement for preventing rotation of piston in relation to barrel of syringe. Also ' within the temperature interval from 15 to 25° C. the temperature correction of these syringes is negligible (less than 0·0001 ml. because the steel rods carrying the adjustable stop were selected to compensate with the change in the glass parts.

Concerning the question of accuracy Krogh claims that with such a syringe pipette any desired volume between 0·1 and 1·5 ml. and an accuracy of ± 0·0001 ml. is easily obtained, and with the most careful working the error may be reduced to ± 0·00002.

Krogh comments that a syringe working between two fixed points is not only much quicker than a pipette of the same volume, but definitely more accurate. The latter part of the statement may now be questioned resulting from the error analyses described in Part III and applied in the use of the simple tube pipette No. 1 in Table III. With this 0·73 ml. was delivered with a probable error of ± 0·00014 ml. There is no difficulty in reaching this accuracy and it may be certainly taken as not representing the obtainable limit. Such a pipette, with some experience, could be constructed in a few minutes, and standardised—mean of five deliveries as recommended in Part III—in five or six minutes. In examining such accuracies the possible maximum was not investigated, it being considered sufficient to show that in the delivery of about 1 ml. the same percentage accuracy could be reached as in delivering 25 ml. with a standard pipette.

In a later chapter it is shown—using the author's burette with graduated tube of much-reduced bore and finely drawn out delivery tip—that 10 c.mm. can be readily pipetted with a probable error of 0·00002 ml.

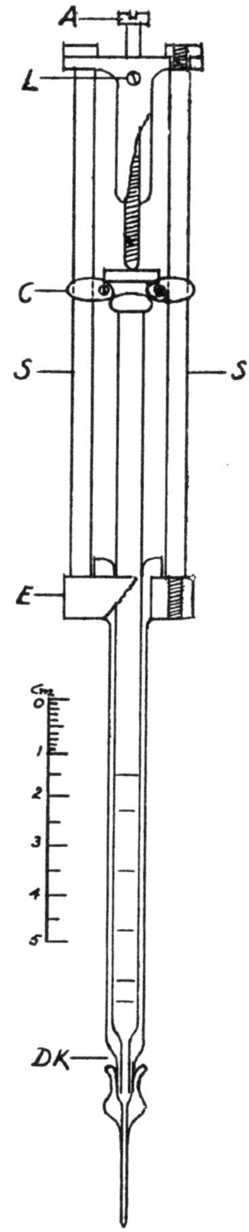

FIG. 12. The Krogh-Keys syringe pipette (from Krogh and Keys, *J. Chem. Soc.*, 1931).

The syringe pipettes are specially valuable for—

(*a*) Rapid and very accurate serial deliveries of the same fluid.

(*b*) Pipetting of viscous fluids.

(*c*) Rapid and very accurate delivery of fluids which should have no or only a minimal exposure to air.

Concerning (*c*) if we are considering, for example, barium hydroxide solution, an ordinary tube pipette delivering between two marks with the lower mark about 2 cms. from the tip will effectively guard against atmospheric action. This arises from the slowness of diffusion in a narrow fluid column and that, at most, only the upper cm. of fluid would be affected during the process, and such fluid will not be delivered. Similarly, the barium hydroxide solution could be taken from a reservoir with soda-lime guard into a narrow test tube and the pipette filled with the tip near the bottom of the tube.

Deliveries into the outer chamber. Deliveries into the outer chamber will consist of the analytical sample plus one or more reagents. The latter in general do not require to be added very accurately and fast delivering straight pipettes or Ostwald pipettes suffice to introduce them.

In pipetting the analytical sample—this may range from 3·0 ml. to 0·1 ml. or less. It will be taken here that the lower limit is 0·1 ml., the more submicro determinations being considered in a subsequent chapter.

The accuracy of delivery of the analytical sample relates to the degree of accuracy that is sought. Standard pipettes from 3 ml. to 1 ml. will deliver with an accuracy ranging from 0·07 to 0·13 as coefficients of variation. Ostwald pipettes—well made—ranging from 0·5 ml. to 0·1 ml. may be expected to deliver with coefficients of variation ranging from 0·2 to 0·4%. If higher accuracies are desired slow delivering straight tube pipettes may be used.

A note on wash-out pipettes. By washing out a pipette one could hope to eliminate the pure delivery error as distinct from the reading error and that arising from variable small adherences around the tip, but it should be remembered that in doing this the fluid should not have been originally sucked to any appreciable extent above the mark. Otherwise one is washing out a variable amount of wall fluid, and may be increasing the error instead of diminishing it. Also, the washing out should be done several times if it is intended to remove the original wall fluid completely.

A special kind of wash-out pipette is very useful for accurate deliveries of small amounts of fluid (order of 0·1 ml. and less) and consists in the use of the Conway burette with a negative hydrostatic pressure caused by sucking up the fluid in the ' braking ' attachment beyond a certain point. After taking in the required amount as noted on the graduated tube, three times this volume is then slowly run out. In this special case the washing out of wall fluid does not arise.

After such deliveries with subsequent washing it must not be forgotten that increasing the volume of the outer chamber increases the time for full absorption, and unless this is allowed for, serious errors can arise.

For a discussion of the effects of drainage time, blowing out of drop, etc., the reader is referred to Part III. Here it may be said that no increased error is introduced when the drop is blown out immediately at the end of delivery—which can be sharply enough determined by watching the meniscus—the point of the pipette being rotated against the glass wall at the instant of delivery.

Standardisation. In standardising any of the above pipettes five deliveries are weighed out into a tared weighing bottle and the mean weight calculated. The temperature of the water is noted and the volume of the delivery for 20° C. is obtained by

$$V = W\left(1 + \frac{s}{100}\right),$$

where s is a correction found in Table XXXII, Part III, W the weight of water, and V the standardised volume. Wash-out pipettes may be standardised with mercury.

Cleaning of pipettes. If pipettes used with the present technique be cleaned with concentrated sulphuric acid and bichromate mixture, they must be very thoroughly rinsed. Attaching to the filter pump and sucking through successively water, alcohol, ether and air will generally be sufficient. This is not readily possible with the 2 ml. Bang burette, which may be treated as exceptional and cleaned with sulphuric acid and bichromate. It is essential after full washing out of the sulphuric acid, that the taps be taken out, well washed, dried and regreased. The fluid for use should also be allowed to remain 24 hours in the burette, which is then emptied and refilled.

When acid of strength less than $N/500$ is used no exception should be made to excluding concentrated sulphuric acid for cleaning purposes.

Storage bottles. With each horizontal burette there is a pyrex storage bottle which holds sufficient fluid for a large number of refills. Otherwise where weak standard acid is being stored, pyrex bottles or paraffin-lined bottles should be used when the acid is of less strength than $N/100$. Bottles may be lined by paraffin by firstly drying well, then heating under the hot water tap and pouring in some melted paraffin (M.P. over 50°C.) with which the internal surface is easily covered, subsequently pouring out excess and allowing to cool in air.

Barium hydroxide solutions—as required for deliveries into the central chamber in carbon dioxide determinations—or other weak standard alkali should be stored in a bottle with soda-lime guard and siphon delivery. The barium hydroxide may be delivered then directly into a reservoir of a 2 ml. burette or into a small tube with subsequent removal of 1 ml. by a long pyrex pipette, the tip of which is inserted to the bottom of the tube.

Effective prevention of the entrance of carbon dioxide from the air can be made either by painting rubber stoppers, etc., with melted paraffin or by tying lightly a paper collar around the neck of the bottle and using it as a support for melted paraffin poured in. On setting an effective seal is formed.

Pipetting into the No. 2 and 2A 'units'. The range here is generally 0·1–0·2 ml. for central chamber and 0·1 ml. (or less) up to 1·0 ml. into the outer chamber. Similar observations apply as for pipetting into the No. 1 'unit'. In addition it may be noted that when pipetting volumes of the order of 0·1 ml., if an accuracy greater than that expressible as a variable error of 0·3% is not required, pipettes such as the long graduated kind supplied with the Warburg manometric apparatus (capacity 0·1 to 0·2 ml.) are often very useful. The volume pipetted is delivered rather rapidly between two marks. This is suitable for such microdiffusion determinations as total CO_2 in 0·1 ml. blood. Alternatively the syringe pipettes would here be very useful and the technique was used by Lehmann (1944) in determining blood bicarbonate by the microdiffusion method.

CHAPTER VI

MICRO-BURETTES (SUITABLE FOR USE WITH THE STANDARD UNIT) AND ERRORS INVOLVED IN THEIR USE

In titrations where we are restricted to minute quantities, it is necessary to keep the final volume of fluid low. In this way we decrease the amount of added reagent necessary to produce an observable change around the end point.

The accurate delivery of small volumes needs special burettes. With the Bang 2 ml. standard burette the percentage variable error in delivering 1 ml. is usually (but not necessarily) greater than the percentage error in delivering 25 ml. with the ordinary 50 ml. burette. That it is not necessarily greater will appear from the considerations in Part III.

With a 2 ml. Bang burette of standard type the variable error in delivering 1 ml. was found to be 0·11% as a coefficient of variation. The variable error in delivering 25 ml. from a standard 50 ml. burette is of the order of 0·05% (including the reading errors in meniscus adjustment). Such errors are determined for the one burette under the same conditions as standard deviations of the single delivery from the mean of a large number. They are conveniently expressed then as percentages of the mean or coefficients of variation.

Owing to the large reduction of volume and provided we are dealing with small quantities of acid (less than 10^{-2} milli-equivalents) increased accuracy over the 50 ml. would be secured with the 2 ml. Bang burette, the advantage being greater the smaller the quantity. At the same time the use of solutions ten times or so more concentrated is an extra advantage.

With micro-titration in general it is occasionally required to titrate quantities so minute that a burette delivering volumes of the order of 0·1 ml. or less is necessary, and such a burette is very suitable for microdiffusion determinations. If the principle of delivery from a vertical burette is retained, adequate control is difficult and drainage errors are likely to become prominent. Such difficulties are removed by delivering from a horizontally placed graduated tube with uniform delivery rate secured by a small hydrostatic pressure.

The error can be made at least as small as with Rehberg's burette (1925), (in which the fluid column is expelled by a mercury column controlled by a screw), but the ease of construction and cleaning is greater.

For such a horizontal micro-burette the variable error in delivering 0·1 ml. (the delivery time taking about ten seconds) was found to be 0·12% or practically the same as that found with the 2 ml. Bang burette delivering 1 ml. Such delivery errors—as discussed in Part III—are very dependent, however, on the time required for the delivery, but in so far as the two burettes examined represented two actual burettes in use and typical of their kind, it will be seen that not only is the horizontal burette more suitable for the titration of small quantities, but may be substituted for the Bang burette over the whole range of micro-titration.

The horizontal burette of present dimensions, as described below, permits the use of fluids ten times more concentrated than with the 2 ml. Bang burette and under better conditions of security from contamination by the carbon dioxide of the air or by alkali from glass surfaces, the delivered fluid having been in contact only with pyrex glass walls.

In the writer's experience the horizontal micro-burette of design described below is the most suitable for routine work with the standard ' unit '. The 2 ml. standard Bang burette is also useful, and no doubt various other types will be found serviceable for the particular conditions, including the Rehberg burette, which has proved of so much value in micro-work and can be readily adapted. These three burettes are described in the ensuing sections.

(*a*) **The Conway micro-burette.** A suitable form of the burette for general purposes is shown in Fig. 13 (and may be procured from Messrs. A. Gallenkamp, Finsbury Square, London ; it is an improvement on the original design, 1934).

A is a piece of thermometer tubing graduated in 0·01 ml. large and 0·001 ml. small divisions. Being of good quality thermometer tubing uniformity of bore is secured, though for the most accurate work calibration is advisable. It is joined by a small piece of rubber pressure tubing to tube *B*. To *B* is attached the delivery tube and tap *C* and also the connecting tube and tap *D*, communicating with the container *E* on a platform which

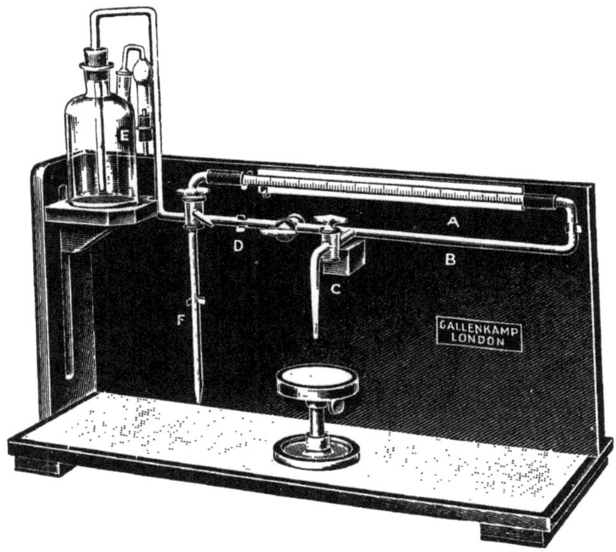

FIG. 13. Conway micro-burette (as supplied by Messrs. A. Gallenkamp & Co., Finsbury Square, London).

can be raised or lowered from the side and fastened in position by a screw. In a recent form, there are supplied, if desirable, tubes leading from the reservoir bottle, which are fitted, with a ball enlargement fitting into a carefully ground socket and held in position by a wooden collar and screw arrangement. In this way the titrating fluid can be readily replaced by another, the burette flushed out with the new fluid and so quickly ready for use with the new solution.

Detachable delivery tips attached by a ground glass joint are an advantage, but add to the price of the instrument. In Fig. 15 a form of this burette is shown with detachable delivery tubes (*E*). All tubing except the graduated thermometer tubing is of pyrex glass. The tube adjoining the graduated tube has twice the internal bore of this latter, so that the volume of the fluid in the graduated tube is comparatively small *and the emerging volume has been in contact only with pyrex glass.**

The cleansing required for a smoothly running meniscus is very easily done by detaching the graduated tube and cleaning by any of the ordinary methods. It may as a rule be cleaned sufficiently well using alcohol and ether. After cleaning it is replaced with the glass ends in contact and washed out with a few drops of fluid from the container. The small hydrostatic pressure ensures a uniform delivery without drainage owing to the horizontal posi-

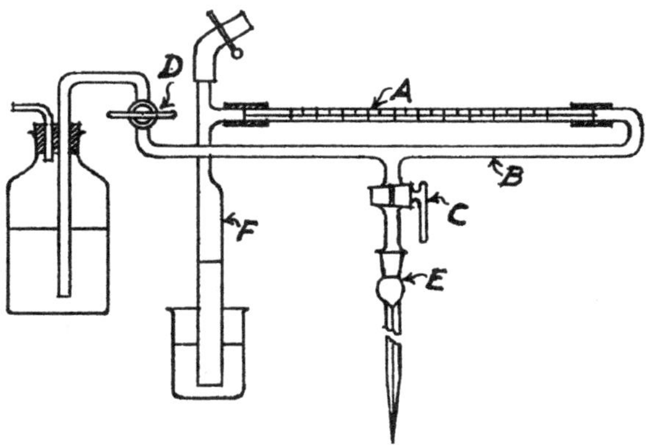

FIG. 14. Diagram of Conway burette (slightly modified after Wilson). Article in Milton and Waters, *Methods of Quantitative Micro-analysis* (Arnold, London).

tion. If the rate of delivery is found inconveniently rapid a simple and efficient method of adequately controlling this tendency consists in attaching a glass tube to the end of the graduated section as in Fig. 14 (*F*). This dips into water below, contained in—say—a 25 ml. beaker. By sucking up some water in this tube and attaching the clip as in Fig. 14, the delivery pressure of the burette is diminished and controlled as desired. The internal diameter of tube (*F*) should be at least 4 mm. The delivery tube should at the same time be finely pointed and not too blunt and the opening sufficiently constricted to give a comparatively slow flow.

In delivering from the burette the tip of the delivery tube may be allowed to touch the fluid for titration. If this method is not

* Hence Kirk's criticism (1950) with respect to rubber contacts is not, in effect, applicable.

used a fine pyrex rod will be found suitable for removing small volumes of fluid from the tip and an amount representing one division on the graduated tube may be controlled in its emergence and transferred to the titrated fluid. Much finer deliveries can be secured by the contact method.

Filling, greasing of taps, etc. The burette is firstly assembled in accordance with the diagram, and the various parts compared with the description. The manner of usage can be readily deduced from the construction, but the following description may prove helpful. The tube *F* is detached from tube *A* and the platform for the ' unit ' also displaced. The pyrex bottle *E* is detached, its supporting platform being lowered, after turning the screw at the side. The bottle *E* is rinsed out with some of the fluid for use, then filled to near the top and replaced in position. It will be necessary at first to suck the fluid across into *A* by attaching a thin rubber tube to the end of *A* and then sucking. The rubber tube is detached and the fluid allowed to drip from *A*, the overflow being caught in a vessel placed suitably underneath. The delivery tap at *C* is opened and the fluid allowed to flow out freely, being, if necessary, gently sucked out at first through a rubber tube attached to the delivery tip. When all the air has been displaced, the taps are turned into the closed position, the platform for holding the ' unit ' is replaced and some titrations carried out to see if the apparatus is in working order. The tube *F* may then be attached and water sucked up if it is thought desirable to have a better control of the meniscus in *A*. When filled and in working order the burette requires no attention for a long period, other than the occasional detaching of tube A for cleaning with alcohol and ether to secure a smoothly running meniscus.

If at any time the taps begin to stick, they are opened after detaching bottle *E*, dried with filter paper, as also their glass sockets after emptying the tubes, then greased in the usual manner and replaced. In greasing such taps a very small quantity of rubber grease is smeared on and then a minute amount of vaseline, which is spread by rotating the tap in socket.

Where extremely weak barium hydroxide solutions are used (of the order of $N/1000$), each day before the series of titrations about 10 ml. of the alkali is run out and discarded ; the cork of reservoir bottle and rubber of soda-lime guard should likewise be sealed from the air with melted paraffin.

In Fig. 15 is shown a modified form (Harman and Webster, 1948) laboratory constructed, in which instead of the ball and socket arrangement commented on above for interchanging

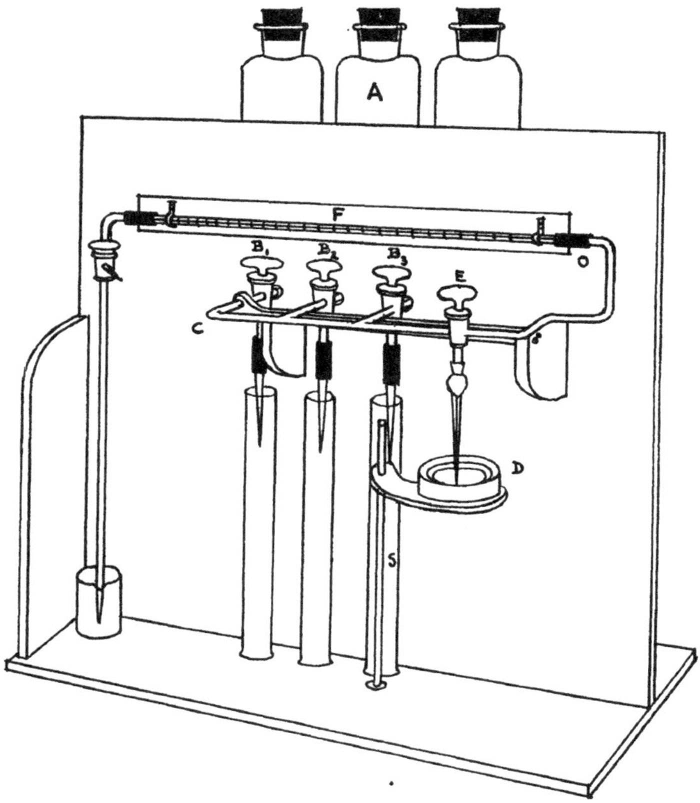

Fig. 15. Modified Conway burette (Harman and Webster, *Amer. Jour. of Clin. Path.*, 1948).

delivery bottles three reservoir bottles are held on an attached platform above and behind in a position behind the burette and feed the delivery tube at E. B_1, B_2, B_3 are three-way stopcocks.

On changing from one reservoir to another the stopcocks of the two inactive reservoirs are adjusted to the drainage position and then the system filled from the active reservoir. After closure of the stopcocks of the 'inactive' reservoirs, the burette tube is filled and emptied through the titrating cock three times, and is then ready for use.

In Fig. 16 are shown simplified forms of the Conway burette by Wilson (1949) and Holt and Callow (1942). The manner of using such burettes appears from the diagrams. Wilson (1949) comments

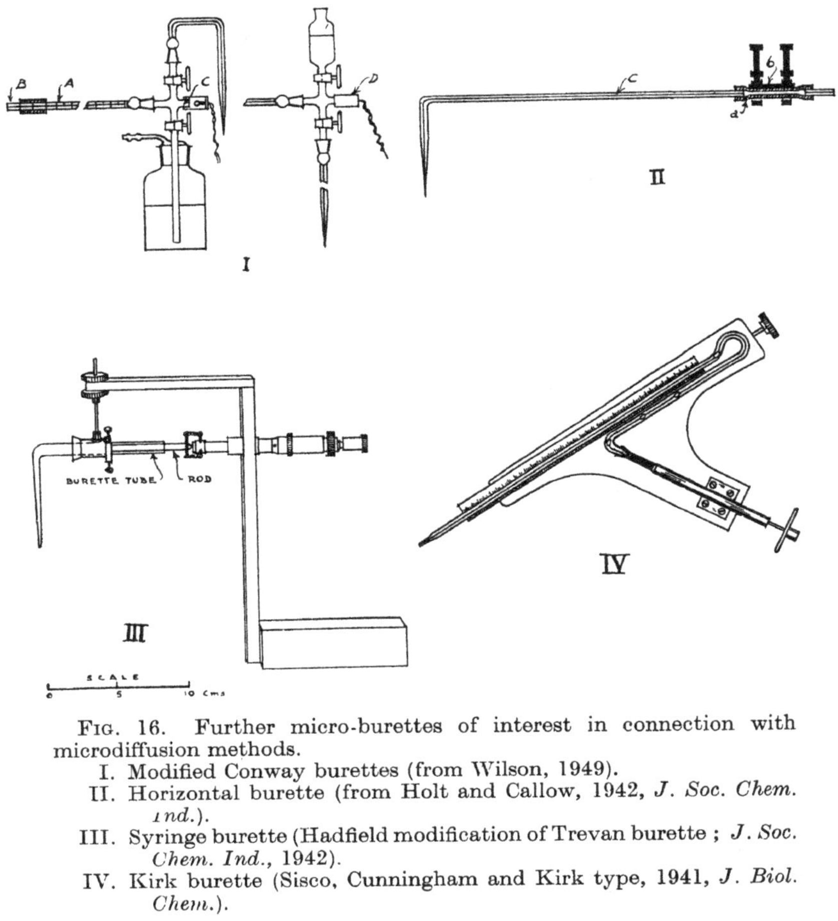

Fig. 16. Further micro-burettes of interest in connection with microdiffusion methods.

I. Modified Conway burettes (from Wilson, 1949).

II. Horizontal burette (from Holt and Callow, 1942, *J. Soc. Chem. Ind.*).

III. Syringe burette (Hadfield modification of Trevan burette ; *J. Soc. Chem. Ind.*, 1942).

IV. Kirk burette (Sisco, Cunningham and Kirk type, 1941, *J. Biol. Chem.*).

on the Holt and Callow burette as difficult to control with a spongy action and very sensitive to temperature variations during the titration.

(*b*) **The Bang burette** (1920). The most suitable burette of this type is the standard 2 ml. with soda-lime guard as shown in Fig. 10. The fluid from the bulb-like reservoir is controlled by the tap below and behind the graduated section. In filling this latter bubbles are generally easily removed by having the tap fully open at first. With a well-made delivery tip the fall of the meniscus down to the 1 ml. level should not take less than ten seconds.

After some time in use the meniscus will not descend evenly and droplets remain on the sides, when the burette must be cleaned. Strong sulphuric acid and dichromate may be used, the burette being filled with this and allowed to stand overnight. It must then be repeatedly rinsed out with water and subsequently with the fluid for titration, being allowed to stand several hours with this and refilled before titrating.

The remarks concerning the suitable greasing of taps in the previous section apply here also, and before using, the new burette should be cleaned and the taps freshly greased.

The end-point emergence from burettes. Before using the horizontal micro-burette for titrations the minimal or end-point emergence for careful routine work should be observed. For the 50 ml. burette this is usually taken as the drop ; for all burettes below this volume the drop is too large and, of course, far too large for the 2 ml. Bang burette and the horizontal burette. With a well-made Bang burette the end-point emergence should not exceed 4 to 5 c.mm. For the present horizontal burette it need not exceed 0·5 c.mm. Such an emergence is transferable with a fine pyrex rod from the tip of the burette delivery tube to the fluid for titration. Finer emergences than this are secured by the contact method of delivery, which has now become rather general in micro work (Emich, 1932). Larger emergences—1·0 or 2·0 c.mm.—may be used when the titration does not call for the finest accuracy.

On turning the tap of a burette carefully until a small amount of fluid emerges it will be noticed that the tiny emergent gush is rather constant for each single burette. It depends on the degree of constriction and the glass surface at the end of the tip, as well as on the volume and shape of hole bored through the movable tap and the extent to which this is greased, minimal greasing giving the best control.

Where in the case of the horizontal burette it is found difficult to control delicately the end-point emergence it will be found of assistance to diminish the hydrostatic pressure by sucking water up in tube F (Fig. 14) until the emerging rate with fully opened tap is going rather slowly. A marked tendency of the meniscus to jump several divisions may be removed by carefully regreasing

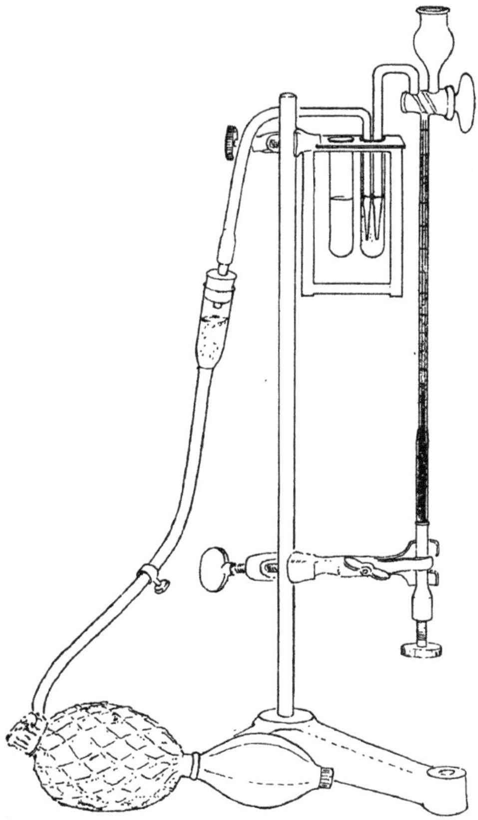

FIG. 17. The Rehberg burette, from Pincussen's *Mikromethodik* (Thieme, Leipzig), showing attachment for bubbling through alkali.

the taps with the least amount of material. If this does not remove the excessive jumping the delivery tip of the burette should be still further constricted.

Precision of reading meniscus. It is of practically no advantage to read the meniscus to less than half the end-point emergence, but it is of advantage in the most accurate work to read it at least so far. This will mean the reading of the 50 ml. and 2 ml. Bang and horizontal burettes to 0·2 of a division. Most frequently it will be necessary to read them only to half or one division.

(*c*) **The Rehberg burette.** In the Rehberg burette the titrating solution is on the top of a mercury column controlled at the bottom by a screw, which is mercury tight. Pincussen's modification (Pincussen, 1930) of this burette is shown in Fig. 17. In this model a greater ease of filling is secured by the insertion of the two-way tap. With this the fluid in the graduated section is connected with the delivery capillary or with the filling reservoir.

The graduated section consists of a thick-walled capillary tube of 0·1 ml. capacity between the terminal graduations, of which there are altogether 100. The reading of the mercury may be easily made to 0·25 of a c.mm. by means of a lens. The delivery tube bent twice at right angles and somewhat finely drawn out is of similar bore to the graduated section, with which it may be connected by suitably turning the tap. Below the graduated section the tube is widened and fits into the screw and plunger arrangement, to which it may be securely fixed by metal spirals as in diagram. Before use the screw is turned so that the plunger is well down when the lower section is filled with mercury and attached securely to the upper part.

The burette is fixed to a stand which also carries an adjustable holder for two small test-tubes backed by a window of milk glass. The apparatus is so arranged that the delivery tube of the burette enters one of these tubes carrying the fluid to be titrated and passes down almost to the end of this. The other tube is used with similar fluid volumes and indicator and buffered to the exact end-point required.

The stirring arrangment as shown in Fig. 17 is carried out by bubbling air through the titration fluid, the air pressure being established by a rubber pump such as is used in ether freezing and controlled by a screw clip. For acidimetric work the air is passed through a soda-lime tube and then through a glass tube bent at right angles and ending rather finely at the tip.

Preparation for titration. After fixing the two parts of the burette, with the plunger low and the metal spirals in position, the two-way tap is turned so that the graduated section connects with the reservoir. The mercury column is raised by twisting the screw until it just appears in the reservoir, which is then filled with the titrating fluid. The mercury is now lowered and the fluid drawn into the graduated section, after which the tap is adjusted, so that the fluid in burette connects with the delivery section and this is filled. The tap is turned again to connect with reservoir and the mercury lowered to a little below the zero reading. The reservoir is then shut off and a little of the fluid is sent out through the delivery tube on carefully turning the screw until the mercury stands at the zero point. Any adhering drop is removed from the tip of the delivery tube by touching with filter paper. The test-tube containing the solution for titration with indicator and companion tube are then placed in position as in Fig. 17, the air current cautiously started and the titration begun.

The use of the Rehberg burette with the standard ' unit '. At the end of the absorption period the fluid in the central chamber is pipetted into the titration test-tube, followed by two washings of distilled water of about 0·5 ml. each, the fluid with washings being then titrated.

The Kirk burette. (Sisco, Cunningham and Kirk, 1941). This is a modification of the Rehberg burette (Fig. 16 IV). The mercury is separated from the standard solution by air. The burette has a capacity of 0·1 ml. and the readings are taken from a scale behind the capillary tube.

Other modifications of the Rehberg burette.

Various other modifications have been described ; e.g. that of Heatley (1935, 1939) with the mercury displacement carried out by means of a levelling tube instead of a screw, and the standard solution is in contact with paraffin oil ; the burette of Scholander *et al* (1943) in which a micrometer screw is used with a spindle advancing into mercury. In the burettes of Rosebury and Heyningen (1942) and of Wyatt (1944) the principle of the Rehberg burette and that of the micrometer syringe burette are combined.

The Trevan Micrometer Syringe Burette (Hadfield modification)

This has already been described under pipettes since the design could serve for both purposes. The following description is of the modification designed by Hadfield (Fig. 16 III). As the piston of the syringe Hadfield used a rod ground to a uniform diameter throughout its length. A rod 0·500 cm. in diameter has a volume of 0·499 ml. for a one-inch length and fits a precision tubing listed as 0·508 cm. bore. The head of the liquid in the burette, which is an important factor in deciding the maximum size of jet and the maximum clearance between the rod and the tube permissible, can be reduced by using the burette horizontally with only the jet vertical. In a similar set-up Wilson uses a 1 ml. ‘ Vim ’ tuberculin syringe and seals a length of capillary tubing to the needle spigot, bent at right angles to give a vertical delivery with the syringe held horizontally.

The micrometer is held by a vertical plate attached to a heavy base. The manner of suspending the syringe tubing, with adjustment of milled nuts and screws and short bored cork extending into only part of the collar appear in his diagram. The attachment of the micrometer to the glass rod or piston (which is necessary) is secured by two tight helical springs carried on a collar near the end of the micrometer, and hooked into loops in a platinum wire ring wound in a circular groove cut in the rod near its end.

Each division on the head corresponds to a volume of 0·0005 ml. With such burettes deliveries of 0·5 ml. are reproducible to at least 3·5 parts in 10,000.

The four types of burette compared.

The principles of accuracy in the use of burettes are formally considered in Part III, and below, certain details are given concerning the accuracies of the horizontal and Bang burettes in routine working. Here, some prominent features of the four will be briefly considered, and the reader is also referred to a summary of these burettes made by Wilson (1949).

Such burettes are to be considered with reference both to convenience and accuracy, for two burettes of equal working accuracy, or which achieve equally the required accuracy, are not of equal value if one needs more attention in use.

The relatively very large reserve capacity of the author's burette (combined with very easy filling out of contact with the atmosphere) and amounting ordinarily to over 1,000 deliveries of

0·2 ml. volumes, may be claimed as an advantage over the other burettes. Also the fact that the fluid delivered has been in contact only with pyrex glass and not with mercury is an advantage over the Rehberg burette with respect to the kind of reagents that may be conveniently used.

The compactness of the micrometer syringe burette is attractive but the micrometer is liable to damage by corrosive agents as noted by Wilson (1949). Also it is somewhat tedious to fill entirely air free (as mentioned in Hadfield's description, 1942).

Concerning accuracy, drainage error limits the use of the Bang burette to about the 2 ml. capacity level, and, for accurate work, it is altogether excluded for deliveries which cannot exceed 0·1–0·2 ml. At the range of about 2 ml. delivery a well constructed burette should give an error only of the order of 0·1% (coefficient of variation) when working carefully, or about the same as is very easily attainable in routine work with the Conway burette delivering about 0·2 ml. from a graduated capacity of 0·25 ml. (In such titrations it is assumed that the end-point is perfectly sharp, which necessitates working above a certain concentration level). From Krogh's data with his syringe pipette this is also about the level of accuracy that may be expected with deliveries of the order of 0·1–0·2 ml. from the micrometer syringe pipette, unless special care is taken and apart from the other burette errors mentioned below.

As discussed in Part III the working accuracy of a burette depends on its manipulation and delivery errors (assuming uniformity of bore or of piston diameter, or alternatively, exact calibration). Under the term manipulation is included the reading error and that of the end-point emergence. The pure delivery error depends on the varying amount of fluid left on the tube wall, and can be practically eliminated to any desired degree by slowing the delivery rate. Such delivery error does not appear in the use of the micrometer syringe burette, which would have the advantage of a more rapid delivery on a similar accuracy level. Reading and end-point emergence errors would be common to all three, and are present or could be minimised to a similar degree depending on the attention given thereto.

The end-point emergence error can be lessened in all four by contact of a fine delivery tip with the titration fluid. It is not altogether eliminated when we consider that the end-point is either being approached at some rate which may slightly overshoot it, or there is added a little extra fluid by turning a tap or a screw some small distance. The average of such minimal additions will produce the error of the end-point emergence, and to this will add in a statistical manner—as later discussed—the reading error.

It may be concluded that the total relative error with all four burettes, working at their appropriate levels, could be reduced towards a similar vanishing degree of error, depending on the attention given, but perhaps with a greater working speed on the highest accuracy levels using the micrometer syringe burette. Where, however, accuracies of 0·1% (coefficient of variation) are considered sufficient in serial titrations the micrometer syringe burette has no longer such advantage.

In the following section, a brief account is given of the errors involved in the routine use of the Conway and the Bang burettes. (The subject is more fully dealt with in Part III.)

The variable error in titrating 1 ml. of standard sulphuric acid with the Conway compared with that for the 2 ml. Bang burette. By variable error is here meant either the standard deviation or the coefficient of variation of the single titration with respect to the mean of a large number (these terms are defined in Chapter LIV, Part III). The error of pipetting out the 1 ml. acid will not be taken into account, the burette error being alone considered.

This may be increased or diminished to a considerable extent according to the degree of care and time expended on the titration. We shall consider what may be taken as typical results of careful working in which the end-point emergence of the two burettes is 4·0 c.mm. and 0·5 c.mm. respectively. (With badly made burettes such emergences may not be obtainable.)

We shall suppose that the reading is brought to 0·2 of a final division, and that the delivery error of each burette is the same, namely 0·12%, as a coefficient of variation.

The sulphuric acid for titration is taken as similar to that used in the ammonia determinations subsequently described and contains 0·0012% methyl red and 0·0002% methylene blue with 20%

alcohol; and that the volumes delivered from each burette are always 2·0 ml. and 0·2 ml. respectively.

From such data the curves in Fig. 18 are constructed and are in accord with the experimental findings, but for a fuller understanding of their nature Part III must be consulted. (The indicator could be used in lesser concentration and 0·0008%

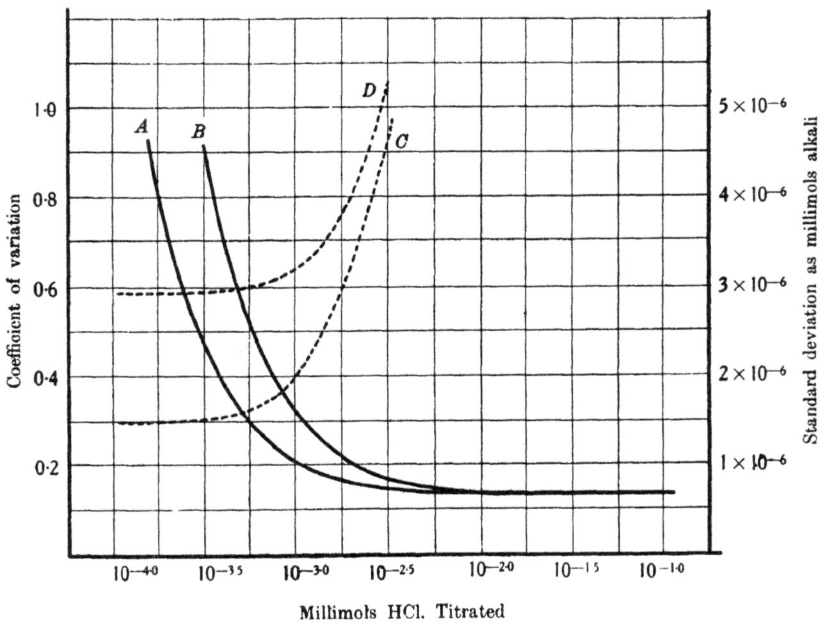

FIG. 18. Curves of variable error in titrating 1 ml. hydrochloric acid with barium hydroxide solution (or carbon dioxide free alkali), using 0·2 ml. from a Conway burette (Curve *A*) and 2·0 ml. from a Bang burette (Curve *B*).

Curves *C* and *D* give the corresponding variable error in millimols of alkali (reckoned as KOH).

methyl red is found suitable for blood ammonia titrations, and even lesser concentrations are workable.)

Two sets of curves are given in the figure, one group (*A*, *B*) representing coefficients of variation, and the other (*C*, *D*) the variable error in millimols. Each group of curves shows two definite regions and an intermediate section. With *A* and *B*, to the right we have pure glass error which when expressed as coefficients of variation (0·14% for each burette) is independent of the quantity of acid titrated until this falls to about $10^{-2 \cdot 5}$, and towards the left we have a region of pure chemical error, which is inversely proportional to the quantity titrated and directly proportional to the end-point volume. This variable chemical error which dominates at the lowest acid levels for titration, arises from the indefinite nature of the end-point change as the alkaline strength decreases.

The variable error of titrating 1 ml. of 0·0002 *N* strong acid illustrating the minimum variable error of acidimetric titration. The object of determining this error is to ascertain the least amount of alkali or acid which can be discriminated in practice by titrating 1 ml. with the burette. If we consider the titration of 1 ml. of $N/10$ acid delivered, for example, from a 1 ml. simple pyrex pipette (drawn out tube, length 17·5 cm.), the standard deviation with the horizontal burette is usually 25×10^{-5} millimols. Using $N/100$ fluids the standard deviation in millimols is ten times less, and further reduction in the strength of the fluids will progressively reduce the standard deviation expressed in absolute units. We reach a point, however, when the standard deviation so expressed becomes constant. This limit of standard deviation sets the limit of our discrimination and is proportional to the total buffering power of the fluid at the end-point multiplied by the indiscernible range of the *p*H figure.

The limit for 1 ml. titration is reached for the Conway burette when the quantity of acid titrated is in the region of $10^{-3\cdot5}$ millimol (as shown in Fig. 18).

For titrations in atmospheric air it is not possible to diminish this figure appreciably for the above volume by titrating closer to a *p*H of 7·0, for any diminution of the buffering power of the water secured thereby is more than counterbalanced by the rapidly increasing buffering power of the carbon dioxide of the air as well as by the instability of the end-point.

The carbon dioxide of the air will be in equilibrium with distilled water of approximately 5·6 *p*H, so that the methyl red indicator has the great advantage that its point of maximum sensitivity lies in the same region.

In a carbon dioxide free atmosphere, and titrating 0·7 ml. acid to a *p*H of 7·0 with the above glass ware and similar fluids and volumes, it is possible to decrease the standard deviation from

$1{\cdot}1 \times 10^{-6}$ to about $0{\cdot}3 \times 10^{-6}$ millimol KOH, provided we substitute for the indicator mixture one equally sensitive at a *p*H of 7·0 and provided also the alkali is perfectly carbon dioxide free.

General uses of the Conway burette. Besides its use in determinations with the standard ' units ' the burette, as already mentioned, may be employed with advantage in general micro-titrations. It can be used to give—and that quite easily—accuracies of the same order as when 10·0 to 15·0 ml. are titrated from the 50 ml. burette, with coefficients of variation of 0·16 to 0·11%. Accuracies of a higher order than this can be obtained—as considered in Part III.

The long delivery tube is of advantage in titrating into centrifuge tubes or small containers.

It has been used here, for example, in the thiosulphate titrations of the Hagedorn-Jensen micro-glucose determinations in blood, using *N*/20 thiosulphate instead of *N*/200, having therefore the advantage of greater permanence in the titrating fluid. Florkin and Bosson (1939) have used it with *N*/200 thiosulphate when titrating minute amounts of iodine in about 1·5 ml. fluid —the final stage of their procedure for determining glucose (2·5–30 μg) in small amounts of blood.

It has also been found advantageous in micro-permanganate titrations of calcium oxalate, and with such, an automatic stirrer may be conveniently set up by fixing a capillary tube connected with an air blower along the delivery tip. The burette, by interchanging the graduated tube with others of different bore, can be converted into one serving wide ranges of delivery. Such interchange with a finer bored tube has been used in dealing with volumes of the order of 20 c.mm. as described in a later chapter.

Levvy (1945) has investigated the use of wider bored tubes than those ordinarily used with the burette. He finds that the meniscus of a horizontal column of water remains vertical in clean glass tubing of 2·5 mm. internal diameter. With tubing of 3·5 mm. internal diameter the meniscus is at an angle of about 45°. For a vertical meniscus the bore diameter should not exceed 3 mm. With the 2·5 mm. tube in place, the burette was found, in a series of twenty-six deliveries, to deliver 1 ml. water with a coefficient of variation of 0·105%, the meniscus moving at 1 cm./sec. With the Bang burette, for similar rate of flow, the coefficient of variation is approximately 0·14%.

Levvy (1943) has also found the burette of advantage in the iodometric micro-determination of arsenic in biological material. He uses a ground-glass joint in place of the rubber stopper, and if

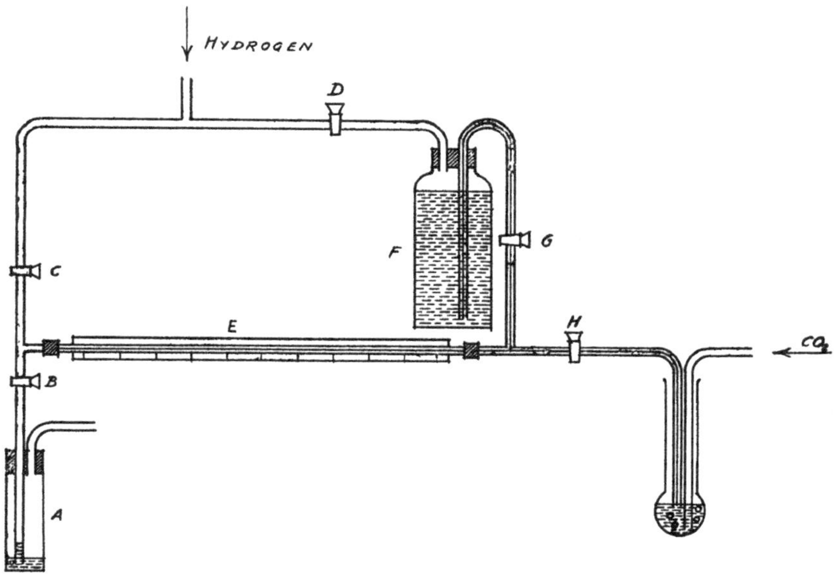

FIG. 19. Modified Conway burette as designed by Ramsay for titrating with titanous sulphate. *A*, pressure-regulating vessel. *B*, *C*, *D*, *G* and *H*, stop-cocks. *E*, the burette, standard bore capillary tubing against mm. scale. 10 cm. ≏ 0·05 ml. *F*, $Ti_2(SO_4)_3$ reservoir, capacity 200–250 ml. *I*, digestion and titration flask. (From the *Biochem. J.*, **38**, 467, 1944.)

the reservoir is shielded from light there is little change in the $N/200$ I_2 over intervals of weeks. He likewise finds the burette of value in the determination of glucuronic acid (private communication), the final procedure turning on a cerium sulphate titration of reduced ferricyanide, with Setopalin C or Lissamine Green as indicator.

A modification of the burette (Fig. 19) has been employed for titrating with titanous sulphate in the micro-determination of iron in biological material by Ramsay (1944). The organic matter is

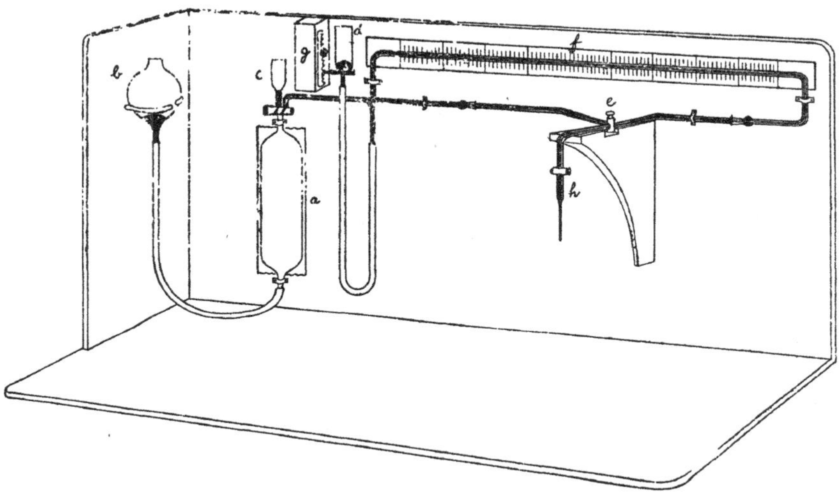

FIG. 20. Modification designed by McFarlane for titrating with titanium solutions. The burette is constructed of soft glass and mounted on a 3-ply board (87 × 48 cm.). Capillary tubing of approximately 1 mm. bore is used for the section, *f*, *e*. The burette proper is 67 mm. long and is backed by a strip of graph paper. At the left it is connected by pressure tubing to a small mercury reservoir (2 × 5 cm.), *d*, and at the right end by a ground-glass joint to a three-way stopcock, *e*. The mercury reservoir, *d*, is adjustable to a height of 3 cm. above and below the level of tube *f* by means of the rack and pinion, *g*. A second ground-glass joint connects the burette to the reagent storage vessel, *a*, of 200 c.c. capacity, joined to a levelling bulb as shown. The storage flask is painted with two coats of black enamel and set into the board by cutting out a strip. The delivery tube, supported by a bracket (13 × 13 cm.), is drawn out to a fine capillary, the tip of which has a clearance of 16 cm. from the base board (28 × 87 cm.) (from *Industrial and Engineering Chemistry*, Anal. Ed., Vol. 8, p. 125).

destroyed by heating with H_2SO_4, $HClO_3$, and HNO_3 in a small pyrex flask (10–12 ml. capacity, internal neck diameter 8 mm., over-all length 80–90 mm.). The digest is diluted and titrated with standard $Ti_2(SO_4)_3$ under hydrogen. KCNS is used as indicator, and CO_2 for stirring (see Fig. 19).

(In a private communication Dr. Ramsay states that he now uses the unmodified burette for such titrations, the hydrogen supply being connected with the reservoir.)

In Fig. 20 is also shown the modification previously designed by McFarlane (1936) for titration with titanium chloride. In this the fluid is expelled by a mercury column.

Its use as a submicro-analytical pipette.

The burette may be also used as a fine analytical pipette not only in the direct delivery of standard solutions, but also in the measurement and delivery of small volumes of fluid for analysis. This depends on the fact that when the water level in the side tube is above a certain mark a suction through the delivery tube of the burette will be produced on opening the tap. Using fine bored graduated tubing, quantities of fluid of the order of 10 c.mm. can be taken up, very accurately measured, and again discharged followed by an equal volume of water or of dilute acid or alkali (if such be used in the burette and do not interfere with the required analyses). An equal volume of discharge suffices to deliver the full amount of the fluid sucked in. Such use of the burette is referred to again in 'drop scale' analyses.

CHAPTER VII

THE MICRODIFFUSION METHOD WITH END-POINT VOLUMES AROUND 20 CUBIC MILLIMETRES

THE use of the method for such small end volumes and those of a similar order in the outer chamber, will rarely arise, and only for special conditions. It is obvious that the standard ' unit ' can be scaled down to any desirable level for use with capillary pipettes and narrower gauge burettes than those already described. With regard to absorption times, from consideration of the principles in Chapter III, and keeping the fluid depths at about 1 mm. and the surfaces of the glass chamber covered, the necessary absorption time on scaling down approaches the limit for clearance from the outer liquid by diffusion. This is 19 minutes at room temperature and about 12 minutes at 38° C. Reducing the depth to 0·5 mm., a possible 5 minutes total absorption time could be obtained at room temperature.

The following gives an account of the procedure developed from a study of determinations of 0·07 μg NH_3–N in 20 c.mm. fluid. The series showed a coefficient of variation of approximately 1%, or a probable error of 0·0005 μg of $NH_3 - N$ (such experiments were carried out in the author's laboratory by Dr. E. O'Malley). Special attention in these analyses was not directed to the reduction of absorption times to their possible limit, but they were complete in less than 30 minutes. It is considered also that greater accuracies are obtainable without undue difficulty.

Pipetting volumes of the order of 10 c.mm.

The standard acid. To the horizontal burette as described in Chapter VI is fitted a graduated tube of approximately 0·36 mm. internal bore, and marked in a similar way in large and small divisions. Each large division on this tube is equivalent to approximately 1 c.mm. The delivery tube of the burette is drawn out to a finer tip than that usually supplied. The burette is filled with the standard acid. To find the variable error of delivery a series of deliveries was made into a small weighing bottle made from tubing 0·5 cm. diameter and fitted with a glass stopper. The coefficient of variation of the weighed deliveries was 0·3% (using a microbalance of the Kuhlmann type) but the actual coefficient was probably somewhat under this value. This figure of 0·3% is equiv-

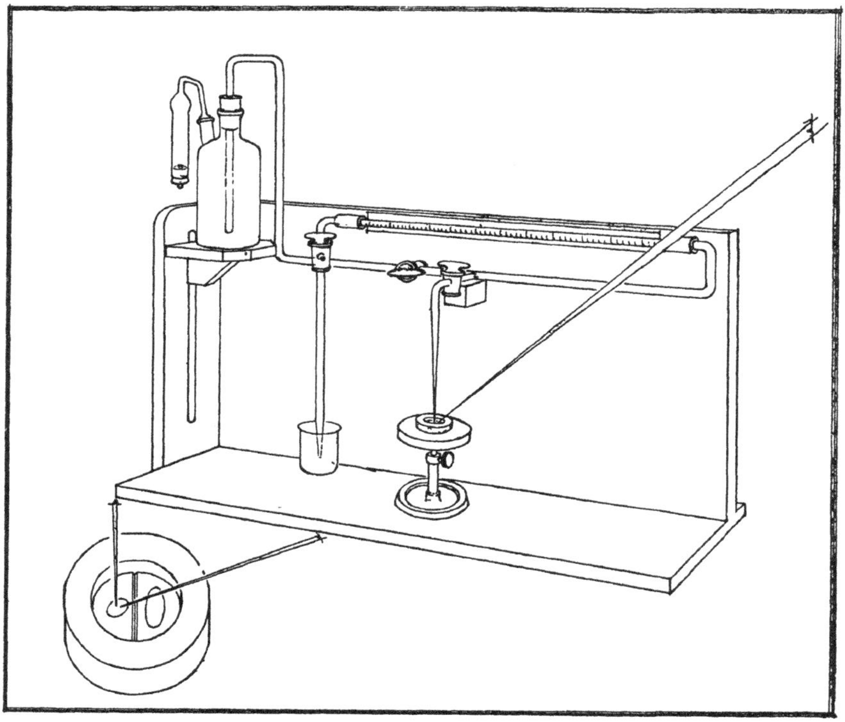

FIG. 21. Titration arrangement and small wax ' unit ' for very minute ammonia determinations (order of 0·1 μg NH_3-N).

alent to a probable error of 0·00002 ml. or the same as that given by Krogh and Keys for their precision syringe pipette.

The standard acid used was $N/1000$ HCl containing 1·32 ml. of 0·1% methyl red and 0·66 ml. of 0·1% bromocresol blue per 100 ml. 10 c.mm. of this are delivered on to one side of the trench in the microdiffusion apparatus constructed out of solid paraffin and shown in Fig. 21.

The sample for analysis. The same burette may then be used for taking up the sample for analysis which is presumed to be of the order of 10 c.mm. for each determination. Prior to the taking up of the sample the graduated tube is emptied at a rate of 10 secs. per large division. The level of fluid in the side tube is then raised so that a suction pressure is developed. The tip of the burette is wiped clean of external fluid and immersed in the sample, and the required amount sucked in. The tip is again wiped and twice the fluid volume delivered. (Before sampling, therefore, the meniscus may be suitably about midway along the graduated tube.) The delivery is made on the other side of the trench in the microdiffusion apparatus (Fig. 20).

A small droplet (about 10 to 20 c.mm.) of saturated potassium carbonate or metaborate is delivered from a simple straight tube pipette (the exact volume is not important) close to the drop from the burette containing the sample for analysis, and the 'unit' closed with a small glass lid. By tapping the apparatus the two drops in the one chamber are made to coalesce and the 'unit' is then left aside for 30 minutes.

The microdiffusion apparatus. When carrying out such drop scale analyses, small microdiffusion 'units' can be constructed out of paraffin wax. The kind of apparatus is shown in Fig. 21, the order of size being evident from its appearance on the burette stand. Across the centre of the microdiffusion 'unit' runs a line or small trench, made with a penknife or pin, and dividing the floor of the 'unit' in two. The 'unit' is covered during the determinations by an ordinary microscope cover-glass, on which a little fixative, to make it air-tight, is smeared.

Titration. After 30 minutes the lid is removed and the acid titrated with $N/500$ $Ba(OH)_2$ from a second burette prepared like the first.

The arrangement shown in Fig. 21 may be used for stirring. In this a long uniform pyrex glass rod drawn out to a point is fixed in an electrical stirring apparatus and kept rotated. Other forms of stirring are likewise applicable, such as that of Wigglesworth, in which a jet of air is arranged to impinge on the side of the drop or the 'electromagnetic flea' of Linderstrøm-Lang and Holter (1931) described in the following Section.

With such technique determinations of 0·07 μg NH_3–N have been carried out. Of 20 such determinations 19 showed an error range of from 0–2%, the coefficient of variation for the 20 being 1%.

Kirk's procedure. Kirk's procedure is illustrated later with his very delicate total nitrogen determinations of 1–10 *μg* NH_3–*N*, conducted with errors less than 1% (s.d. = 0·3%).

Technique of Linderstrøm-Lang and Holter (1931, 1933). The method adopted by Linderstrøm-Lang and Holter for ammonia absorptions and titrations (1933), and appearing shortly after the description of the standard microdiffusion apparatus or 'unit', indicates the lines on which the principle may be applied in a different way to drop scale determinations. The method adopted by them for the absorption is of an ingenious kind but not ideal, since they required 5 hours for the full absorption even at 40°, arising from the stratum depth below the diffusing surface. Their 'micro-distillation' arrangement is of the following kind. At the bottom of a small tube (25/3·8 mm., 6 mm. outer diameter) lined with paraffin is placed a drop of fluid for analysis (about 14 c.mm.). Into the tube, held horizontally, is then introduced approximately 7 c.mm. of 2*N* NaOH about 5 mm. above the bottom of tube, and 35 c.mm. of water about 5 mm. from the top. On slanting the tube the drops are run together, the water drop preventing the escape of ammonia, until 7 c.mm. of hydrochloric acid is introduced towards the upper part of tube. The tube is then ready for the ammonia diffusion and equipped with a cap as in Fig. 21, *C*, being introduced into its holder in a thermostat at 40° C. After about five hours the absorption is complete and the acid is titrated with *N*/10 NaOH or *N*/100 borate from a burette on the Rehberg principle (with Widmark and Ørskov's modification, 1928), to a *p*H of 5·9, the end-point being judged by the presence of a companion tube as in Fig. 22, *E*, with acid containing the indicator, buffered with phosphate at 5·9, and introduced in a similar way.

In the above procedure it is necessary that both the alkalinised fluid and the titrating fluid be stirred, and an interesting procedure has been used for the purpose. The stirrers ('electromagnetic fleas') consist of small capillary tubes (1–1·5 mm. long) filled with Ferrum Redactum and fused to form small pear-shaped bodies. One of these is introduced (after testing) into the fluid and moved by a magnet (*A*, Fig. 21) externally placed, the magnet current being broken periodically by a metronome arrangement.

The stirrers are tested after making by boiling in dilute nitric acid, and subsequently testing this with potassium thiocyanate.

In the above procedure it will be seen that the fluid volume at the bottom of the tube is altogether about 56 c.mm, giving with an internal diameter of less than 3·8 mm. (some of the 3·8 mm. diameter being taken up by the paraffin lining) a depth of about 5 mm. From the principles of Chapter IV this should give from the equation

$$t = 1230a^2$$

307 minutes or 5 hours which corresponds to the time allotted by Linderstrøm-Lang and Holter. With respect to the exclusion of atmospheric carbon dioxide and the use of bromocresol purple, this is unnecessary. for the kind of titrations used. As shown in Fig. 18, the glass error on titrating with methyl red with full exposure to carbon dioxide begins to be affected by the variable chemical error only in the region of $N/1000$. The constant error of the titration to a *p*H of 5·5 need be of no consequence in itself, the ammonia absorbed being determined by a subtraction of two titrations to the same end-point. The variable error in judging the end-point colour may be of importance, and is reflected in variations of titrating alkali related to the total buffering of hydrogen ion, of indicator and atmospheric carbon dioxide at the end-point. This is fully discussed in Part III. (In a later paper by Brüel, Holter, Linderstrøm-Lang and Rositz (1946), it would appear that the set up above described has been altered so that the ammonia diffuses from only about one-half the fluid volume and the time would be reduced to about one-fourth. This corresponds approximately to the time now given—approximately 1·5 hours. Also bromocresol green is used as the indicator instead of bromocresol purple. After the absorption into acid 18 c.mm. $M/150$ Na_2HPO_4 with indicator is added, and $M/100$ HCl is used to titrate to a *p*H of 4·6, comparison being made with a standard. (Some criticism of this whole procedure is given later when considering nitrogen determination).

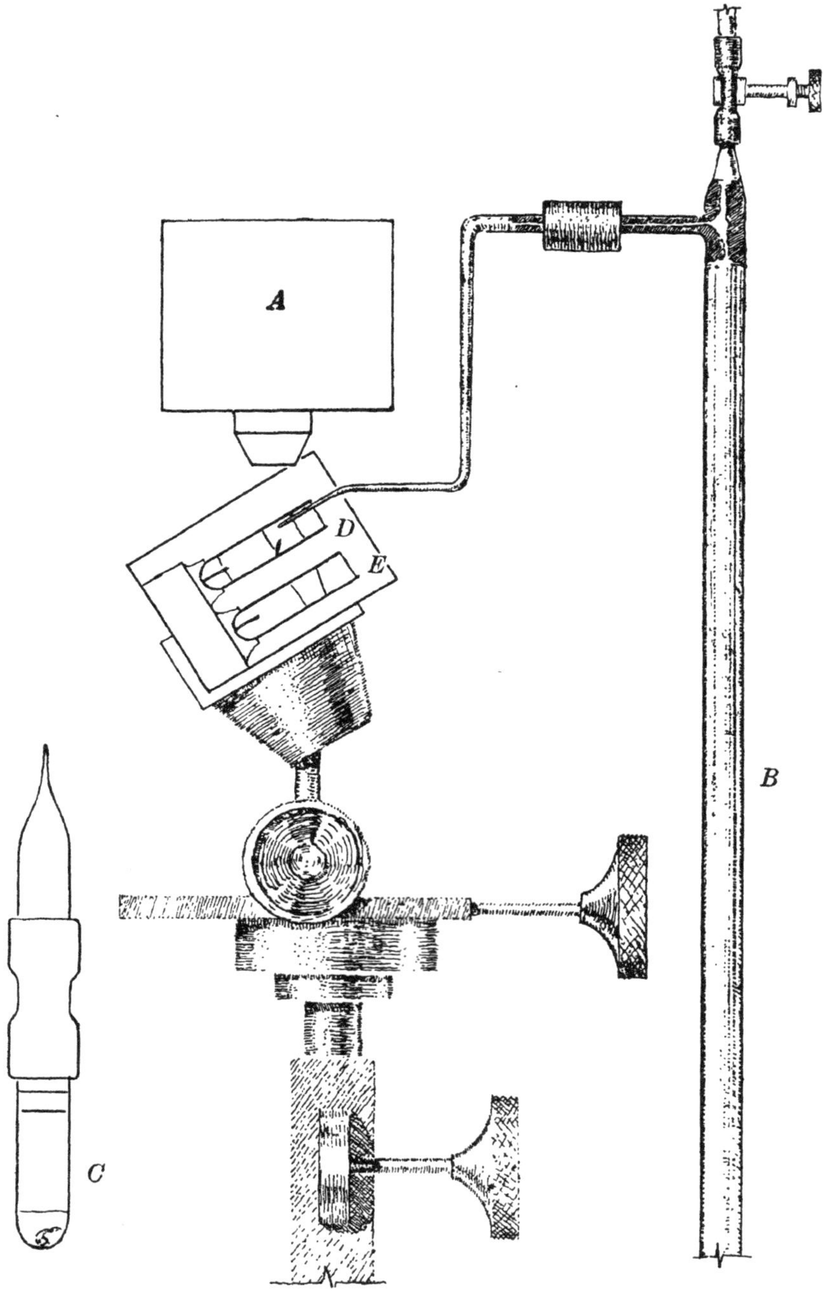

FIG. 22. The completed absorption tube (*C*) and titration arrangement from Linderstrøm-Lang and Holter (*Z. physiol. Chem.*, **220**, 5). The magnet *A* is used to move the stirrer in the drop of acid in the upper portion of tube *D*. *Vide* text.

Pipettes used. The kind of pipette used by Linderstrøm-Lang and Holter (1931) for introducing small drops into the tubes already described, is shown in Fig. 23. The pipette itself (*A*) is a capillary tube drawn out to a fine tip with a slight bend at the end, which serves the purpose of delivery against a wall surface without fear of contact by the body of the pipette. The pipette, the end of which dips into the fluid for analysis, is filled somewhat above the mark by sucking through tube, the clip being then replaced. The meniscus falls slowly and is observed by the reading

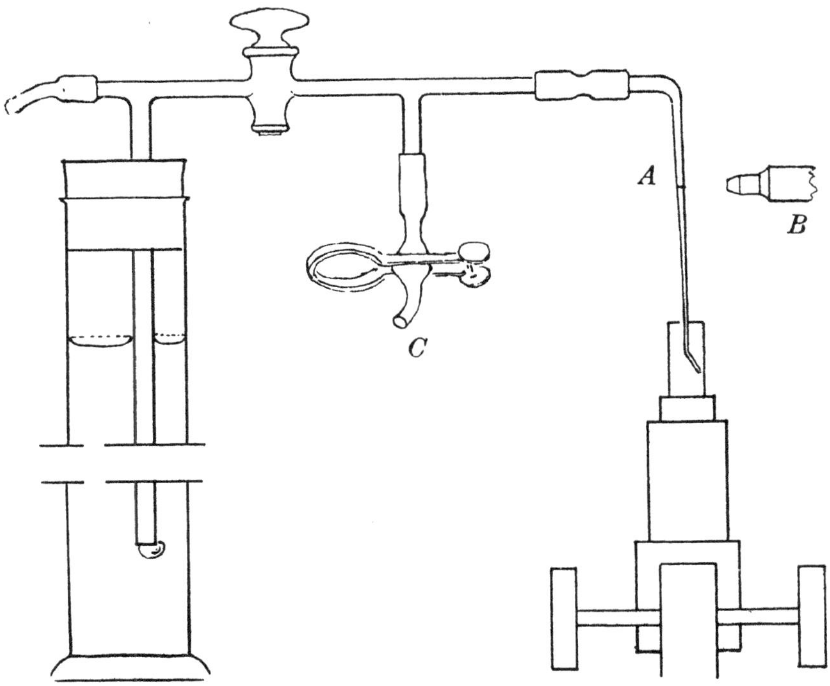

FIG. 23. Arrangement for fine pipetting (order of 10 c.mm.) by Linderstrøm-Lang and Holter (from *Zeitschr. für physiologische Chemie*, **201**, 1931).

microscope at side (*B*); when the meniscus has exactly reached the mark the support of the supply tube is suddenly lowered by twisting on its screw and the fluid is then held in the pipette by capillary forces. The contents of the pipette are ejected by the arrangement shown in Fig. 23, the tap being opened and a pressure of about 20 cm. water allowed to expel the drop, within about

5 seconds. Such pipettes may be standardised on a micro-balance, but the wall fluid should be allowed for, and this may be done by the formula in Chapter V.

Alternatively they could be standardised by a titration procedure, in which a delivery of *N* HCl is titrated with *N*/100 or *N*/200 alkali, from a micro-burette of an ordinary type.

The burette. The burette used is of the Rehberg type with the modification of Widmark and Ørskov (1928), in which the volume of fluid expelled is measured by the amount of turn of the screw, which is graduated for this purpose.

CHAPTER VIII

COLORIMETRY IN THE MICRODIFFUSION METHODS

THE colorimetric methods (for minute halogen concentrations) described in Part II may be suitably modified for any kind of accurate micro-colorimetric technique in which volumes down to about 1 ml. are dealt with. It has been found, however, most suitable to use the spectrophotometric procedure.

This has also been employed by Borsook (1935) in the determination of minute amounts of ammonia and total nitrogen, using an absorbing ' unit ' similar to that described in Chapter II, and developing the colour with the Berthelot reaction.

Crismer (1937) has likewise used this reaction—which he has modified by the use of chloramine—with the ammonia absorbed in the central chamber of the standard ' unit ' (No. 1), the Pulfrich photometer being used for measuring the colour produced. Crismer determined 3–50μg $NH_3 - N$ in this manner to within 1·3% error.

Russell (1945) has recently improved the Berthelot reaction in depth of colour and stability by dissolving the phenol in much weaker alkali and using manganous ions to catalyse the colour development. She uses it likewise in conjunction with the micro-diffusion technique, determining the depth of colour by a photoelectric colorimeter.

Other examples of spectrophotometric or photoelectric measurements in the methods described in Part II are the measurement of the remaining unoxidised Fast Green in the determination of chloride (Gordon, 1952) ; acetaldehyde-semicarbazide compound in the determination of acetaldehyde in blood (Burbridge, Hine and Schick, 1950) ; the reaction with chromotropic acid in the determination of formaldehyde (Boos, 1948 ; as used by Feldstein and Klendshoj, 1954) ; and various other colour reactions with volatile substances.

In this chapter it is proposed to review briefly the principles of such colorimetry and spectrophotometry as has been found useful in the analytical procedures following microdiffusions.

Principles of colorimetry, including spectrophotometry. In all the simpler colorimetric methods of analysis the concentration of a coloured substance in solution is determined by comparing the depth with that of a known concentration of the substance which gives to the eye the same colour intensity. The instrument or colorimeter in which this comparison is made is now most generally constructed on the Duboscq principle, illustrated in Fig. 24. In this two cylindrical vessels contain the solutions for comparison and have the under surface of plane glass. Through variation in the depths of the cylindrical glass plungers P and P_1—the positions of which are indicated by a scale at the back—the light reflected from a mirror is made to pass through different strata of the coloured fluids. Passing up through the plungers, this light enters the prisms p and p_1, thence through a lens system into the eye of the observer, who sees the light from each coloured stratum as a corresponding half of a circular field of vision. On varying the position of the plungers a series of matches can be made in which both halves of the field of vision appear of equal intensity.

In any such match we may suppose that the light on each side has passed through the same number of molecules of the coloured substance. Then, if the layer of the unknown concentration is 10 mm. and that of the known 20 mm., it may be assumed that the concentration in the solution for analysis is twice that of the standard solution, or generalising,

$$\frac{c}{c_1}=\frac{d_1}{d} \quad \text{or} \quad c=\frac{c_1 d_1}{d},$$

in which c is the concentration of the observed solution and c_1 that of the standard, d_1 and d being the corresponding stratum depths giving equal light intensities.

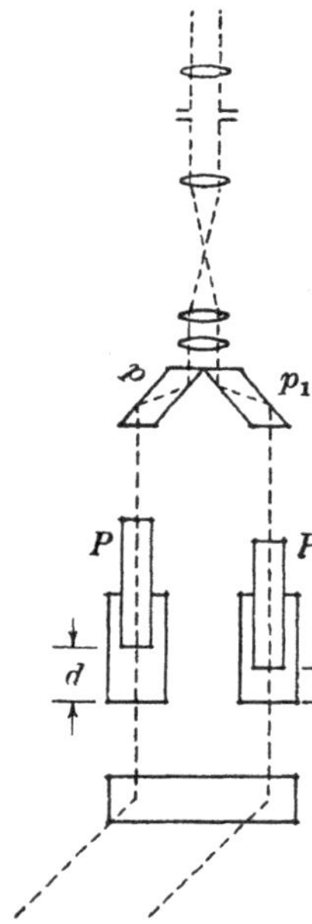

FIG. 24. Construction principle of the Duboscq colorimeter (from Pincussen's *Mikromethodik*, Thieme, Leipzig).

Actually in the above equation we make four different assumptions, at least three of which need not necessarily be true. It is supposed :

(1) That the absorption of light per unit distance through a stratum of uniform concentration and temperature is proportional to the light intensity at any level of the stratum. The truth of such an assumption is expressed as Lambert's law. If it were incorrect the ratio of the strata giving equal light intensity would not be as 2 : 1, with a concentration ratio of 1 : 2.

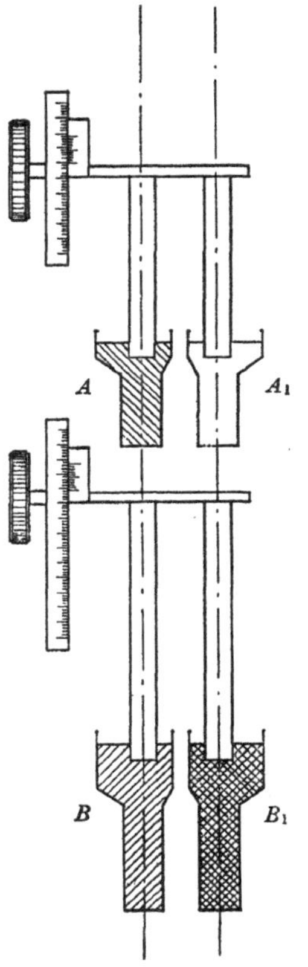

FIG. 25. The figure shows the manner of compensation for coloured solvent as used in the Leitz instruments. If a coloured substance is dissolved in varying concentration in a coloured solvent of constant or known colour value, then the above arrangement in which B is the coloured solvent, B_1 the same containing the coloured substance, A a standard solution of the substance and A_1 distilled water, the effect of the solvent is eliminated. This follows from the fact that in any colour balance the same depth of coloured solvent is looked through on each side.

(2) It is assumed that the absorption of light through any stratum is proportional to the concentraton of the absorbing or coloured substance. The truth of this assumption is expressed as Beer's law, or both (1) and (2) may be grouped together as the Lambert-Beer law.

It not unfrequently happens that Beer's law does not hold strictly, and occasionally the deviation therefrom is gross.

(3) It is supposed that the coloured substance is the only substance dissolved in exactly similar solvent in both solutions, or that any extra material does not affect the optical properties of the coloured substance or solvent. This is seldom strictly fulfilled, but often sufficiently for the most accurate analysis. Gross errors can arise by assuming the similarity of solvents when in fact suspended material or a second coloured substance may be present in the solution investigated.

(4) It is supposed also that the solvent is not itself coloured and that under the colorimetric conditions it does not absorb any appreciable amount of light.

Difficulties arising from points (3) and (4) are generally removable by the procedure of compensation. How this is applied may be seen from Fig. 25, showing the principle as used in the Leitz colorimeters. In B is placed the solvent only and in B_1 the solvent containing the coloured substance for analysis.

A contains a solution of the substance in water of known concentration and A_1 contains water. The plungers in A and A_1 move together as also those in B and B_1. When a match is made it is obvious that a compensation is likewise effected, the light passing through the same amount of water and of solvent on both sides.

Deviations from the Lambert-Beer law can only be avoided in ordinary colorimetry by having the concentration of the standard solution close to that of the observed solution, and in general it is found inadvisable to have it exceeding the range of 0·3 to 3 times.

Such difficulties or restrictions are removed in the practice of spectrophotometry.

Monochromatic light and spectral filters. Monochromatic light may be used in ordinary colorimetric work with advantage. Such light could be introduced at the light source, but it is simpler and equally effective to remove all but the given wave band by a spectral filter inserted, as in Fig. 27, in the path of the mixed rays after absorption and before entering the eye. The spectral filter is a small circular disc of coloured glass with uniform thickness. A number of such (about 10–12) covering the visible spectrum are mounted on a disc, which can be moved alternately into position. These filters allow light to pass of maximum intensity at some particular wave-length and with diminishing intensity on each side, so that at a range of the order of 40 $m\mu$ the intensity is about one-tenth that at the centre.

In order to secure a readily measurable absorption of light, a filter is chosen which has a colour towards the opposite end of the spectrum to that of the coloured solution, or more specifically one which corresponds to the region of maximum absorption.

Even with ordinary colorimetry the use of monochromatic light secures the following advantages. It simplifies the visual judgment, since on moving the plunger up and down the colour observed is darker or brighter, the judgment being concerned only with luminosity, though of a particular colour. With ordinary lighting as the stratum depth increases the total luminosity decreases, but the colour observed appears to grow in intensity. It also allows a certain range of colours to judge from, and we can change a comparison of yellow lighting to one in the blue or green region of the spectrum. Furthermore, it will occasionally permit the elimination of an interfering coloured substance by the use of a wave band in which this shows no absorption.

Beer's law will be also more strictly fulfilled using monochromatic light.

Points in the use of the colorimeter. Before use the cups should be well cleaned, particularly the underneath surfaces, and the plungers, etc. freed from dust. In the more modern instruments the prisms are inclosed and protected, as also the front of the colorimeter when not in use or when a comparison is being made. The requisite cups are then filled with the observed solution and the standard, the arrangement in Fig 25. being adopted when the compensating device is used. All cups not containing standard, observed solution or solvent, should be removed if this can be done similarly on each side or else contain distilled water in which the plungers are immersed. The plungers in the observed solution and standard are now turned down until they touch the glass. If the scale readings are not exactly zero they should be adjusted thereto. With the plungers at zero the eyepiece is focused so that the dividing line is made as sharp as possible and the lighting arranged until equal luminosity appears in both halves of the field. The adjustment is made with the front of the instrument protected with its shutter or alternatively in an evenly illuminated part of the room. The lighting is very easy to regulate in the newer instruments with the lamp incased, fixed to the instrument and provided with centering or other adjusting devices.

In making the comparison, depths below 5·0 mm. should be avoided if possible. There appears to be a difference of opinion as to the most suitable number of readings to take with the one solution, and the author prefers to take at least five. When the reading of one solution is finished it is essential that the lighting be again adjusted with the plungers at the zero position.

In using micro-cups (1 ml.) a small air-bubble is occasionally trapped underneath the plungers and can give rise to very erroneous readings. This should be borne in mind when lowering the plungers into the solutions.

Spectrophotometry. This is the measurement of the relative diminution in intensity of monochromatic light on passing through a coloured or absorbing stratum.

We may consider in the colorimetry previously described that a spectral filter is in use, then if instead of the standard solution and plunger on one side we had some means of progressively and measurably diminishing the light until a match is obtained with a given stratum depth on the other side, we could obtain a measure of the ratio in which the light intensity for a given wave band has been diminished by the coloured solution. This would exemplify the principle of spectrophotometry, and different methods have been used for producing and measuring the light diminution. In the simpler types of spectrophotometers in use to-day for analytical work the light intensity is diminished by a progressive constriction of the light orifice, though this in turn allows a variety of applications. The use of polarised light and the rotation of a Nicol prism has also been used in physical investigations. Apart from such devices for diminishing monochromatic light in a measurable way over the visible spectrum, the same effect could be produced by retaining the Duboscq principle but using instead of standard solutions of the substances investigated, one which has a constant absorbing power over the whole spectral range. Such a universal standard is approximated to by the *grey solution* of Thiel.

Instead of judging visually a similarity of illumination in each half of a field of vision we may suppose the light on passing through the layer of coloured fluid to impinge on a selenium cell and that this is then balanced to a null point on a galvanometer by a second cell (illuminated for convenience by similar monochromatic light not passing through the coloured layer, but modifiable in intensity by an adjustable diaphragm).

On allowing the beam illuminating the first cell to pass through a similar layer of water instead of the coloured solution, the degree to which it must be reduced to balance the first cell by a diaphragm controlled by a drum, will give a measure of the diminution in intensity produced by the coloured layer.

In this lies the principle of such photoelectric apparatus as that of the Spekker Absorptiometer.

Using these spectrophotometric methods deviations in the application of the Lambert-Beer law are of no consequence, when after an initial observation of the light diminution in unit stratum the concentration is read off from a curve originally constructed for solutions of known concentrations. We not only secure this advantage, but after such a curve is drawn, any further labour in the making of standards is removed.

In the actual plotting of such curves the relative diminution of light intensity is not used directly, but rather a function of this termed the extinction coefficient, for it is the extinction coefficient and not the light intensity that goes in linear proportion to the concentration when the Lambert-Beer law applies. How these quantities are related is considered in the following account.

Relations of light intensity, extinction coefficient and stratum depth. If we suppose radiant energy such as monochromatic light passing through a coloured fluid, as in a colorimeter cup, with the incident rays perpendicular to the fluid surface, then we may write

$$-\frac{di}{dl}=ik, \qquad (16)$$

where 'i' is the intensity of the incident light, 'l' the length of the absorbing layer and 'k' a constant expressing the absorptive property of the substance in solution.

Lambert's law is expressed by this equation. That the absorption per unit length (measured over a very small or infinitesimal layer) should be proportional to the amount of radiation entering the layer may be interpreted by supposing the radiation made up of units which are absorbed when they impinge directly on molecules, total absorption being therefore proportional at any time to the 'concentration' of the radiation.

The relation may be further expanded to

$$-\frac{di}{dl}=ick', \qquad (17)$$

in which form it may be termed the Lambert-Beer law. Here the absorption per unit length is not only proportional to the radiant energy entering, but also to the concentration of the absorbing molecules, which should apparently follow from our first supposition of moving molecules being struck by the radiation units, since doubling the molecular concentration doubles the number of molecules struck. This at least forms a simple if figurative interpretation, though as noted, the Lambert-Beer law is not always strictly obeyed (for example, it is not valid in the presence of proteins, as in blood serum).

From the first equation we obtain on integrating

$$-\log \frac{i}{i_0} = lk \times 0{\cdot}434, \qquad (18)$$

and if the intensity of the incident light is regarded as unity then

$$-\log i = lk \times 0{\cdot}434$$

$$\text{or} \qquad i = 10^{-lE}, \qquad (19)$$

(where $E = k \times 0{\cdot}434$).

For different concentrations of the same substance and with incident light of unit intensity, then when the depths are adjusted to produce the same intensities, from equation (19)

$$\frac{E_1}{E_2} = \frac{l_2}{l_1}. \qquad (20)$$

From equations (16) and (17) and writing E again for $k \times 0{\cdot}434$, as in equation (19), we have $0{\cdot}434ck' = 0{\cdot}434k = E$, from which and equation (20)

$$\frac{E_1}{E_2} = \frac{c_1}{c_2} = \frac{l_2}{l_1}, \qquad (21)$$

which is the fundamental relation of colorimetry and spectrophotometry. E is termed the *extinction coefficient* and $\frac{E}{c}$ (or E_0) the *specific extinction coefficient.* The extinction coefficient was originally defined by Bunsen as the reciprocal of the depth which absorbed 90% of the incident radiation (with 90% absorption we may write

$$i = 10^{-1} = 10^{-lE}$$

$$\text{or} \quad E = 1/l,$$

which accords therefore with the original definition).

It is convenient to define $E \times l$, the extinction coefficient multiplied by the stratum layer, as E' or the 'extinction'; E is then given by $\frac{E'}{l}$. It is the E' values or 'extinctions' that are marked on the drum of the Pulfrich photometer.

It will be seen that *we can associate such extinctions or extinction coefficients directly with the concentrations to which they are simply proportional, whereas this is not so with light intensities.*

Determination of concentrations by measurement of extinction coefficients. The relation of the extinction (as defined above), the extinction coefficient and the specific extinction coefficient may be written

$$\frac{E'}{l} = E = E_0 c. \tag{22}$$

Since the specific coefficient is a constant for a particular substance and given wave-length and may be determined once and for all, it follows that any method or instrument whereby we can determine E', or E for light of fixed wave-length, gives us a means of determining concentrations without standards.

Because of the detailed literature now supplied by the makers of various photometers or spectrophotometers it is not proposed to describe or compare such instruments here in any detail, apart from the principles used in the Pulfrich, Spekker and Beckman types of instrument.

Determination of extinctions by the Pulfrich Photometer. A special principle of the Pulfrich photometer may be seen from Fig. 26, in which by means of two opposed screws controlled on the one axis by the movement of a graduated drum, the diagonal of a square opening may be progressively diminished and in direct relation to the angular rotation of the drum.

If similar monochromatic light passes through water and a coloured fluid contained in opposite cells, then the emergent light will appear of the same colour, but of different intensities. If this emergent light passes through two such similar square apertures,

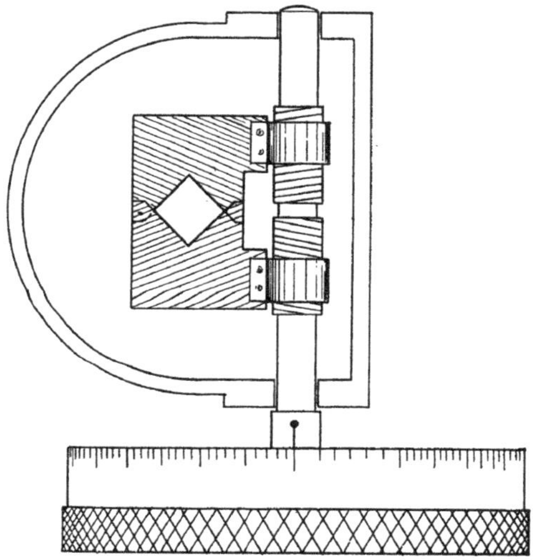

FIG. 26. Diagram illustrating an essential principle of the Pulfrich photometer.

it can be reduced on either side to form equal intensities. On the side containing the coloured fluid the aperture may be fully open or at 360° angle on drum, the required rotation of the other drum

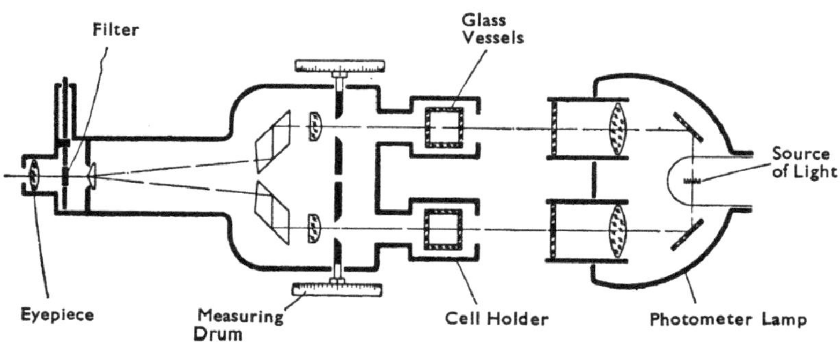

FIG. 27. Diagram of optical arrangement in the Pulfrich photometer.

giving a measure of the light absorption and hence the extinction of the coloured fluid.

Fig. 27 shows how this principle is rendered optically effective in the Pulfrich photometer, the optical components of which comprise a monocular double telescope set at infinity with an axial separation of 70 millimetres. The emergent light from the two cells and the drum-controlled aperture passes through two object lenses, the two beams being afterwards combined into one field of view in the eyepiece by the two separate prisms and the single biprism. The setting is arranged so that the optical axes of both telescopes and hence the aerial images of the square apertures of entry coincide. The images of both the apertures of entry are seen, on looking into the inspection eyepiece placed in front of the ordinary eyepiece, as bright square apertures. Looking through the ordinary eyepiece the two images are received by the pupils of the eye in such a way that two semicircles appear representing each half of the field of vision with a sharp line of demarcation on focusing. The light in each half-field represents the emergent light from the corresponding absorption cell and the spectrum filter.

Relation of drum rotation to light intensity and extinction coefficient. The intensity of the emerging light from the square aperture is proportional to the surface, hence to the square of the diagonal and consequently to the square of the angular rotation, since each angle of this corresponds to an equal increment of aperture diagonal. We have, therefore,

$$i=\left(\frac{a}{360}\right)^2\times\text{a constant.}\tag{23}$$

We can regard the intensity as passing from 100% to zero as the aperture is fully opened and closed, so that when a is 360°, i is 100, and therefore

$$i=\left(\frac{a}{360}\right)^2\times 100.\tag{24}$$

This relation of intensity to a^2 rather than a ensures much greater drum movements for equal concentration increments at the lower levels. The drum is graduated in percentage intensities, and in the newer models in extinctions also. The extinction value is given by

$$\begin{aligned}\log\frac{100}{i}&=E'\\&=El\\&=E_0cl,\end{aligned}\tag{25}$$

where E is the extinction coefficient, l the stratum layer, E_0 the specific extinction coefficient and c the concentration.

Reading E' directly on the drum we obtain E_0cl

$$\text{or} \quad \frac{E'}{E_0l} = c \text{ (as g./litre).} \tag{26}$$

When Beer's law is not obeyed the concentration must be determined from a curve of extinctions against concentrations. Such curves may be used generally and are particularly useful for large serial determinations.

Determination of extinctions by photoelectric apparatus. The optical arrangements and the photoelectric circuit are shown in Fig. 28.

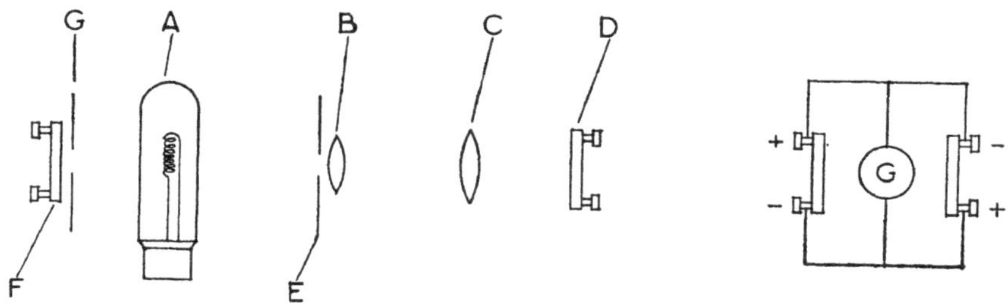

FIG. 28. The optical arrangement and photoelectric circuit of the Spekker Absorptiometer.

A 100-watt projector lamp, A, in the cylindrical lamphouse in the middle of the instrument, and run from the electric mains supply, is the source of light. A lens, B, mounted in the housing to the right of the lamp forms a parallel beam of light ; the beam passes on through the specimen and then falls on another lens, C, which forms an image of the lamp filament on a photocell, D, mounted at the right hand of the instrument. A calibrated variable aperture, E, is mounted immediately in front of the lens system, and enables the intensity of the light falling on the photocell to be varied by known amounts. Since there is an image of the filaments on the cell, there is no change in the cell area illuminated when the aperture alters. The scale associated with the aperture is so calibrated that it reads in ' extinctions ' as already treated.

Light from the lamp also falls on a photocell, F, mounted in the housing to the left of the lamp. The amount of light falling on this photocell (which is protected by 1 cm. water cell) can be varied, by means of an iris diaphragm, *G*, mounted in front of it. The two photocells are connected in opposition across a galvanometer so that, when the photoelectric currents given by the cells are equal, the galvanometer shows zero deflection.

The coloured solution contained in a glass cell is placed in position between the lenses, *B* and *C*, and two similar coloured filters (the colour being about complementary to that examined) are arranged in position on each side of the lamp. With the drum controlling the aperture, *E*, adjusted to zero the iris diaphragm, *G*, is arranged to give no deflection of the galvanometer. A similar cell containing water or the fluid from a ' blank ' analysis is then substituted for the first cell and the drum turned to give again zero deflection. The reading on the drum gives the required extinction.

Such in general is the procedure with apparatus of the kind. For further details the makers' catalogues and instructions sheets should be consulted.

The Beckman Spectrophotometer (Model DU). The main principle in this instrument is the use of a quartz prism to produce a spectrum from which a suitable band is isolated by moving a narrow slit across the spectrum. The resulting light, which is almost monochromatic, passes through the analytical solution and falls on a photoelectric cell. The narrow band of light selected results in better agreement with Beer's law than is possible with filter photometers and also gives a much wider choice of wavelength. The determination of the complete absorption spectrum of the solution is also possible. A considerable advantage of this type of instrument lies in its use for absorptions in the ultraviolet and infra-red, and with the aid of special attachments a wave-length range of 220 to 1000 millimicrons can be covered. Various cells for holding the sample solutions are available, giving stratum depths ranging from 1 to 100 mm. Other accessories include a fluorescence unit and flame photometric attachment.

The range of error in the determination of concentrations using the Pulfrich and Spekker instruments For such absolute colorimetry or photometry, using the Pulfrich photometer, a difference of intensity in the two halves of the field of vision can be detected, with some experience, when one field is about 1 to 2% lighter than the other.

From this we may calculate the following table :

TABLE V

Intensity of transmitted light. i ($= \%$ *of* i_0).	*Extinction E'* $\left(=\log \frac{100}{i}\right.$ *and proportional to the concentration*).	*Percentage error in concentration with an error in light discrimination of*	
		1%	2%
5	1·301	0·3	0·7
10	1·000	0·4	0·9
20	0·699	0·6	1·2
50	0·301	1·5	2·9
80	0·097	4·4	8·9

(A somewhat similar table has been compiled by Weigert (1927), as quoted by Heilmeyer (1929).)

The most efficient range lies between 10% intensity (90% absorption) and 50%. Below 10% the field is becoming too dark, and error from this source increases more rapidly than that of the lessening of the relative percentage error. Higher than 50% the percentage error is too high for accurate working.

These intensities correspond to extinctions of 1·0 and 0·3 respectively, and to 114° and 255° with the Pulfrich photometer, which is roughly one-third to two-thirds the whole drum rotation.

With the Spekker absorptiometer, the major error source would seem on a first consideration to lie only in the reading of the drum. Apart from this drum reading, there is a possibility that the lamp may radiate unequally over the time for adjusting to zero deflection with the two fluids examined.

Concerning the accuracy of the drum reading, Messrs. Hilger state ‘ the reading can be read to an accuracy of 0·005 at a density reading of 0·5. This corresponds with an accuracy of measuring of concentrations of 1% when the concentration and filters are chosen to give this optimum density value.’ Their word density is here equivalent to ‘ extinction ’ as used above.

From an inspection of the drum calibration it would appear that in the region of 0·5 extinction readings could be made to about 0·2 of a division instead of half but the 1% error probably corresponds to the makers’ opinion of the working performance of the instrument as a whole. To test this matter of accuracy for a Spekker instrument and compare it with that obtained for the same stratum layer and fluid in the Pulfrich photometer, a series of readings on the one solution with a specific extinction of near 0·5 was taken using blue filters in each case, and 22 single readings. The coefficient of variation found with the Spekker instrument was 1·0% and for the Pulfrich 1·2%. Two readings on the Pulfrich could be expected to give $1{\cdot}2/\sqrt{2}$ or 0·9% and thus be about equal to the single reading with the Spekker ; but the more objective nature of the results with the electrical instrument may well be preferred, and the fact that numerous readings can be carried out without any sense of visual strain or fatigue.

At the same time it is clear that, at least with this one Spekker instrument, which may presumably be taken as typical, the error of the drum reading is not the major error involved.

PART II

DESCRIPTION OF METHODS WITH THE STANDARD UNITS

CHAPTER IX

AMMONIA GROUP

AMMONIA. GENERAL METHOD USING STANDARD ACID AS ABSORBENT

THE ammonia group of determinations includes not only ammonia itself, but substances which yield ammonia after the action of ferments easily procurable and relatively stable. Such are urea, adenosinetriphosphoric acid, adenosine and adenylic acid. The method is also suitable for micro-Kjeldahl analyses and for macro-Kjeldahl determinations of total nitrogen, and can give an accuracy for the ammonia determination not less than that with the usual 25 ml. technique. There should be no difficulty experienced in attaining accuracies of the order of 0·5% as a coefficient of variation.

General method for ammonia determination (acidimetric procedure). The method which the author has investigated most extensively for minute amounts, and found very satisfactory, consists in the absorption of the ammonia in hydrochloric acid and subsequent titration with barium hydroxide, which latter is always standardised against the acid by preliminary titration. For this titration the methyl red with methylene blue indicator is very sensitive.

For quantities of ammonia *N* greater than about 10 μg (with the standard ' unit ', No. 1, which is always understood in these descriptions, unless the others are specially mentioned) the procedure of absorption in boric acid or other such buffered absorbent, may be preferred for reasons given in the next chapter.

With this latter procedure, and using bromocresol green with methyl red indicator, the titration end-point, though quite satisfactory, has not the same sharpness of colour change as with the first method. Both methods will be described here for the whole ranges to which they may be suitably applied. The following then gives an account of the $Ba(OH)_2 - HCl$ procedure.

The same description applies with minor differences to all ranges of ammonia determination down to 0·1 μg (0·0001 mg.) ammonia nitrogen with the standard ' unit ' (No. 1). The range is suitably covered by three standard acid and alkali solutions whereby 90 to 0·1 γN may be determined with the maximum available accuracy for end-point volumes of the order of 1 ml.

Where the amount exceeds 90 γ per ml. the fluid may be diluted or a smaller volume taken, since there is practically no advantage from the point of accuracy in the determination of larger quantities.

Delivery of the acid. Into the inner chamber of a ' unit ' cleaned according to the procedure described on page 11, is run 1 ml. of the standard acid containing indicator (Solution 1). The acid is delivered from an Ostwald, simple pyrex tube pipette, graduated 1 ml. pipette, or the 2 ml. standard burette, as described in Chapter V. For the finest work the simple tube pipette with slow delivery and comparatively fine point is recommended. Where less than $N/500$ acid is required for the central chamber, the simple pyrex pipette should be used, as previously mentioned. Help in the choice of suitable acid and alkali strengths will be obtained from Table VI.

It is essential that the acid should cover the full surface of the central chamber, but no difficulty will be found here if the ' units ' have been properly cleaned.

Delivery into the outer chamber. For delivery into the outer chamber, volumes may range from 0·1 to 2·0 ml. with the standard ' unit ' (No. 1). These may be introduced as described in the previous chapter, i.e. either by the Ostwald or the graduated pipette or by a simple tube pipette, and here again the simple tube pipette is to be preferred for the finest work. To prevent spattering vigorous blowing should be avoided. Washing out of pipette is to be avoided except for the smallest deliveries (0·1 to 0·2), and then only once or twice, owing to the effect of volume increase on absorption times.

Covering with lid. The lid is smeared with the fixative (described below) and placed on the ' unit '. It is just as convenient to smear the whole lid surface as part thereof.

TABLE VI. *Suitable solution strengths in determining ammonia with the hydrochloric-barium-hydroxide procedure*

Range.	*Acid in central ch.*	*Solution in burette.*	*Each large div. on burette (0·01 ml.) as μg. N.*	*Max. ammonia μg. N.*	*Applications.*
I	1 ml. *N*/150 hydrochlor. ac. (*Vide* p. 95.)	35·7 ml. *N*/10 bar. hyd. diluted to 250 ml.	2·0	93·4	All accurate ammonia determs. beyond about 14 γ *N*. Useful for urine ammonia.
II	1 ml. *N*/1000 hydrochlor. ac. (*Vide* p. 95.)	21·4 ml. *N*/10 bar. hyd. diluted to 1000 ml.	0·30	14·0	Various tissue ammonia determs. Micro-Kjeldahls, etc.
III	0·7 ml. of *N*/5000 hydrochlor. ac. (*Vide* p. 95)	Sols. of Range II diluted five times	0·05	1·96	Blood and tissue ammonia.

The above solution strengths are given as general guides, though many others will no doubt be found useful for the particular purposes in view.

The acid introduced into the central chamber is very conveniently made up for use as described in the text.

When the ammonia to be determined lies beyond 0·09 mg. *N*, the solution may be diluted to fall within range I. No appreciable increase in accuracy will be obtained by using stronger solutions.

Liberation of the ammonia with alkali. The ' unit ' is tilted by resting on a spare lid and the lid displaced just sufficiently to allow the introduction of the tip of a pipette. 1 ml. of saturated potassium carbonate is then usually introduced by a simple tube pipette with rapid delivery. The delivery of the saturated carbonate need be only very approximate, since if less than 1 ml. is delivered there is less volume in the outer chamber, and if greater the mixture is more concentrated in the carbonate, the effects tending to balance out. At the same time it is essential that the carbonate solution be saturated or very close thereto and it is always advisable to have some solid carbonate in evidence in the stock solution. *Sodium carbonate cannot be substituted for the potassium salt unless much longer times are allowed for the absorption.* Alkaline fluids other than potassium carbonate may be found more suitable for special determinations—particularly where

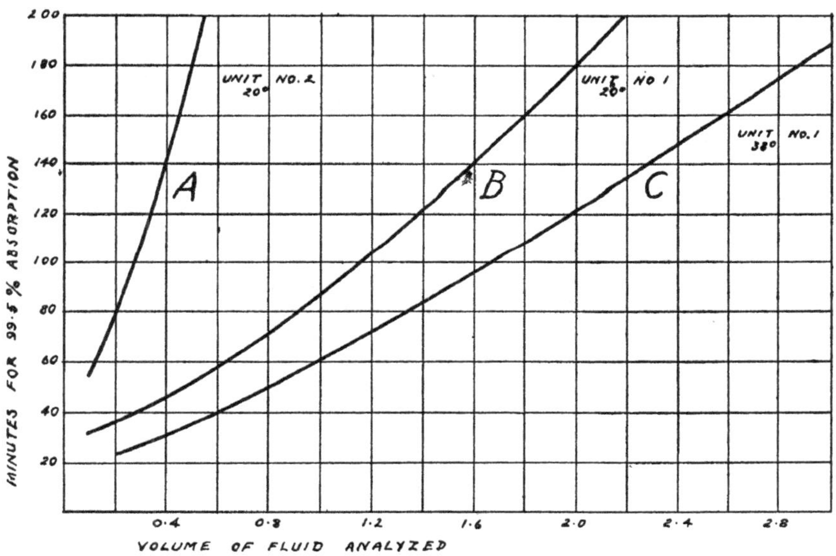

FIG. 29. Times for full (99·5%) absorption in ' unit '.
A.—For ' unit ' No. 2 or 2A. 0·5 ml. sat. K_2CO_3 solution always added to outer chamber.
B.—For standard ' unit ' (No. 1) at 20° C. 1 ml. of sat. K_2CO_3 solution always added to outer chamber.
C.—Same as for *B* but temp. of 38° C.

the ammonia-containing fluid is already strongly acidic, in which case saturated potassium metaborate or 50% KOH may be used.

The absorption period. This varies with the fluid volume that is present before introducing the 1 ml. saturated carbonate as shown in Fig. 29. The figure gives the full absorption periods. (In the original account minimal absorption times were given from equation 3.)

Titration. The acid in the central chamber of the ' unit ' is now titrated with one of the three barium hydroxide solutions of

Table V, according to the range. The end-point may be taken as the complete disappearance of the reddish colour, or the first appearance of a faintly perceptible green (pH$=5 \cdot 5$). A thin pyrex glass rod should be used for stirring. It is essential that a blank determination be done for the smallest ranges since minute quantities of ammonia may be present in the fixative, etc.

Calculation of results. If acid and alkali strengths are those of Table V the calculation of the results becomes very simple, the number of large divisions on the burette being multiplied or divided by simple numerals.

An example may be useful to illustrate the calculation. In this example—as throughout for similar calculations—the ' unit ' containing the fluid and all the required reagents is termed the ' analytical unit,' in contrast to the ' blank unit ' set up as described in the special determinations.

EXAMPLE. Solutions of Range II used (Table V).
1 ml. of $N/1000$ acid in central chamber.
0·5 ml. of fluid for analysis in outer chamber.

Titrations.

Analytical ' unit ' - - -	36·75 large divs.
Blank ' unit ' - - - -	55·30 ,, ,,

Calculation.

$$(55 \cdot 30 - 36 \cdot 75) \times 0 \cdot 30 = 18 \cdot 55 \times 0 \cdot 30 = 5 \cdot 56\,\mu g\ N,$$

$$\mu g \text{ ammonia } N/100 \text{ ml.} = \frac{5 \cdot 56}{0 \cdot 5} \times 100 = 1112$$

or $$\text{mg. ammonia } N/100 \text{ ml.} = 1 \cdot 112.$$

The result must be corrected for pipette and alkali factors when these differ from the nominal volume or strength.

Solutions required.

(1) *Standard hydrochloric acid with indicator.* To cover all ranges three strengths of standard acid are made up (see Table V). One of these is used depending on the range expected. The three acids are $N/150$, $N/1000$ and $N/5000$ hydrochloric acid. (Sulphuric acid could also be used with no appreciable difference at these dilutions.) In preparing these stock acids $N/10$ and $N/100$ should be firstly made up from standard normal acid. In preparing the standard acids with indicator, we will suppose that 500 ml. is the quantity made up in each case.

N/150 HCl *with indicator*. 5 ml. of Tashiro's reagent are run into a 500 ml. flask and 100 ml. of absolute alcohol are added. Distilled water is then run into about three-quarters the volume of the flask and the indicator brought to the neutral point, which usually means dropping in a little dilute alkali until the red colour is discharged. 33·3 ml. of the stock *N*/10 are now run in from a burette and the mixture brought to the mark, etc. The mixture is stored in a pyrex bottle or an ordinary bottle lined with solid paraffin. It will keep practically indefinitely. The distilled water used should be tested for its reaction by some of the Tashiro indicator and—at least for the two weaker acids—should be neutralised to the end-point if necessary. Tashiro's reagent (1932) is made for the above purpose by adding to 200 ml. of 0·1% alcoholic solution of methyl red, 50 ml. of a similar solution of methylene blue (0·1%). Stored in a brown bottle it keeps indefinitely.

N/1000 HCl *with indicator*. The same procedure is followed exactly except that 5 ml. of the stock *N*/10 acid from a pipette is used instead of 33·3 ml.

N/5000 HCl *with indicator*. Here again the same procedure is adopted except that 10 ml. of the stock *N*/100 acid is used.

(2) *Standard barium hydroxide solutions*. These are prepared from the stock *N*/10 barium hydroxide as supplied by the British Drug Houses, and in accordance with the dilutions in Table V. The barium hydroxide stock solution may be run in from a standard 50 ml. burette. All undue exposure to the atmosphere should be avoided, and after making up to the mark and mixing, the solution should be transferred to the reservoir bottle of the burette. Before a titration series the alkali is always standardised against the acid.

For the weaker solutions it is advisable to paint the cork of the reservoir bottle with melted paraffin wax, including the connections with the soda-lime tube, to prevent slow diffusion of atmospheric carbon dioxide into the solution.

(3) *Saturated potassium carbonate*. A very large amount of potassium carbonate is required to saturate water (53·2 g. of $K_2CO_32H_2O$ being contained in 100 g. of the solution at 25° and 60·9 g. at 100°). Excess should be added, and it is convenient to

weigh out 110 g. per 100 ml. water. After dissolving in a large beaker and cooling (it heats on mixing with the water) the saturated solution is decanted into a pyrex conical flask with pyrex stopper, some glass beads are added and the solution boiled for about ten minutes at first vigorously and then gently. The stopper is then placed in position with a small fold of filter paper inserted between it and the flask to allow the air to enter in cooling and prevent the stopper from sticking. After cooling, a little solid carbonate should be in evidence. This boiling of the carbonate solution is necessary to free it completely of ammonia, for potassium carbonate generally contains some small quantity of free ammonia.

Other alkali solutions besides K_2CO_3 may be used and found advantageous for special determinations. Thus, saturated potassium metaborate (69 g. KBO_2 may be allowed to 100 c.c. water) or 40% KOH may be used. With potassium metaborate about the same times are to be allowed as with potassium carbonate, but with KOH 20% increase in these times is advisable.

The fixative. A suitable fixative for smearing on the lid, when using the 'unit' with the $N/150$ acid at room temperature exclusively, is a good quality vaseline. If incubations are carried out at 38° the mixture is hardened by melting 3 parts vaseline with 1 part of paraffin wax (M.P. 55°) and allowing to cool. When using the $N/1000$ and $N/5000$ acids, it is advisable to use a fixative containing only a mixture of pure solid and liquid paraffin melted together and allowed to cool. The proportion may be as follows: 50 g. of solid paraffin (M.P. about 49°) in 80 ml. liquid. This mixture may be varied in accordance with the prevailing laboratory temperature, and is not suitable for incubation.

Apart from the above Kawerau (1941) has found liquid paraffin very serviceable, at room temperature, a few drops being placed on the centre of one glass slide and another slide is placed on the top and drawn across so that two are prepared together. For ordinary room temperatures results with this fixative have been found by Kawerau the same as with vaseline.

Silicone fixative. Silicone fixative (e.g. Dow Corning silicone stopcock grease) has been recently employed in many of the microdiffusion methods for example, by Pirt and Chain (1952) in microhalogen determinations ; also by Feldstein and Klendshoj in the determination of ethanol in biological fluids (1954), etc.

Notes on water-soluble fixatives.

While the above fixatives have been found in practice very satisfactory, a water-soluble fixative may be preferred because of the advantage of easier cleaning. The following is a description of water-soluble fixatives which may be tried :

(*a*) *Water glass fixative.* This has been mentioned to the author by Dr. Silbermann, who has found it very satisfactory. In using it the ordinary commercial water glass for egg preserving may be used. This diluted by adding about one-third its volume of water forms a suitable consistency, and can be applied by a glass rod or small brush around the upper surface of a unit and the lid then placed in position. It is inadvisable to smear it over the whole surface of the 'unit'.

(*b*) *Gum tragacanth fixative.* In preparing this 3 parts of the powdered gum are ground with water which is gradually added until 34 parts are finally included as well as 15 parts of glycerol and 8 parts of saturated K_2CO_3 (all parts being added by volume except the gum, which is added by weight). The whole forms a soft jelly which may be easily spread on the lids.

(*c*) *Gum arabic fixative.* To 10 parts by weight of powdered gum acacia (also termed gum arabic) are added 15 vol. water, 5 vol. glycerol and 5 vol. saturated K_2CO_3. In preparing this fixative, water is added slowly to the powder, the mixture being ground in a large mortar to hasten the solution of the gum. The glycerol and carbonate are then added and stirred in. The mixture may then be used or poured into a large separating funnel and allowed to settle overnight, when the lower fluid can be separated easily from a somewhat frothy upper layer and stored in a stoppered vessel. In using a few drops are smeared over a glass lid, or may be applied as for (*a*) with a glass rod around the rim of the 'unit'.

(*d*) *Glycerol fixative.* Very recently Jørgensen (1956) describes as very satisfactory a mixture of glycerol and $2N$ HCl, 9·1 v/v for CO_2 analysis and glycerol ; water, 9·1 v/v saturated with K_2CO_3 for urea determinations.

CHAPTER X

AMMONIA. GENERAL METHOD

(USING THE BORIC-HCl PROCEDURE)

THE use of boric acid as an ammonia absorbent was first introduced by Winkler (1913–15) and used with the microdiffusion procedure by Abelin (1938). This procedure was further studied by Kawerau (1941) and later by Conway and O'Malley (1942). Boric acid is only one of a number of such buffer absorbents, the theory of which is discussed by the author (1942). It would appear on the whole to be the most useful, though up to about 300 μg NH_3—N half saturated magnesium chloride is about as serviceable. Where the absorbent is titrated electrometrically, such substances as potassium acid phosphate could be employed (Hawes and Skavinski, 1942).

The chief advantage of this buffer absorption with subsequent titration with HCl or H_2SO_4 is that the ammonia is finally determined by a direct titration, and not by a difference in titration. The ammonia therefore can always be measured after absorption, whereas with the other procedure it may happen occasionally that the titre of the absorbing acid is exceeded by the ammonia and the analysis must be repeated.

It is also an advantage that the volume of the absorbing fluid need not be exactly measured, and that it is perfectly stable.

On the other hand the end-point of the titration with the boric or other such absorption is not quite so sharp as in the method previously described, and this difference becomes relatively greater at the lower levels (below 10 μg) when the titrating acid is likewise reduced in strength. It is true that the absorbing fluid could likewise be reduced in strength, but even so at the lowest ammonia levels the barium hydroxide absorption with acidic titration, using the Tashiro indicator is preferable, and some may prefer it even over the whole range of ammonia determinations.

Concerning Abelin's criticism (1938) that there may be a loss of ammonia on titrating with alkali into the central chamber it may be said that this does not occur in the actual practice of a titration, and can only arise with a large rush of the titrating alkali without stirring.

Procedure. The ammonia liberated in the outer chamber by alkali addition in the usual way is absorbed in boric acid mixture plus indicator in the central chamber (slightly reddish), the *p*H of this mixture being approximately 5·0 before absorption. During the ammonia absorption the *p*H of the mixture rises, and to upwards of 8·0 (the strength of the mixture being such as to prevent it rising further). After the required period the fluid in the central chamber is titrated to a faint permanent reddish tint with standard acid. The indicator incorporated in the solution is a mixture of bromocresol green and methylene blue, this being found more advantageous for such titrations than the methylene blue—methyl red indicator, as found by a number of observers ; e.g. Hahnel (1935) ; Prater, Cowles and Straka (1942), Hawes and Skavinski (1942).

At the end of the absorption period the contents of the central chamber is titrated with standard acid from the Conway burette, the end-point being a colour change of green to red (*p*H of about 5·0 to 5·1).

Using $N/50$ acid in the burette, each large division is equivalent to 2·8 μg NH_3—N.

Solutions.

Boric acid. The strength depends somewhat on the range expected. Up to 300 μg NH_3—N, 1·0% suffices. To make a litre of the reagent, 10 g. of the purest boric acid are introduced into a 1 litre flask, 200 ml. alcohol are added, then about 700 ml. distilled water. The boric acid is brought into solution and 10 ml. of the mixed indicator (bromocresol green and methyl red) are added. On mixing, the whole is brought to the desired end-point colour of faint reddish, which usually requires the addition of a little alkali, and the mixture made up to the mark.

Mixed indicator. This contains bromocresol green, 0·033% and methyl red, 0·066% in alcohol. It keeps indefinitely.

Standard acid. $N/50$ is suitable in the horizontal burette for ammonia N up to 300 μg.

Notes on the method.

(1) The lower the boric acid concentration the more delicate the end-point, but when the strength is reduced so also is the absorbing power.

The following equation shows the relation obtaining between the fraction of ammonia not absorbed (U), the volumes in the outer and inner chambers, and the hydrogen ion concentration to which the central fluid is raised :

$$U=\frac{bs}{a}\times\frac{10^{-9\cdot4}}{10^{-9\cdot4}+[\dot{H}]} \qquad (27)$$

where a and b are the volumes in the inner and outer chambers $[\dot{H}]$ the equilibrium hydrogen ion content of the absorbent and s the relative solubility (outer/inner) of NH_3 gas.

Where b and a are equal and s approximately 0·3, as happens with half saturated potassium carbonate in the outer chamber, then 0·01 of the ammonia remains there at equilibrium, 99% being absorbed. As a guide to the conditions of absorption, a pH of about 8·0 is produced by the absorption of 1 ml. of $N/50$ NH_3 in 1·0 c.c. of 1% boric acid.

(2) Where the ammonia absorption is considerable and the titration prolonged, the end-point when reached fades a little. This is due to the evolution of carbon dioxide gas absorbed during the titration. Leaving aside for a few minutes, the true end-point is obtained on titrating a little further.

CHAPTER XI

SPECIAL FACTORS INFLUENCING THE RATE OF AMMONIA ABSORPTION

It is not necessary to know of these factors when carrying out the above determinations in accordance with the instructions; but it is essential when it is required to vary the quantities and conditions in the outer chamber. The influence of fluid volume in outer chamber, of temperature and other factors have been already discussed in Chapter III. Here the special factors of the *p*H value and saline content arise for consideration.

***p*H influence.** If for a given quantity of ammonium salt we consider the rate of ammonia absorption (A_1) at high *p*H values and the rate (A_2) at a *p*H where only a fraction exists as gas, the ratio of these two rates will be the ratio of the tensions of ammonia gas developed. This, however, is clearly the ratio of the degrees of dissociation of the ammonium ions, if we regard these for convenience as dissociating into ammonia gas and hydrogen ion, so that we have

$$A_1 = \alpha A_2. \tag{28}$$

Thus, when the ammonium ion is half dissociated, A_1 is half the full absorption rate A_2. The degree of dissociation or the value of α is given from the equation

$$pH = 9{\cdot}4 + \log \frac{\alpha}{1-\alpha}. \tag{29}$$

From this it will be seen that when $\alpha = 0{\cdot}5$, $\log \frac{\alpha}{1-\alpha} = \log 1{\cdot}0 = 0$ and the *p*H of the solution is 9·4. At *p*H 9·4, therefore, the ammonia will be absorbed at half the full absorption rate (considering such absorption rates without any appreciable salt effect). Similarly at *p*H 8·4 it will be absorbed at 9% of the full absorption rate. Little advantage will be obtained in itself by pushing the *p*H beyond 10·4, when 91% of the full absorption rate is occurring.

It may be noted that the *p*H of normal ammonia solution is approximately 11·6, 99·5% of the total ammonia being present as the free gas, and the *p*H of *N*/10 ammonia is approximately 11·1, with 98·5% as free gas. Such standard ammonia solutions containing the free ammonia may therefore be used in any investigations of absorption rates and various influences thereon without the need of making standard ammonium salt solutions and adding alkali thereto.

The effect of salt addition on the ammonia absorption. Apart from the *p*H effect, the addition of certain salts to ammonia solutions can very greatly increase the tension of the ammonia gas therein, and therefore the absorption rate in the 'units'. Such effects of salt addition on ammonia tensions have been exhaustively investigated by Gaus (1900) and by Abegg and Riesenfeld (1902). The potassium salts were found to be very much more effective than the sodium, and amongst the salts of potassium, the carbonate and the metaborate have the most marked action. The effect of the cations goes in the following series :

$$Mg—Ca—Li—NH_4—Na—K,$$

increasing from left to right, and the first three of these actually produce a *decrease* in ammonia tension. The effect of the anions goes in the following sequence :

$$I—Br—Cl—NO_3—(C_2O_4, CrO_4, SO_3)—HPO_4—F$$
$$—OH—BO_2—CO_3.$$

Table VII gives a comparison of the absorption rates at 38° C. in the 'unit' after the addition of various salts, 1 ml. of the saturated or standard salt solution being added to 1 ml. of a pure ammonia solution. The rate from pure ammonia solutions (2 ml. in outer chamber) is taken as the standard for column 4.

It will be seen that the absorption rate can be trebled by the addition of an equal volume of saturated potassium carbonate or metaborate to the solution in the outer chamber. Potassium fluoride is also very effective.

It may be taken that for every 1% of the saturation value of K_2CO_3 in the mixture there is an approximate increase of 4·4% in the absorption rate, and that at the same time for mixtures of equal concentration the time of clearance from the outer chamber is inversely proportional to the volume therein. From these data it will appear that a rough calculation of absorption times can be made for any concentration of potassium carbonate in the outer chamber and—within limits—for any volume of the mixed fluids.

Potassium metaborate is particularly effective for adding to rather concentrated acid solutions containing ammonia. Here carbonate would cause a marked evolution of carbon dioxide.

The error of the ammonia determinations. The possible lower limit of variable error in such determinations consists in the summation of that involved in the pipetting of 1 ml. acid into the central chamber and of the fluid into outer chamber, and the titration of the acid residue after absorption. A large number of such ammonia determinations (barium hydroxide titration) were carried out with the Bang burette before the horizontal burette was elaborated and gave a standard deviation or coefficient of variation of 0·5%. The results are given in Table VIII, and exceed the possible lower limit of error.* Titrations with the horizontal burette, and using the more refined pipetting already described, would give considerably reduced errors.

TABLE VII

Substance investigated.	*Concentration of substance in mixed fluid at 38°.*	*Absorption rate at 38° (A_1 value, see equation 1 in text).*	*Calculated % increase in ammonia tension in outer chamber.*	*% increase in NH_3 tension found by Abegg and Riesenfeld using 0·5 N sols. of the substances in col. 1.*
KBO_2	½ sat.	0·054	218	24
K_2CO_3	,,	0·050	194	27
KF	,,	0·045	165	19
K_2HPO_4	,,	0·035	106	17
$K_2C_2O_4$	,,	0·024	41	15
KOH	20%	0·041	141	20
NaOH	,,	0·037	118	13
NaOH	0·5 *N*	0·019	12	13
Water	—	0·017	0	0
Glucose	25%	0·014	− 17	—

* 0·2–0·3% under the analytical conditions.

TABLE VIII

Quantity of ammonia-nitrogen added to outer chamber, mg.	*Percentage 'recovered' by absorption in inner chamber.*	*Standard deviation of individual estimation as mg. ammonia-nitrogen.*	*Coefficient of variation of individual estimation.*	*Number of estimations.*
1·40	99·5	$7{\cdot}0 \times 10^{-3}$	0·50	48
0·140	99·5	$6{\cdot}7 \times 10^{-4}$	0·48	33
0·014	99·5	$1{\cdot}0 \times 10^{-4}$	0·72	24

CHAPTER XII

OTHER METHODS FOR DETERMINING THE ABSORBED AMMONIA IN THE MICRODIFFUSION PROCEDURE

Bromometric. Rappaport and Glaser (1932) used such a method for exact blood urea determinations after distilling off the ammonia in a Parnas-Wagner apparatus (1921). Following this Rappaport and Gutman (1935) adapted their procedure to the microdiffusion method subsequent to a reading of the paper by Gibbs and Kirk (1934) on minute urea determinations with a reduced model of the 'unit'. They found it superior to the distillation method.

As used by Rappaport and Gutman, the ammonia is first absorbed by diffusion into 40% sulphuric acid in the detachable chamber of a Widmark alcohol apparatus. This is not nearly as efficient as the standard 'unit', and it is unnecessary to use so strong an acid solution. After the absorption the chamber with acid is removed to a Hagedorn-Jensen tube in which it is neutralised, 5 c.c. of a hypobromite mixture added, and then a tiny crystal of KI and 2 c.c. of a hydrochloric acid solution made by diluting fuming hydrochloric acid with an equal volume of water. The free iodine formed is titrated with $N/100$ thiosulphate in the usual fashion with starch.

Since a very much weaker acid would have sufficed for the ammonia absorption, the subsequent and rather awkward neutralisation procedure described by them could be dispensed with and is not here given.

The essential reactions in the above method are (subsequent to absorption and neutralisation of the acid), according to Rappaport and Gutman :

$$NaOBr + NaBr + 2HCl = Br_2 + 2NaCl + H_2O$$
$$2NH_3 + 3Br_2 + 6NaOH = N_2 + 6H_2O + 6NaBr$$
$$Br_2 + 2KI = I_2 + 2KBr.$$

The reaction between ammonia and hypobromite is, however, more usually given as

$$2NH_3 + 3NaOBr = N_2 + 3NaBr + 3H_2O$$

(e.g. Mellor, 1928).

Such a method would in no way increase the accuracy of the methods here described for solutions of Ranges I and II (HCl absorption), and would only lengthen the procedure and the making of solutions. When using solutions of Range III it may possibly be of use, but owing to the necessary increase in volume and the use of many more reagents the writer would prefer the direct barium hydroxide titration with the methyl red–methylene blue reagent for such minute ammonia absorptions.

The hypobromite solution used in the above method is made as follows :

(*a*) *Bromine solution.* 20 g. KBr are dissolved in 100 c.c. of H_2SO_4, then 8 g. bromine (2·5 c.c.) are added, the solution shaken and made to mark with water. It keeps for a long time in the cold and stored in a brown bottle.

(*b*) *Buffer solution.* 84·5 g. boric acid and 14·6 g. sodium hydroxide are dissolved in about 800 c.c. distilled water, boiled for 30 minutes to drive off any ammonia present and after cooling filled to the 1000 c.c. mark.

Before the experiment, to 10 c.c. of the bromine solution in a 50 c.c. volumetric flask, dilute sodium hydroxide is added till the colour changes from brown to lemon yellow ; then about 25 c.c. of borate buffer are added and distilled water to the mark. This hypobromite solution should be made up fresh before using.

Colorimetric or spectrophotometric. After the ammonia absorption in dilute acid the procedure may be readily adapted to colorimetric or photometric examination. The titration method is much preferable on grounds of accuracy, and also of convenience when the procedure has been going as a routine. For the minutest amounts (Range III), however, the spectrophotometric method may be preferred by some to the titration procedure, since it is not so dependent on a high grade cleaning of the ' units '.

The spectrophotometric procedure has in this way been used by Borsook. He used a ' unit ' similar in dimensions to the standard design, and the phenate hypochlorite reagent as described by Van Slyke and Hiller (1933), instead of the Nessler reagent.

In using these procedures, the problem is merely one of determining the ammonia in a pure weak acidic solution. The acid for absorption may be chosen as $N/100$ or $N/200$ sulphuric acid, without indicator. The Nessler reagent, if used, should not be directly added to the acid in the inner chamber, owing to possible loss of ammonia. It may be pipetted out and made up with washings to 10 ml. in a small cylinder and 0·5 ml. Nessler reagent added, the reading with the Pulfrich photometer being then made with filter *S* 43, and using Figs. 14 and 15 in Krebs' description of Cleghorn and Jendrassik's method (1934) for residual nitrogen in blood.

The extinction figure on the instrument is divided by the stratum depth used and the abscissa corresponding in mg. % residual nitrogen in these diagrams divided by 2 gives the microgramme ammonia *N* in the 1 ml. of fluid examined.

For the very lowest ranges the central fluid may be 1·0 of pure acidic solution of $N/1000$ strength or less. 0·9 ml. is carefully pipetted out in the end into a small tube and 0·15 ml. Nessler added. With this mixed fluid the micro Pulfrich cells of 1 ml. capacity and 50 cm. stratum depth are filled and the extinction determined and compared with that for standard solutions similarly treated. In this way ammonia *N* down to 0·1 to 0·2 γ/ml. should be determinable with an accuracy at about the 5% level. Borsook has in fact estimated 0·5 γ with an error not greater than ±2% (1935).

Apart from such spectrophotometric determinations, ordinary colorimetry can of course be used in which the Nesslerised solution is directly compared with a similarly treated standard.

Van Slyke and Hiller have revived the use of the blue colour produced by ammonia with an alkaline solution of phenol and hypochlorite for use in blood ammonia determinations, and it has been used by Borsook (1935) for micro-ammonia determinations.

For quantitative determinations, it was first used by Thomas (1912, 1913), who attributed its discovery to Berthelot.

Van Slyke considers it to be more delicate than the Nessler

reagent, and its greater suitability for minute quantities has been confirmed by other workers.

Crismer (1937) has modified the Van Slyke and Hiller procedure by using chloramine instead of hypochlorite. Russell (1945) has deepened the colour of the reaction considerably by conducting it at a lower *p*H and catalysing with manganese ions.

Preparation and use of the Nessler and phenate-hypochlorite reagents ; Van Slyke and Hiller procedure.

The Nessler reagent. This may be prepared by grinding 10 g. mercuric iodide (HgI_2) in a porcelain mortar with a little water, the mixture being rinsed into a dark bottle and the remainder of the water (100 ml. in all being used) in which are dissolved 5 g. potassium iodide and 20 g. pure NaOH, prepared from metal, added. The precipitate will settle completely within 24 hours, so that the perfectly clear solution can be decanted off. The reagent should be kept in a well-stoppered dark bottle.

The phenate-hypochlorite reagent.

Van Slyke and Hiller procedure (1933).

Sodium phenate reagent. This is made by dissolving 25 g. of phenol in a small quantity of water, adding 50 ml. of 40% NaOH and diluting to 100 ml. with ammonia-free water.

The hypochlorite solution (*Javel water*), containing 1 g. of Cl per 100 ml. Dissolve 50 g. of calcium hypochlorite, having 56·5% available Cl, in about 500 c.c. of hot water. Mix this solution with one containing 50 g. of anhydrous potassium carbonate in 200 c.c. of cold water. Make the whole to a volume of 1 litre. Test the clear solution for excess calcium with K_2CO_3 and add more of the latter if necessary to precipitate all the calcium. This reagent should be kept in the ice box in a number of small bottles, so that the chief source of supply is not opened each day, and so that there is a minimum of replacement of solution by air as the reagent is used. The active chlorine is estimated before use and at intervals of a few weeks as follows : To 5 c.c. of the Javel water are added 25 c.c. of water, 2·5 c.c. of 10% potassium iodide, and 2 c.c. of glacial acetic acid. The mixture is then titrated with 0·1 *N* thiosulphate which has been freshly standardised against 0·1 *N* potassium iodate solution.

In using the reagent, to each 5 c.c. of dilute acid containing the ammonia 1 c.c. of the sodium phenate solution is added and 0·5 c.c. of the Javel water. Mix, place in a boiling water bath for 3 minutes, cool to room temperature and compare with standard in a colorimeter.

The heating in above procedure was modified by Borsook (1935) to 50 minutes at 37° C. Crismer (1937) used 5% chloramine instead of the hypochlorite solution, combining 2 ml. of the solution for examination with 1 ml. of the phenol solution and 3 ml. of the chloramine and heating in a boiling water bath for 10 minutes. After cooling, readings were taken with the Pulfrich photometer using filter S 61.

Russell's procedure with the phenate-hypochlorite reagent (1945).

1. *Alkaline phenate solution.* 25% phenol in 2·7 *N* NaOH. Mix 25 g. of crystalline phenol with water, add, with stirring, 54 ml. of 5·0 *N* NaOH to make to 100 ml. Preserve in brown bottle in refrigerator.

2. *Hypochlorite solution.* Grind and sift 25 g. of Ca hypochlorite (bleaching powder, U.S.P.) and dissolve it as far as possible in 300 ml. of hot water. Add with stirring 135 ml. of potassium carbonate solution (20 g. of anhydrous salt in 100 ml. solution, previously boiled to free it of ammonia). Mix thoroughly, heat briefly to about 90° C., cool, and make to 500 ml. Filter a small portion of the mixture and test for Ca ions ; if the test is positive, add more carbonate to the mixture until a negative test is obtained. Filter the mixture and store in the refrigerator. (This solution should be water clear and contain 1·30 to 1·40 g. of free Cl per 100 ml.)

Test for Ca *ions.* To 1 ml. of the solution add a little of the K_2CO_3 solution and heat in boiling water a few minutes. The solution should remain crystal clear in the absence of Ca ions.

Test for free Cl. To 2·00 ml. of the hypochlorite add 10 ml. water, 2 ml. 5% KI and 1 ml. glacial acetic acid. Titrate with 0·1 *N* sodium thiosulphate (starch indicator) ; 7·5 to 8·0 ml. of thiosulphite should be required. Re-test the strength of the solution occasionally.

3. *Manganese solution.* Manganous chloride or sulphate 0·003 *M*.

Colour development.

Into a calibrated test tube or colorimeter tube is introduced 1·5 ml. of the sample containing 0·5 to 6 μg NH_3-N, one drop of the manganous salt solution, 1 ml. of the alkaline phenate reagent and 0·5 ml. of hypochlorite solution.

The two latter solutions should be cold when added to the sample. Loss of ammonia from the alkaline solution may be further reduced by keeping the sample tubes in an ice bath during their preparation. The contents of the tubes are mixed by gentle rotation, and placed immediately in a briskly boiling water bath for about 5 minutes, then cooled and made up to a convenient volume (e.g.—6 or 10 ml.). Russell then recommends reading in a photoelectric colorimeter, with a filter having an absorption maximum near 625 $m\mu$. Fe, Cr and Mn ions catalyse the reaction but the Mn^{++} ions are best for the purpose.

It may be noted that 0·1 μg $NH_3 - N$ in the sample gives a detectable colour, and 0·5–0·6 μg $NH_3 - N$ may be accurately determined.

The colour so developed has a smooth reproducible absorption curve, with a maximum at approximately 625 $m\mu$. It is three times as intense as the colour developed with similar ammonia quantities using previous modifications of the Berthelot reaction.

The sample for pipetting may be increased to large volumes if more ammonia is present than considered above, the reagents being increased in proportion. Also, the sample should be either neutral or with an acidity not greater than 0·01 or 0·02 N.

When the procedure is used with the microdiffusion technique described here, the contents of the inner chambers are convenently removed with a bulb pipette into the graduated tubes.

CHAPTER XIII

AMMONIA. BIOLOGICAL DETERMINATIONS

(1) Ammonia in urine.

0·5 to 1·0 g. of ammonia nitrogen are excreted per day by the normal human subject, but as high as 12 g. have been observed with diabetics. Ammonia must be formed by the kidney since, as shown by the author, the amount in the blood of the normal resting subject is below the analytical level (1935). The rate of its excretion is independent of the urine volume, but is closely related to the excretion of acid and also related to the total nitrogen output. This has been expressed in the following form by Rafflin (1926).

$$\frac{p\mathrm{H}}{4} + \log(\mathrm{NH_3}) - \log(\text{total N}) = \mathrm{K}.$$

Hasselbalch (1916) considers the urinary ammonia nitrogen expressed as a percentage of the total nitrogen at a *p*H of 5·8 (' reduced ammonia coefficient ') to be a reliable method of measuring acidosis and the ability of the individual to combat acidosis. The percentage at *p*H of 5·8 varies from 2·2 to 5·5, but is much more constant for the individual.

Principle of determination. The ammonia concentration of urine is comparatively high (but goes inversely as the rate at which urine is excreted by the kidneys). The method with the boric reagent is very suitable, as it gives results over the widest ranges.

Procedure. Into the inner chamber 1 ml. of the boric reagent with indicator is roughly pipetted, and 0·2 ml. of urine is carefully pipetted into the outer chamber. The lid smeared with fixative is placed in position. 1 ml. of saturated potassium carbonate is then introduced in the usual way. The ' unit ' is left on the bench for about 40 minutes when the contents of the central chamber are titrated with *N*/50 HCl from the Conway burette.

Calculation. Each large division on the burette corresponds to

2·8 μg NH_3 – N, so that multiplying the number of large divisions by 1·4 gives the milligrammes NH_3 – N per 100 ml. urine.

If the urine is very dilute, 0·5 ml. may be used in the outer chamber—and the ' units ' left about ¾ to 1 hour before titrating.

(2) The Blood Ammonia.

Principle and introduction. The determination of the free ammonia of normal circulating blood has long been regarded as particularly difficult. The early literature, dating from the researches of Nencki and co-workers, has been reviewed by Parnas and Heller (1924), Parnas and Klisiecki (1926), and the subject has been reviewed by Schneller (1928), in the *Ergebnisse*, and by Stanoyevitch (1938).

Folin and Denis (1912) were the first to find values under 0·1 mg. ammonia N/100 ml. If the special conditions that affect determination be not taken into consideration it is possible to obtain results as high as 2·0 mg./100 ml., and many published figures are of the order of 0·3 mg./100 ml. Table IX gives a brief summary of the data published over the ten years prior to the microdiffusion determinations, which are included at the end of the table.

TABLE IX

Observers.	*Date.*	*Human blood ammonia (Venous) µg. ammonia* N/100 *ml.*
Parnas	1925	30
Van Slyke *et al.*	1925–26	0–100
Parnas and Klisiecki	1926	26–15
Labbé, Nepveux and Hejda	1929	30–72 mean 47
Stanoyevitch	1931	19–42 mean 26 } men 17–30 mean 23 } women
Folin	1932	60–116 (after 30 mins.) 75–171 (after 1 hour)
Fuld	1933	28–60 mean 42
Kalk and Bonis	1933	10–45
Conway; Conway and Cooke	1935–39	0 (Immediate) 44 (mean, 5 mins. after shedding) 70 (mean, 60 mins. after shedding)

It will be seen that with the microdiffusion method there would appear to be no free ammonia in normal circulating blood, which is to be interpreted in the sense that the amount is below the analytical level.

In the microdiffusion method all aeration and distillation are avoided. Numbers of determinations can be rapidly carried out so that the true curve of ammonia formation from zero time of shedding can be readily obtained. From zero up to five minutes after shedding (see Fig. 31) the ammonia content of human blood rises in a steep curve up to a mean of 46 μg ammonia $N/100$ ml. Fig. 31 (*a*, *b*, *c*) illustrates the course of ammonia development in single samples (man, rabbit and fowl). The first stage in ammonia development in shed blood has been termed the ' alpha ' ammonia by the writer and forms on the liberation of carbon dioxide. From five minutes onwards there is a slow formation of ammonia (Fig. 31) representing at room temperature (16°) about 0·43 μg $NH_3-N/100$ ml. blood per minute for the human subject. The rate of formation is considerably affected by temperature, being 0·97 μg $NH_3-N/100$ ml. per minute at 24° C. This ' beta ' ammonia is very largely accounted for by the breakdown of adenylpyrophosphate, mainly through adenosine in shed blood with escape of CO_2, but mainly from adenylic acid if CO_2 tension be maintained. A third form derives from a precursor in plasma and corpuscles and its formation is best studied in separated plasma in which no ammonia derives from A.T.P. or adenylic acid after shedding. Evidence was given (Conway and Cooke, 1939) that this ammonia derived from adenosine after dephosphorylation of a substance different from muscle adenylic acid and which appeared to be vegetable adenylic acid. It is of interest that Archibald (1944) has shown that part of this ammonia moiety is due to glutamine.

In recent years 1952–56, a number of papers have appeared in various clinical journals in which blood ammonia levels were investigated in relation to hepatic disease, the microdiffusion procedure being used for the ammonia determinations.

White, Phear, Summershill and Sherlock (1954) from observations on 21 control subjects without hepatic disease (including 15 normals) support the author's conclusion that no measurable free ammonium exists in normal circulating blood, but that free ammonium appears rapidly after shedding in air. After three minutes they find an average of 0·79 μg/ml. ammonium *N* with 2 control patients. This is somewhat higher than we obtained, but their procedure was a little different and allowance for ammonia formation by the alkali action *per se* does not appear to have been made. For comparison with levels in hepatic disease, this is of no importance.

Using CO_2 to inhibit ammonia formation as described here, Singh, Barclay and Cooke (1954) find the mean normal blood ammonia level to be $10{\cdot}16 \pm 3{\cdot}15$ μg $NH_3 - N$ per 100 ml.

The blood ammonia determinations for clinical purposes are now usually carried out on blood after shedding in air, or without a CO_2 atmosphere provided, the 3 or 5 minute period after shedding being selected. Thus it is the ' alpha ' ammonia level that is determined. This is generally accepted as being of the order of 50 micrograms *N* per 100 ml. (McDermott, Adams and Riddell, 1954.)

It is reasonable to assume that a significant elevation of the blood ammonia over the normal ' alpha ' level is due to so much free ammonia in the circulating blood (and this could be readily certified) and small amounts of ammonia added to shed blood are quantitatively recoverable.

The ' alpha ' ammonia level is strictly measurable after extrapolating back the line of relatively slow ascent of ammonia formation in the ' beta ' stage, and the ' alpha ' level is that where this line cuts the ordinate, but in fact the level after 3 or 5 minutes is so close to this value that no practical difference arises on taking such to represent the ' alpha ' or first stage ammonia. It may be also pointed out that the extra free ammonia formed during the determination by the action of the alkali *per se* on blood constituents should be subtracted (in accordance with the procedure described later) if a true estimation of this ammonia moiety is required but this is not important when difference in levels between normal and diseased subjects is being considered.

Koprowski and Uninski (1939), using the microdiffusion technique described here, have shown that the ammonia content of dog's blood and its development after shedding are similar to that for the human subject. Florkin and co-workers (1938–1940) also with the same technique have made observations of biological interest on invertebrate blood. They have found a relatively high blood ammonia in the lobster, crayfish and snail (0·7–2·0 mg. $NH_3 - N$/100 ml.) which remained steady for 30–80 minutes after shedding. There was much less ammonia in the blood of the mussel (0·05 to 0·07 mg./100 ml.) and its value after shedding shows a curious augmentation and diminution. The blood of insects (dytiscus, hydrophilus) like that of man, dog, etc., had no measurable ammonia content. After shedding there was a rapid rise to high levels (2–3 mg. $NH_3 - N$/100 ml.).

In the formation of the ' beta ' ammonia after shedding a curious phasic formation is evident for human blood at 38° (Conway and Cooke, 1939) and this phasic formation occurs for the rabbit's blood at room temperature, the rate of formation being here much more rapid than for man. The phasic formation appears generally characteristic as it is evident also in fowl's blood.

That preformed ammonia is present in measurable amounts after exercise and in certain diseased conditions is very likely, but has not been investigated so far by the microdiffusion technique. Whatever stage of formation it is desired to examine or whatever conditions with regard to their influence on the preformed ammonia, the procedure here described will be found suitable, with slight alterations according to the conditions.

Referring briefly to other methods of blood ammonia determinations, it may be said that these are either developments of the original Folin technique (e.g. Nash and Benedict (1921), using special absorbers) or a steam distillation method elaborated by Parnas and Wagner (1921). The ammonia for 5 ml. blood—or 2 ml. as with the Parnas-Wagner method—is finally present in about 4 or 5 ml. fluid. It is then colorimetrically determined after adding Nessler's reagent. Considerable differences exist in the reports of the delicacy of this reagent for such a purpose. Van Slyke and Hiller (1933) state that 0·001 mg. ammonia nitrogen (the ammonia from 5 ml. blood containing 0·020 mg. ammonia N/100 ml.) in 5 ml. fluid is not visible at all with the Nessler reagent. Parnas and Heller, on the other hand (1924), claim to be able to detect the difference between 0·001 and 0·0009 mg.

ammonia N in 5 ml. (corresponding to the ammonia from 2 ml. blood containing 0·050 mg./100 ml.). This latter would in any case seem quite workable in general with the spectrophotometer.

Van Slyke and Hiller (1933) made use later of the phenate hypochlorite reagent as already described.

In the method to be described (*following the general ammonia determinations of Range III*), using the hydrochloric-barium-hydroxide procedure, the ammonia is finally present in 0·7 ml. of a 0·0002 N hydrochloric acid solution in which is incorporated the indicator as described. The titration error here sets the limit of delicacy for the determination since no appreciable error is introduced by the ammonia absorption itself. The titration error of 0·7 ml. $N/5000$ acid (with indicator, see below) with 0·9 ml. final fluid corresponds to 0·02 $\mu g.$ ammonia N as a standard deviation and for 1 ml. of original fluid investigated is equivalent to 0·002 mg./100 ml. When blood is used and the absorption time is limited the error is 0·003–0·004 mg./100 ml. This is approximately five times more delicate than the method of Van Slyke and Hiller (1933) and permits the same accuracy using only 1 ml. blood or 0·5 ml.—as will appear—instead of five.

In the ordinary use of the ' unit ' the absorption at room temperature takes two hours under such conditions, which is too long a period for blood ammonia determinations as there is a steady formation of ammonia in the presence of the alkali. *When we consider, however, that each ' unit ' is turned out from the same mould with the same dimensions and surface relations, and further, that since after the first few seconds the fluid surfaces are perfectly still, a uniform passage of ammonia may be expected and in fact occurs. An error of less than 5% is introduced by cutting short the absorption to about the half period, but less than 1–2% when allowed to go to 83% absorption (as with 0·5 ml. blood and 0·5 ml. saturated carbonate for 20 minutes).* Whatever accuracy may be lost here is more than compensated for in comparison with other methods by the entire freedom from contamination during the absorption period. In 10 minutes approximately 42% of the ammonia is absorbed at room temperature with 1 ml. blood plus 1 ml.

carbonate (83% in 20 minutes with 0·5 ml. blood and 0·5 ml. carbonate). The ammonia formation in the presence of the alkali over this period is comparatively small.

Procedure.

Preparation of 'units'. The whole success of the method depends on the 'units' being properly cleaned in the manner already described. The reagents of Range III (Table V) are used with this method and the number of 'units' required are prepared up to the point of introducing the blood into the outer chamber and left aside for about twenty minutes before using.

The saturated carbonate may be introduced before or after the introduction of the blood into the outer chamber, depending on the speed with which the determinations are required after shedding. If required immediately after the blood is shed it is convenient to have the 'units' prepared with all additions—acid and carbonate and lid in position—thus allowing for a very rapid blood determination. (If the determination is not required until five or ten minutes have elapsed after shedding, it is better to add the blood first to be followed by carbonate.) Allowing twenty minutes after the 'units' are prepared and before blood introduction will ensure the absorption of small traces of ammonia that may possibly exist in carbonate and fixative.

To obtain the maximum possible accuracy, and to ensure completely that no contamination of any kind will interfere during the absorption period, a number of 'units' may be prepared with every addition except the blood and left aside for twenty minutes or half an hour. At the end of this period the lids are detached in succession, the central acid being titrated to neutrality, the mixture being then sucked out—using a filter pump and pyrex tube—a refill of acid made, and the lid replaced.

The acid is introduced as 0·7 ml. from a pyrex tube pipette. This amount will be found to cover the surface of the internal chamber, when the 'units' have been carefully cleaned.

Collection and introduction of the blood. The manner of collection depends on the object of the determination. When human venous blood from the normal resting subject is received into alveolar air there is no apparent free ammonia (Fig. 30).

If the blood is allowed to drop through the air into a container ammonia forms, as already described, to reach in about five minutes (Fig. 31) a level of 0·46 'micrils' NH_3-N, rising then slowly from this point onwards.

If the ammonia content after three or five minutes from shedding is required—and this will correspond to the other methods at present in use—a few ml. of blood are taken from an arm vein in a

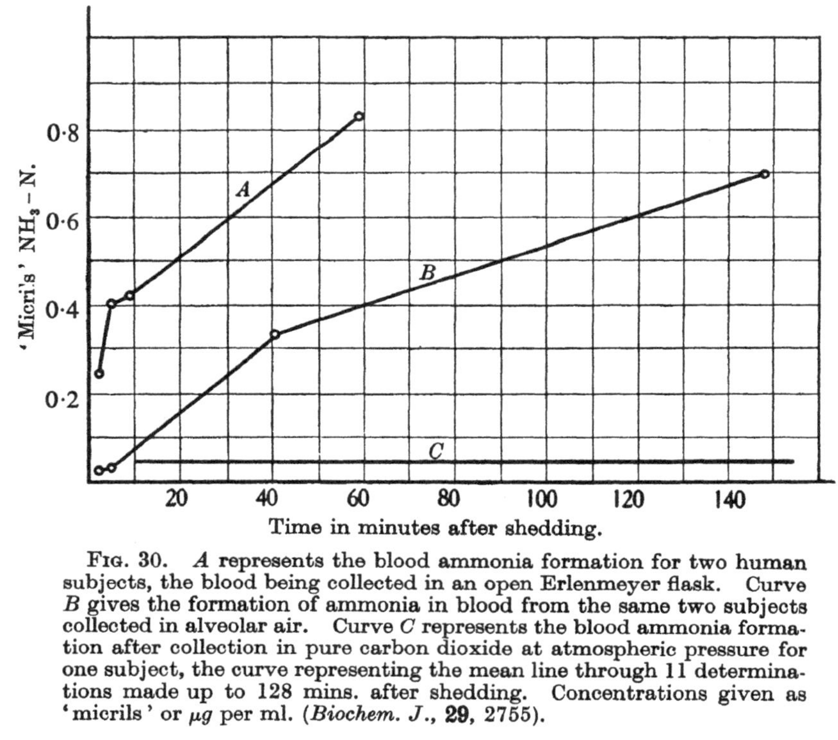

FIG. 30. *A* represents the blood ammonia formation for two human subjects, the blood being collected in an open Erlenmeyer flask. Curve *B* gives the formation of ammonia in blood from the same two subjects collected in alveolar air. Curve *C* represents the blood ammonia formation after collection in pure carbon dioxide at atmospheric pressure for one subject, the curve representing the mean line through 11 determinations made up to 128 mins. after shedding. Concentrations given as 'micrils' or μg per ml. (*Biochem. J.*, **29**, 2755).

syringe specially cleaned, and then this blood is introduced into a small tube or flask with a little ammonia-free potassium oxalate.

When a large number of determinations are in prospect on the blood sample and 10–25 ml. are required, the blood may be allowed to flow directly from the vein through a hypodermic needle (the syringe being removed) into a small flask containing a little oxalate. A rubber tourniquet tied around the arm above the elbow will assist in distending the veins and securing the necessary pressure, the tourniquet being immediately loosened when enough is taken and the arm flexed on a piece of cotton wool soaked in alcohol and pressed on the site of the puncture.

If the blood is being received into alveolar air, a tonometer of about 200 ml. capacity or larger and containing a little oxalate is filled with alveolar air, and the hypodermic needle to be intro-

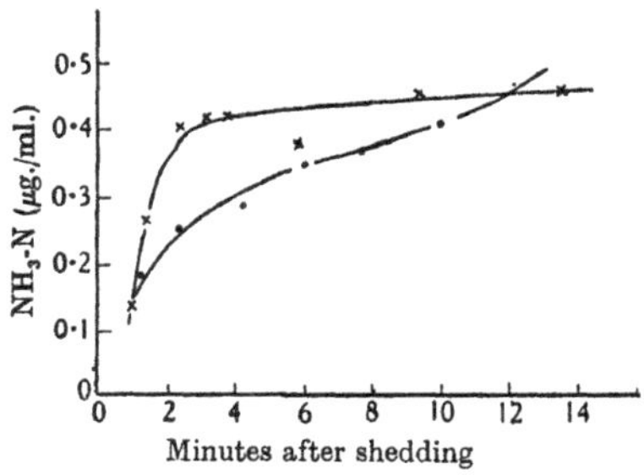

FIG. 31.

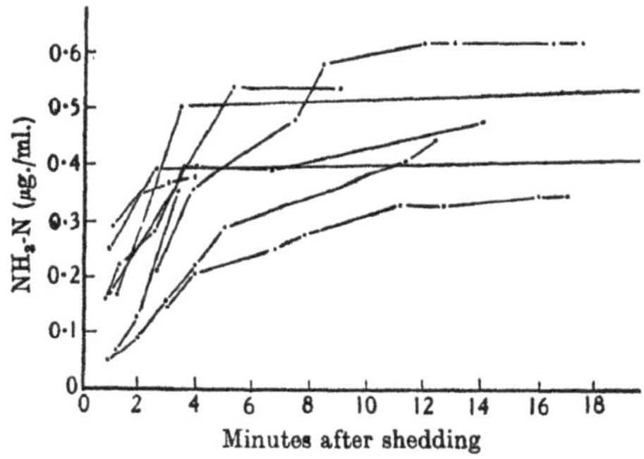

FIG. 31*a*.

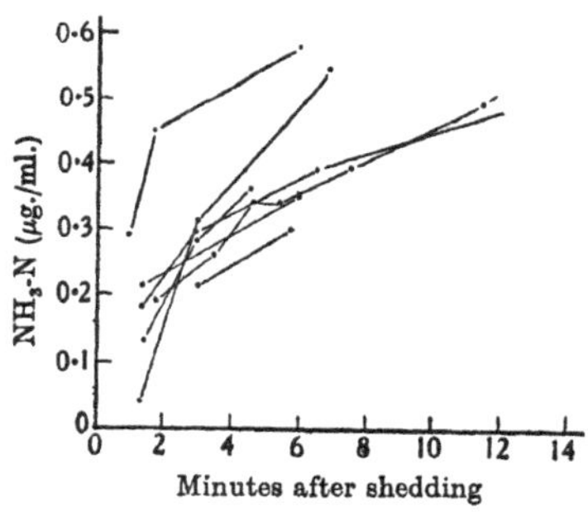

FIG. 31*b*.

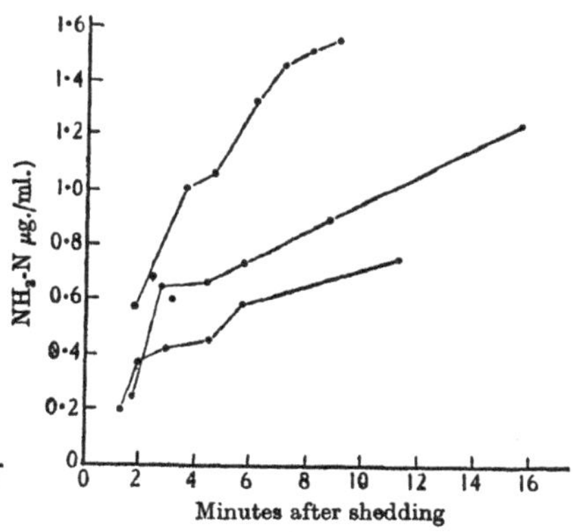

FIG. 31*c*.

FIG. 31. Characteristic curves of the blood ammonia for man (crosses) and rabbit (dots) after shedding into an open flask. Room temperature. Each point is the median of consecutive samples of 11 observations in order of time after shedding.

FIG. 31*a*. Serial curves of blood ammonia for man.

FIG. 31*b*. Serial curves of blood ammonia for rabbit.

FIG. 31*c*. Serial curves of blood ammonia for fowl.

duced is attached to a long tube (best coated internally with paraffin), the syringe not being used here for assistance in entering the vein. The first drops of blood may be discarded and then the tube inserted down into the tonometer, the stopper of which has been removed immediately before.

Introduction of the blood after collecting, etc. The 'units' being previously completely prepared, with the carbonate already present in the outer chamber, 1 ml. of the blood is pipetted therein, the time being noted and the contents mixed by rotation 15 times in about as many seconds. After allowing the absorption to proceed for ten minutes the lid is removed and the content of inner chamber titrated. At least one blank 'unit' is titrated and one control set up in a similar way. The control may suitably consist of 1 ml. of standard solution containing 2 'micrils' NH_3-N.

The titration is carried out in the ordinary way, using a fine pyrex glass rod. A colour change will be evident at the end-point with an addition corresponding to 2 or at most 3 small divisions. Accuracy in determining the end-point will be increased with a little experience. A very faint green appearing in an almost colourless solution when titrating with the barium hydroxide solution is preferred by the writer as the end-point to be aimed at.

Solutions required.

The absorption acid and titrating alkali are those of Range III (Table V, p. 92, under general ammonia method).

The only additional material required is :

Oxalate (ammonia-free). About 50 ml. of a strong solution of potassium oxalate are evaporated in an open pyrex dish after bringing to a *p*H of 10 or 11 by adding a little alkali, the reaction being tested by a universal indicator. The water volume is maintained by occasional addition. The mixture is kept boiling for about 10 minutes and then evaporated until the solid separates ; after cooling, the crystals are filtered off on a small Buchner funnel and preserved in a desiccator.

The fixative. The same fixative is used here as for chloride determinations (*4*) : 40–50 g. paraffin (M.P. about 40°) are melted with 80 ml. of pure liquid paraffin and cooled. Vaseline should not be used, as it contains traces of ammonia.

2 '*micrils*' (μg/ml.) NH_3-N *control solution.* 0·471 g. of pure $(NH_4)_2SO_4$ is dissolved and made up to a litre with distilled water. 2 ml. of this stock is diluted to 100 ml.

Calculation of the blood ammonia. This is based on the result of the control ammonia. The number of large divisions on the burette corresponding to the blood ammonia is divided by the number for the control and the result multiplied by 2·0. This must at the same time be multiplied also by 1·15 (where ten minutes are allowed for the absorption) owing to the fact that the ammonia comes over somewhat more rapidly from a water-carbonate mixture than from a blood-carbonate mixture. (These rates were at first considered to be the same but a fuller investigation under improved conditions gave the factors shown in Table IX.)

TABLE X

Absorption time. Mins.	*Ratio of ammonia absorbed from water-carbonate to that from blood-carbonate mixture.*	*The absorption as an approx. percentage of the total ammonia in the aqueous solution.* %	*Allowances in the final result for the special deaminating action of the alkali. μg. ammonia N/100 ml. blood at* 25°	20°	15°	10°	0° C.
10	1·15	52	11	7	4	3	1
12	1·14	57	14	8	5	3	1
14	1·12	61·5	16	10	6	4	2
16	1·11	65·5	19	12	7	5	2
18	1·11	69	22	14	9	5	2
20	1·11	72	25	16	10	6	2

In the above table made out for the pyrex ' unit ' as made by Messrs. A. Gallenkamp, London, it is understood that 1 ml. blood or water is added to 1 ml. of saturated potassium carbonate in the outer chamber from pipettes exactly calibrated for this purpose, and that the mixing is carried out with a fixed routine of fifteen rotations in approximately 15 seconds. The allowances in the fourth column for the special deaminating action of the alkali have a temperature coefficient of 10% per degree (from the *Biochem. J.* Vol. XXXIII, p. 457, 1939).

An allowance must also be made for the ammonia formed by the action of the alkali on the blood and absorbed in the given time. Such allowances are given in Table X, and are listed for different room temperatures. An example will show clearly how the calculation is made.

EXAMPLE. (All titrations carried out in duplicate, and the control solution containing 2 ' micrils ' (μg./ml.) $NH_3 - N$.)

Titration of ' blank ' unit - - - -	35·6 large divs.
Titration of ' analytical ' unit - - -	31·3 ,, ,,
Titration of ' control ' unit - - -	14·2 ,, ,,
Time of absorption - - - - -	10 minutes.
Room temperature - - - - -	15° C.

The amount of the absorption in the analytical unit corresponds to 35·6 − 31·3 or to 4·3, and in the control of 21·4 large divisions (each large division of the burette corresponds to 0·01 ml. of the solution of Range III).

Calculation : $\frac{4\cdot3}{21\cdot4} \times 1\cdot15 \times 2\cdot0 = 0\ 462$ 'micrils' NH_3-N. The factor derives from Table IX and 2·0 represents the control.

From the gross value of 0·462 'micrils' NH_3-N 0·04 is subtracted to allow for the special deaminating action of the carbonate as given in Table IX, column 4.

Before titrating with the barium hydroxide it is advisable to run out about 7–10 ml. from the burette, and also to have the cork of the reservoir bottle and of the soda-lime guard painted and sealed with melted paraffin. It will be seen that in the use of the method the exact strength of the barium hydroxide is of no consequence.

Some modifications of the above procedure used by other workers.

1. McDermott and Adams (1954) used 1% boric acid with indicator in the central chamber, and titrated with 0·003 N HCl after 10 mins. absorption.

Reasonably good results appear to have been obtained even with about 0·2 $\mu g\ NH_3 - N$/ml. The boric acid procedure in general is described in Chapter X, but we have always considered it preferable to use the hydrochloric-barium hydroxide procedure below 10 $\mu g\ N$ owing to the greater sharpness of end-point, due to the comparative absence of buffering.

The boric acid procedure has in general the two advantages of an indefinite upper limit and a stable titration fluid (standard HCl solution).

The disadvantages of relative lack of sharpness in the end-point could be largely eliminated by considerably lowering the boric acid concentration since only minute amounts of ammonia are absorbed in such determinations with freshly shed blood. From equation (27) and with $\frac{bs}{a}$ approximately 0·6 (with 1·0 ml. boric reagent in central chamber and 1 ml. blood + 1 ml. saturated potassium carbonate in outer chamber) then if [H] at the end point is $10^{-8}N$ it follows that the proportion of ammonia remaining after maximum absorption is 2·3% of the original amount.

With one ml. of 1% boric acid it takes 280/$\mu g\ NH_3 - N$ to produce a pH of 8·0 so that with only 3$\mu g\ NH_3 - N$ absorbed (about the highest to be expected in the 10 minute absorption period, even in the ammonium tolerance test) the boric acid concentration need be only 0·01 % resulting in a much greater delicacy of end-point.

2. Here it may be noted that in fact Reif (1960) has used 0·1 % boric acid absorbent in a number of research investigations on normal blood ammonia levels with apparently excellent results. A Rehberg micro-burette of 0·225 ml. capacity (Macalaster Bicknell No. 3095) was used to titrate the 1 ml. boric acid containing indicator in the central chamber. This paper of Reif deals in a very thorough manner with present evidence with respect to the normal blood ammonia levels. It may be mentioned that after considering a number of factors which could give high and low ammonia levels he considers the *upper level* of the free ammonium ion concentration of fresh normal resting human plasma lies below 20 μg NH_3N/100 ml.

Reif also states 'The corresponding value for whole blood lies below 10 μg NH_3N/100 ml. as determined by Conway's procedure. However, severe clumping queries the completeness of ammonia recovery from red cells. Nevertheless, coincidence of ammonia values, whether determined on whole blood or on plasma, make Conway's method a convenient tool '.

The present author is, however, of the opinion that such clumping of the red cells will in fact produce little relative effect, as owing to the small distances to be traversed by the diffusion of the strong carbonate solution it may be taken that the differential permeability of the red cell membrane will be rapidly destroyed and the diffusion of the ammonium ions very little affected. In Reif's paper recent work on the exact levels of the normal blood ammonia (e.g. Seligson *et al.*, 1951, 1957; Brown *et al.*, 1957; Jacquez *et al.*, 1959; Calkins, 1956; and DeGroote and Vandenbroucke, 1958) is reviewed.

3. White, Phear, Summershill and Sherlock (1954) used the procedure described in the text, with the following variations: 0·5 ml. blood is used instead of 1·0 ml. and 15 mins. are allowed for the absorption. This has the advantage of using lesser volumes of blood and secures a greater fractional absorption of the ammonia. This in turn is offset by the increased new ammonia formation due to action of the alkali over a longer time and in greater relative strength.

These workers do not appear to allow for this latter ammonia production—but for comparison with diseased conditions this is not important. For blood shed in air (or without CO_2 atmosphere) they obtained a mean value for NH_3-N of 0·79/μg/ml. and s.d. of 0·09. If allowance were made for the extra ammonia under their conditions, it may be assumed that the value would lie in the region of 0·4–0·5 μg/ml.

The blood samples were taken from fasting subjects and three minutes after shedding.

Other modifications of the microdiffusion procedure have been used.* For example, 60 min. absorption periods have been allowed, with mechanical devices for turning, etc. All this only increases the inaccuracy and the labour of the determination.

Ammonia levels in the peripheral, hepatic, portal and renal veins in normal subjects and those with hepatic disease.

The following diagram is taken from the paper on ‘ Ammonia Metabolism in Man ’ by McDermott, Adams and Riddell (1954) and illustrates the normal levels.

Determinations of the ammonia level in the portal vein, inferior or vena cava, renal artery and renal vein were obtained from patients undergoing surgical procedures in which these vessels were accessible to vena puncture. These workers also stress the fact that the samples of portal blood were obtained from fasting patients under anaesthesia.

It will be seen that the amount of ammonia entering from the intestinal tract through the portal vein is considerable and when expressed in probable quantities per day would amount to some 20 to 25% of the total ammonia production in the body—even measured in this way on the fasting subject.

The greater fraction of such ammonia production occurs of course in the liver itself in the deamination of amino acids.

It may be expected from the diagram that elevated blood ammonia values would occur with advanced liver disease and especially where there is anastomotic circulation between the portal system and the inferior vena cava.

This corresponds with the general results obtained.

In cirrhosis without mental symptoms the level of NH_3-N in

* See note at end of chapter.

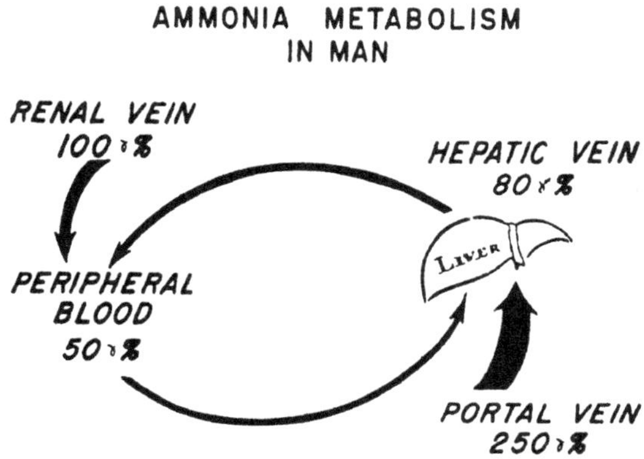

FIG. 32. A diagramatic representation of the extra-cellular circulation of ammonia based on representative samples (McDermott, Adams and Riddell, *Anal. Surgery*, *140*, 4, 1954).

the peripheral veins is in the average some 20μg/100 ml. higher than the normal, but with mental symptoms the difference is usually much greater.

The question of a direct relation between mental symptoms and elevated blood ammonia has been much discussed. McDermott, speaking in the discussion after the reading of his paper with Adams and Riddell (1954) states the following: 'There is no question but that a large percentage of these cases represents true ammonia intoxication, but obviously the liver with its many complex metabolic phases, cannot be expected to show only one sign of degeneration at any given time. The problem of hyponatremia, hypoalkemia and often hypoglycemia and other facets of liver disease, will bring about the death of a patient in the terminal phases of liver disease without of necessity this being related to ammonia intoxication.'

It may be noted that Adams and Foley (1953) used the term 'hepatic coma' to include the entire syndrome of disturbed central nervous system function associated with liver disease. 'This term may be misleading, for in many instances coma does not occur' (Sherlock *et al.*, 1954).

Sherlock *et al.* (1954) mention with regard to their results, 'In all patients with neurological complication, the fasting level of ammonium in the periphereal blood was above mean levels found in patients with liver disease uncomplicated by nervous disorder.'

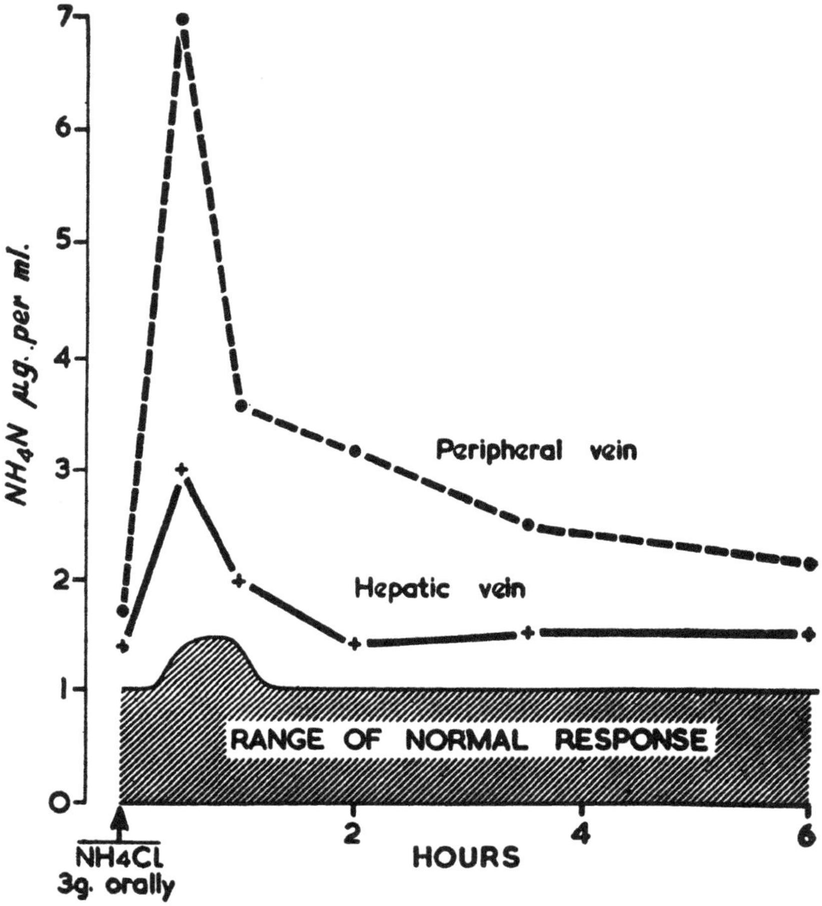

FIG. 32*a*. Blood-ammonium curves following administration of 3g. ammonium chloride by mouth in a patient with cirrhosis of the liver and extensive portal collateral circulation (Sherlock *et al.*, *Lancet*, Sept., 1954)

They also write, 'We also confirmed that the elevation of peripheral blood ammonium could not always be correlated with the neurological state' (Traeger *et al.*, 1954).

Ammonium Tolerance Test

White *et al.* (1955) describe the results of a large number of tests, in which 3·0 g. of ammonium chloride were given by mouth in gelatine capsules. ('This almost invariably provoked nausea, rarely vomiting.') Preliminary accounts of this test were given by

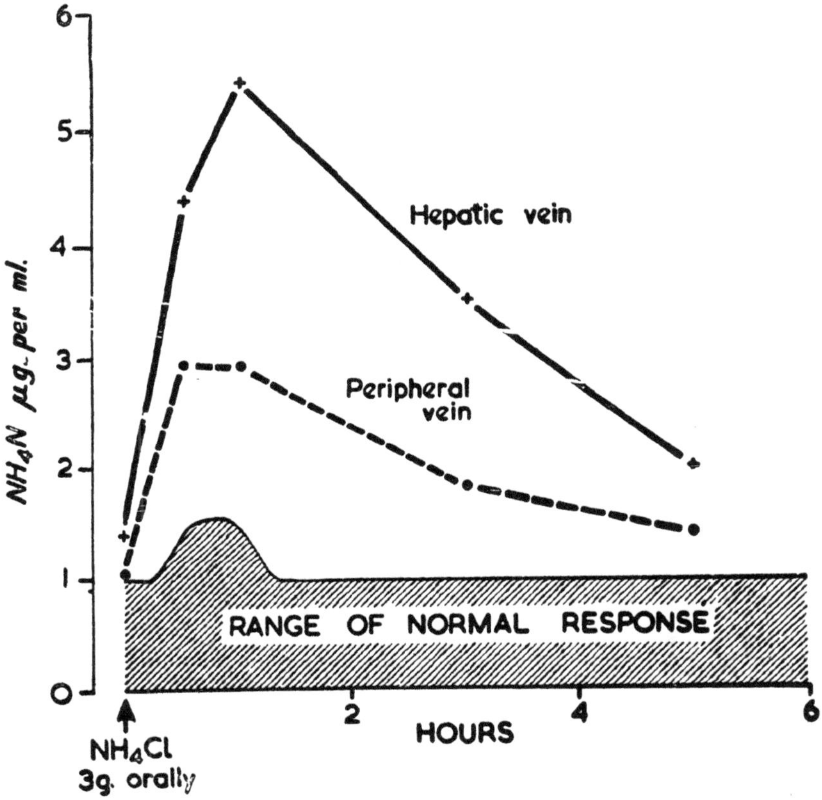

FIG. 32*b*. Simultaneous samples from peripheral and hepatic veins showing abnormal blood-ammonium curves following administration of 3 g. ammonium chloride by mouth in a patient with infective hepatitis (Sherlock *et al.*, *Lancet*, Sept., 1954)

Sherlock *et al.* (1954). Blood samples were taken from an antecubital vein and through a catheter passed under fluoroscopic control into a hepatic vein. Fig. 3 of their paper, Sherlock *et al.* (1954) summarises the results obtained in cirrhosis and having portal collateral circulation, and Fig. 2 the results with hepatitis. (These are represented here as Figs. 32*a* and 32*b*.)

The effects on the blood ammonia in such conditions, are such that halving the dose above mentioned may be suggested.

About the same time as such a test was described, Eisenman (in the discussion after the paper by McDermott, Adams and Riddell, 1954) mentioned his use of a similar test using ammonium citrate, and described how in a cirrhotic patient the tolerance curve is slightly elevated but where there is present a portocaval shunt it is markedly elevated and remains so for a long period. 'In brief, we believe that the oral ammonium citrate tolerance test is a reliable method for determining the potency of a portocaval shunt.'

(3) Determination of the total ammonia formed after 24 hours in sterile blood at room temperature or after some hours action at 38°. After 24 hours at room temperature the amount in human blood is approximately 0·7 mg. N/100 ml., deriving mainly from the breakdown of adenylpyrophosphoric acid (it is approximately 1·8 mg. N/100 ml. in rabbit's blood). The ammonia formed by the action of the alkali may be neglected. Using the method of suspended absorption, the mean of a duplicate determination may be relied on to within 2–3% error. Much smaller volumes of the blood may be used with 1 ml. of saturated carbonate and enough time allowed for the total absorption. Solutions of Range II would then be appropriate.

A suitable procedure here would consist in using 0·2 ml. of the blood and 1 ml. saturated carbonate and allowing the absorption to proceed for 45 minutes, in which time with these volumes the absorption is complete. Solutions of Range II could be used with respect to the acid and alkali.

Alternatively 0·1 ml. blood and 1 ml. carbonate may be used with solutions of Range III. (*Vide* p. 92).

(4) The preformed ammonia in muscle.

From the moment of its excision from the animal ammonia begins to form in muscle, and if the muscle be ground with water or saline the rapidity of the formation is enormously accelerated, passing from an apparent value of approximately 7 to 40–60 'micrils' NH_3-N within five minutes. Its ammonia content also increases during activity as is now well known from the work of Embden (1926, 1927) and, later, of Parnas (1927). As it is necessary to have the muscle finely divided, chemical or physical measures must be taken to prevent the formation during this division. In the method of Embden (1928), who discovered adenylic acid in muscle (now known to be combined with phosphoric acid) and showed it to be at least the chief mother substance of the formed ammonia, the muscle is firstly frozen by immersion in liquid air contained in a Dewar flask, the muscle being held under the surface of the liquid by a forceps to ensure rapid freezing, after which it is ground in a fold of linen or muslin in a mortar—which may in turn be cooled with some liquid air.

The powder is then introduced into weighed glass-stoppered vessels containing a measured amount of 4 hydrochloric acid (somewhat modified in later experiments). This fluid is considered to stop any further ammonia formation, prior to the ammonia determination.

The ammonia method used by Embden is a modification of the old method of Nencki and Zaleski, in which distillation was carried out *in vacuo* after the addition of excess magnesium oxide at a temperature of 30° (immersion in water bath). In the modification ground glass joints are used and the distillate (with air current sufficient to produce a positive pressure of about 11 mm.) received into 5 ml. or so of standard acid containing some Tashiro's reagent. The distillation procedure is continued for 35 minutes. It may also be noted that the magnesium oxide is added as 15 ml. of a suspension containing 2 g. of the light magnesium oxide.

The method of Embden may now be considered as chiefly of historical interest.

In the method of Parnas (1921, 1924, 1927, 1934), instead of freezing in liquid air the tissue is rubbed with a cold saturated borax solution. In the procedure a rough, deep glass mortar and a very large pestle are used. At the bottom of the mortar is a little quartz sand, sharp and granulated, and a few c.c. of cold saturated borax solution, previously boiled and ammonia-free. The muscle is thrown in and squashed as quickly as possible with a strong pressing and rubbing motion. Experience must be acquired, but offers no difficulty. The method is suitable only for small muscles. Note that the ammonia content in grinding in the ordinary way averages 4 to 6 mg.% and with the borax solution one finds 0·4 to 1·0 mg.%, mostly 0·6 to 0·7 mg.%. Similar results are got after freezing in liquid air and then proceeding to the borax grinding. The subsequent ammonia is then determined by the well-known Parnas method as modified in later adaptations.

Microdiffusion method for determining the preformed tissue ammonia.

Preparation of ' units ', etc. These are prepared as for determining the ammonia content of blood, the solution of the Range III, Table VI, being used in the burette. The saturated carbonate may be introduced beforehand into the outer chamber or added after the introduction of the muscle-phosphate mixture. As with the blood ammonia, where the highest accuracy is required the ' units ' may be firstly prepared, the central chamber titrated after 30 minutes and the contents sucked out, being replaced by a fresh acid delivery. Such a procedure, however, is unnecessary, if the fixative is absolutely free of ammonia and if the ' units ' have been carefully cleaned.

Preparation of tissue suspension and ammonia determination. Here use is made of the fact first discovered by Parnas *et al.* (1934, 1935), that phosphate suspends the ammonia formation in muscle extracts for upwards of half an hour, but the method is also applicable, with suitable modification, to tissues prepared by freezing in liquid air. Taking the gastrocnemius of the *Rana temporaria*, for example, this is excised immediately after the killing of the animal or subsequent to whatever conditions have been allowed to act on it. It is quickly weighed on a tared watch glass and introduced into 2·0 ml. of $M/10$ phosphate and some fine quartz sand (Merck's purest) in a small clean mortar. The phosphate solution is formed by mixing equal volumes of $M/10$ alkaline and acid phosphate. The *p*H of this mixture is approximately 6·8. Immediately after introducing the muscle into the phosphate it is ground with as large a pestle as convenient, the grinding being conducted as rapidly and efficiently as possible. When the muscle is thoroughly ground the mixture is poured into a small clean tube and allowed to settle for about one minute, when duplicate 0·5 ml. volumes, either directly or after clearing by centrifuging, are pipetted off into the outer chamber of prepared ' units ' already containing 1 ml. saturated potassium carbonate (though this may be added subsequently).

The ' units ' after mixing contents of outer chamber, are left aside on bench for 1 hour and then titrated. *Alternatively, if a quick result is desired, the absorption is suspended after 20 minutes, when 93% of the ammonia will have come over.* The suspension will here occur at a region of the absorption curve when the change with time is small.

Possibilities of the method with the ammonia absorption and titration. Concerning the ammonia absorption, it is clear from accounts of previous determinations that this offers no difficulty, but proceeds automatically and after 1 hour is complete under the conditions (93% after 20 minutes).

The titration possibilities are such that 1 μg ammonia N can be determined to a standard deviation of 0·02 μg. 1 μg ammonia N corresponds to about 20 large burette divisions of Range III solution and is contained in about 0·1 g. resting muscle of frog. If 0·5 ml. extract is pipetted into the outer chamber and contains 1 μg ammonia N, this means a dilution of about 1 in 5. The dilution in the above procedure will be about this order—usually somewhat greater.

Results obtained for muscle with the above method. The average of 25 determinations, using phosphate as above, was 0·72 mg.% NH_3-N with a standard deviation of 0·05. Later series gave higher mean values—about 1·2 mg.%. The mean of a large number of determinations by Parnas and co-workers was 0·85 mg.% NH_3-N. There was no significant difference found by him using cold saturated borax solution and a previous freezing in liquid air.

Further studies on the ammonia content of fresh resting muscle made here (O'Brien, 1941) show that the fresh resting muscle of the rabbit (the animal being instantly killed) gives an average gross value of 0·39 mg./100 g., but as will be described elsewhere, much of this would appear due to ammonia formed in grinding the mammalian muscle. This has been estimated at 0·25 mg./100 g. so that 0·14 mg. (or possibly less) exist as free ammonia in the normal resting rabbit muscle.

If in rabbit muscle the ammonium ion is concentrated similarly to the potassium ion *in vivo*, or about 30 times, then we could expect a blood ammonia figure of about 0·004 mg./100 g. or at about the level analytically determinable. This is in reasonable agreement with the direct estimations of the blood ammonia.

Alternative method for preformed ammonia in muscle. The following method for the ammonia in frogs' sartorii was investigated and found so far very suitable (using solutions of Range III). The sartorii muscles (1 or 2) after weighing quickly are transferred to the outer chamber of a prepared ' unit ' as above, and the lid placed in position. The 2 ml. of saturated potassium carbonate are run into the outer chamber and the contents mixed by rotation. Such ' units ' are rotated after each 10 minutes and titrated after 30 minutes. The saturated carbonate rapidly breaks through the membrane system of the muscle fibres liberating the ammonia, and owing to the relatively high proportion of saturated carbonate to muscle fluid, the ammonia is rapidly given off and absorbed. For this very simple and accurate method it is advisable that the size of the tissue be not more than about 50–60 mg.

(5) Determination of the total ammonia formation in muscle. (See subsequently the adenylpyrophosphate determination.) This follows simply from the previous description, the ammonia content of the muscle here being upwards of ten times greater. A considerably greater dilution of the muscle suspension may be used to keep within Range III or the same dilution with Range II solutions, the latter giving more accurate results.

We can proceed in a variety of ways, remembering that each g. of muscle will give about 50 μg. ammonia N. Thus with a 1/5 extract 0·5 ml. in the outer chamber will give 5 μg. or 16·7 large divisions with solutions of Range II. With 0·5 ml. in the outer chamber with 1 ml. carbonate, 1 hour may be given for complete absorption at room temperature.

The absorption rate of ammonia from muscle extract is the same as from water.

(6) Ammonia content of natural waters (e.g. water from rivers, springs, etc.). Table XI gives amount in different kinds of natural water, and the present method may be used when a titration procedure is preferred to Nesslerisation.

For these ammonia determinations the solutions for the third or lowest Range are used, each large division on the burette being equivalent to 0·050 μg. ammonia N, or when 1 ml. of the water is being used for the analyses to 0·05 ' micrils '. Blank determinations are carried out at the same time.

TABLE XI

Water.	*Ammonia content* ($\mu g./ml.$) *N.*	*References.*
Rain water -	0·3 – 4·2	Boussingault (1853); Smith (1872).
River water -	0·01 – 4·0	Boussingault (1854).
	0·58 – 1·9	Knop (1868).
Sea water -	0·13 – 0·9 (11.4 observed in Adriatic)	Audoynaud (1875); Dieulafait (1878); Thorpe and Morton (1870); Vierthaler (1867).
	0·00 – 0·10 (Irish Sea off Dun Laoghaire)	Conway (1944).
Spring water -	Generally lower than river or sea water.	Mellor (1928).

Into the outer chamber is introduced 1 ml. of the natural water, then 1 ml. carbonate, mixed and left aside for 2 hours. If a rapid result is desired the absorption is cut short after 10 minutes by removing the lid and titrating, the ammonia found being then multiplied by 1·95 at 18° C., with a decrease of 2% in factor for every degree increase in temperature. This latter procedure involves at most a 5% error in the ammonia, which for the minute quantities in this determination may be considered negligible.

Calculation. Each large division on burette is equivalent to 0·05 μg. ammonia N. This value will need to be corrected by a factor found by titrating one or two acid deliveries. When 1 ml. of the water is used each large burette division represents 0·05 'micrils,' mg. per litre or parts per million.

Solutions. Range III solutions (Table VI, p. 92).

* Referring to foot-note on p. 123, Hulme and Cooper (1956) have published a paper on 'Blood-Ammonium estimation as a routine laboratory procedure" (from the Netherne Hospital, Coulsden, Surrey, England), using the micro-diffusion method largely as described by the author, with various details suitable for routine work. They include a table of determinations on 60 normal subjects.

CHAPTER XIV (*Aa* 6)

TOTAL NITROGEN. (OVER 100 μg N)

When large numbers of determinations of total nitrogen are being carried out either of macro or micro kind, the microdiffusion principle, in conjunction with the Kjeldahl digestion, is of value in considerably diminishing the work in such serial determinations, and also for all micro ranges increases the accuracy of the procedure (e.g. Bentley and Kirk, 1936).

The volumetric accuracy need not fall below that for the traditional full macro determination, as shown in Part III, and it is evident at the same time that the elimination of the usual distillation removes an important source of error. With quantities as low as 1 μg N, Tompkins and Kirk (1942) have reported errors of only 0·3%.

In this chapter some methods of practical value for nitrogen quantities above 100 μg N will be described.

In the incineration process the use of certain catalysts in excess should be avoided—for example, copper sulphate, since it slows the absorption considerably when in comparatively high concentration. It may still be used when in small quantities of 0·1% solution. Copper is not so good a catalyst as mercury, selenium, tellurium, titanium, or iron, as shown by Osborn and Wilkie (1933) in an extensive analysis of the catalytic action of various metals, etc., in Kjeldahl incinerations. They found mercury to be the best catalyst. In previous editions mercury was considered inconvenient for the microdiffusion procedure, owing to its removal as sulphide before distillation. Recently Weil-Malherbe and Green (1955) have shown that it can be used very successfully by the procedure they adopt, as described later.

Kirk (1947) has reviewed the extensive literature on the use of catalysts in the Kjeldahl digestion.

It is now evident from the findings of many investigators that clearing time for the digest is no criterion of complete digestion, and the tendency is to allow much longer digestion times than heretofore. Chibnall, Rees and Williams (1943) recommend from 8 to 16 hours' digestion after clearing.

Kirk concludes that the 'available evidence certainly indicates the superiority of selenium over other catalysts as a clearing agent, and it probably is superior also in reducing over-all digestion time'. He adds that mercury is as good if not better for obtaining maximum nitrogen recovery.

There would appear to be an upper limit for the selenium catalyst, beyond which nitrogen recoveries decrease. Bradstreet (1940) places this limit at 0·25 g. Se for a macrosample.

For the alkalinisation of the digest or aliquot part thereof, saturated potassium metaborate or 40% KOH would appear the most suitable. With 40% KOH the times for diffusion are about 25–30% longer than with metaborate.

Saturated potassium metaborate and saturated potassium carbonate have practically the same effect in raising the ammonia tension of solutions, and hence increasing the absorption rate, but the carbonate cannot be used here owing to the high acidity of the digest.

The following account gives examples of incinerations and ammonia determinations at various levels, but they may of course be suitably changed for special conditions.

Macro-Kjeldahl determinations.

EXAMPLE 1. *Wheaten flour.*

Incineration. 0·5 g. of flour is weighed into a 300 ml. Kjeldahl incinerating flask. 10 ml. of the digest mixture are added and the mixture incinerated in the usual way with a small funnel in the top of the incinerating flask, somewhat longer than the usual clearing time of the fluid being allowed.

Digest mixture. The digest mixture is made of equal parts of concentrated H_2SO_4 and saturated $KHSO_4$ and contains about 0·2 g. SeO_2 per 100 ml. (Copper sulphate may be also included, but not more than 0·1 g. per 100 ml., as it tends to inhibit the rate of the subsequent ammonia release.)

Ammonia determinations. Subsequently the mixture is cooled, 25 ml. water added cautiously and cooled under the tap. The mixture is now made up to 50 ml. in a volumetric flask, and 0·5 ml.

in duplicate is pipetted into the outer chamber of two or three 'units' (No. 1). The pipetting may be very accurately and suitably done by means of a straight tube pipette taking about 10–20 seconds to deliver, and the final fluid blown out immediately after delivery with the tip touching the glass (see Chapter V).

After pipetting the fluid, the 'units' are allowed to remain on the bench for about five minutes (to allow escape of any volatile acid formed in the incineration), 1 ml. of boric acid reagent which need be only roughly measured (see p. 98) is run into the central chambers of the 'units'. 2 ml. of saturated potassium metaborate or 1 ml. of 40% KOH are added in the usual way after placing the lid with fixative in position, and the contents of the outer chamber mixed by rotation and left on the bench for $1\frac{1}{2}$ hours. The contents of the central chambers are then titrated with $N/50$ HCl from the Conway burette.

Calculation. The number of large divisions (each equal to 0·01 ml.) required in the titration, is multiplied by 0·056 to give the nitrogen as g.% in the sample of flour.

EXAMPLE 2. *Urine.*

Incineration. 2 ml. of urine are pipetted into the incinerating flask from a straight tube pipette (a standard 2 ml. pipette is also suitable). The pipette, about 15 to 20 cm. long, should take about 10–20 seconds to deliver, the final amount being blown out gently with the tip against glass and rotating while blowing. Remaining procedure as in Ex. 1.

Ammonia determinations, etc. After the incineration as above the dilution is made to 200 ml. and the ammonia determination made in the same way. The calculation is the same as for Ex. 1.

Accuracy of the above macro-Kjeldahl determinations. The accuracy of such procedures on the purely volumetric (including titration) side is analysed in Part III. It is shown there that the diffusion technique can give an accuracy as great as the macro-Kjeldhal conducted at the 25 to 50 ml. titration level. (Recently, Hallam (1955) has confirmed this statement.) With regard to the passage of the ammonia from the Kjeldahl digest to the acid, it is obvious that the diffusion procedure is much less liable to error, and goes automatically without any observation being required during the process.

Micro-Kjeldahl determinations.

The above determinations were described as macro, though usually somewhat more material is taken in the typical macro procedure. The N incinerated was of the order of 10 mg. or more. In the following account much smaller amounts of N (down to 100 μg) are present in the samples.

EXAMPLE 1. *Determination of albumin and globulin in a sample of serum.*

1 c.c. volumes of the serum are pipetted into two conical flasks of about 50 c.c. capacity. Into the first is run 25 c.c. of 0·9 or 0·6% NaCl.

Into the second flask is run 25 c.c. of warm 23% (stored at 37° C.) sodium sulphate (anhydrous). The stoppered flasks with the contents mixed are placed in the incubator at 37° C. for 1½ hours.

The contents of flask 2 are filtered. 5 c.c. volumes of each are introduced into 100 c.c. incinerating flasks, and 2·5 c.c. of the digest mixture added (from a graduated 5 c.c. pipette). Incinerate in the usual manner, with small funnel in neck of flask.

After incinerating, cool and add 25 c.c. water from a pipette to each and mix well.

Ammonia determination. 1 c.c. boric acid reagent is added to the central chambers of microdiffusion 'units' and 1 c.c. of the final mixtures in the outer chambers. The subsequent procedure is as before, using 1 c.c. of 40% KOH as the alkali. After 2 hrs. on the bench titrate with $N/50$ HCl from the Conway burette.

Calculation. Multiply the number of large divisions (minus blank) for the first mixture (without sodium sulphate) by 0·217 which gives the total albumin and globulin as g./100 c.c.

The similar reading for the second mixture is multiplied by 0·209 which gives the albumin as g./100 c.c.

Solutions required.

23% sodium sulphate (anhydrous) solution maintained in the incubator.

0·6 or 0·9% NaCl.

Digest mixture (see page 135).

Usual solutions for the microdiffusion procedure, boric acid procedure, and using 40% KOH as alkali for the outer chamber and *N*/50 HCl to titrate.

EXAMPLE 2. *Total nitrogen in brain tissue. Weil-Malherbe and Green's procedure using a mercury catalyst* (1955).

To the sample containing 5–50 μg.N is added 1 ml. of a solution containing 0·625% (w/v) HgO in 9 N—H_2SO_4. After the water has been boiled off, 100 mg. K_2SO_4 are added. Digestion is continued for 2 hrs. after clearing. The cooled digestion mixture is diluted to 5 ml. and a sample of 2 ml. is placed in the outer chamber of a Conway unit, followed by 1 ml., of a solution containing 10% (w/v) $Na_2S_2O_35H_2O$ in 60% (w/v) KOH. The inner chamber contains 1 ml. 0·01 *N* HCl. When the diffusion of ammonia is complete the contents of the inner chamber are removed with a fine-tipped teat pipette and transferred to a test tube followed by two washings with 1 and 0·5 ml. water respectively; 0·25 ml. Nessler reagent is then added. Readings are taken in colorimeter cells with capacity 3·5 ml. and diam. 1 cm.

The method gave theoretical recoveries with various purines and amino acids.

CHAPTER XV (Aa 7)

TOTAL NITROGEN. (UNDER 100 μg N—FURTHER PROCEDURE)

WITH quantities as low as 5 μg. ammonia N, Bentley and Kirk (1936) obtained an error of only 0·7% (standard deviation) for the full volumetric procedure, and of 1·2% for the complete process with thiobarbituric acid, *p*-amino-benzoic acid and *p*-brom-acetanilide in amounts giving from 3 to 10 μg. ammonia N.

Borsook and Dubnoff (1939) using special small tubes for the Kjeldahl digestion and a microdiffusion unit somewhat similar in

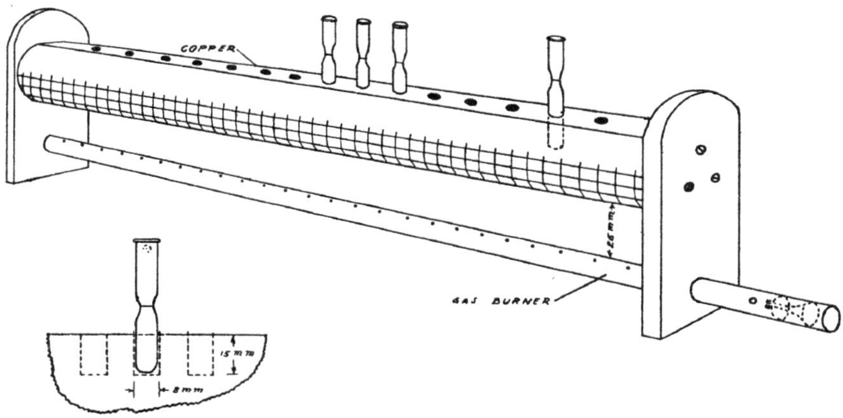

FIG. 32c. Digestion rack and Kjeldahl tubes as used by Borsook and Dubnoff (from the *J. of Biol. Chem.*, **131**, p. 165, 1939).

dimensions to the present standard 'unit' (No. 2) obtained similar accuracies with 5 and 10 μg. total nitrogen.

In these methods digestion and diffusion were carried out in separate vessels. More recently Needham and Boell (1939) developed a procedure in which both processes were carried through in the one piece of apparatus (Fig. 34), and proved very useful for total N determinations of the order of 1 to 10 μg. N. Tompkins and Kirk (1942) described a special small digestion-diffusion vessel (Fig. 35), which could also be adapted for evacuation prior to the ammonia diffusion, the rate of which it hastens considerably.

With this apparatus the diffusion time was reduced from the 12 hrs. for Needham and Boell's technique to 5 hrs., and to about $1\frac{1}{2}$ hrs. *in vacuo*. The accuracy achieved was striking, being only 0·3% as a standard deviation from the mean when dealing with quantities as low as 1 μg. N.

Hawes and Skavinski (1942) have also described a method for total N determination of the order of 10 to 100 μg. N, using the microdiffusion procedure. They too carry out the digestion and diffusion in the one vessel. A unique feature in their procedure is the use of a plain test tube (or conical centrifuge tube, if desired) which serves both as a digestion and a diffusion vessel. The diffusion receiver consists of a helix of platinum wire carrying a drop of absorbing solution held in the tube by means of a grooved rubber stopper (Fig. 36) Hawes and Skavinski obtained a standard error of about 0·5% for 100 μg. N and 0·98% for 11·3 μg. N. Brüel, Holter, Linderstrøm-Lang and Rositz (1946) described a total nitrogen method dealing with quantities on similar levels. This method has been recently referred to as the method of choice by Glick (1949) largely on the grounds that the 'precision' is of the order of 0·005 g. N. This seems to overlook Kirk's results, as well as those for ammonia given by the author (1947). As seen in Chapter VII 0·07 μg. has been determined with a probable error of 0·0005 μg NH_3–N. Even this is certainly not the limit by the microdiffusion procedure. (This matter has also been referred to in the Preface to this edition.)

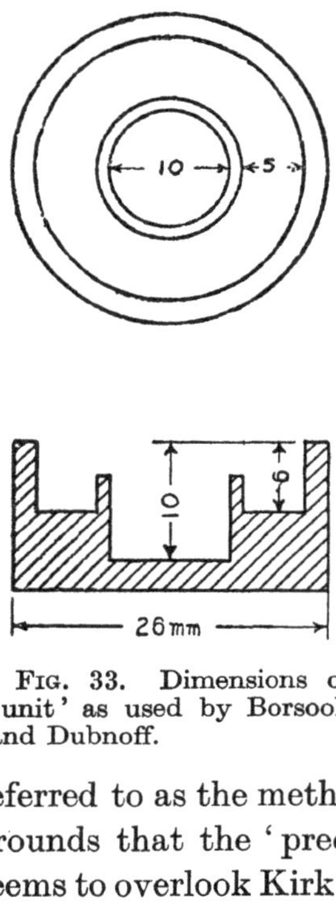

FIG. 33. Dimensions of 'unit' as used by Borsook and Dubnoff.

In the following account a description is given of the procedures mentioned above.

Digestion.

(1) *Borsook and Dubnoff's procedure.*

0·1 or 0·2 ml. aliquots of the solution followed by 0·1 ml. of the digestion mixture are pipetted into a digestion tube shaped as in the diagram (Fig. 32*c*) and with a mark around the constriction representing a volume of 0·45–0·50 ml. All the tubes so set up are placed in an air oven at about 105° C. until most of the water has been evaporated off and the solution has become dark brown.

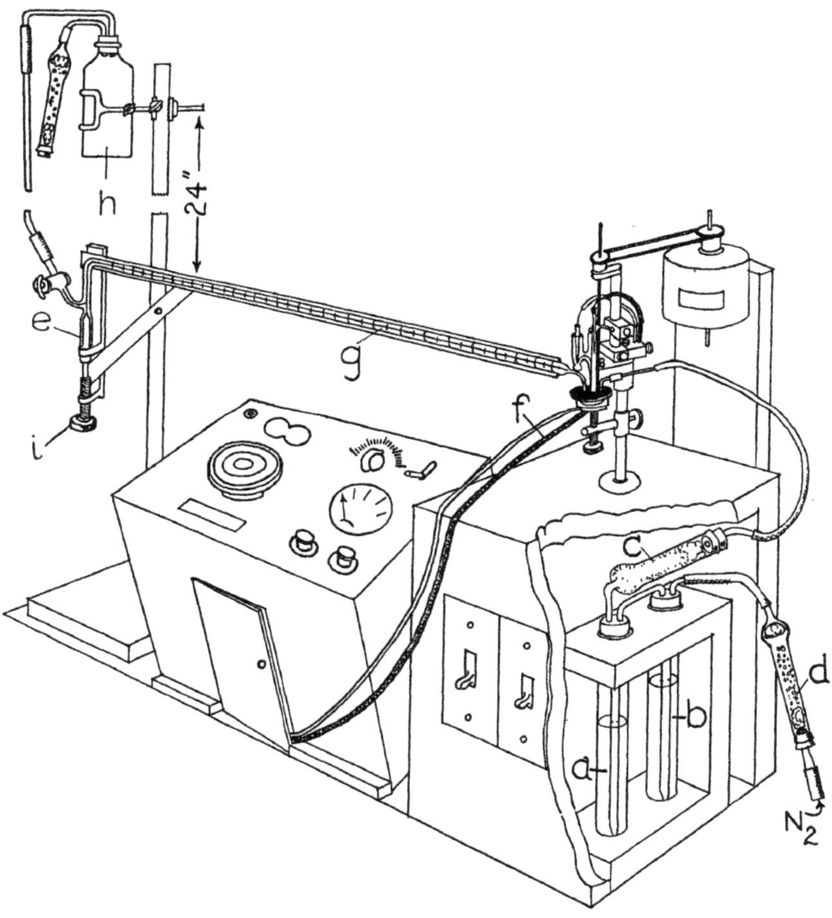

Fig. 33*a*. Titration apparatus used by Borsook and Dubnoff (from the *J. Biol. Chem.*, **131**, p. 168, 1939).

c, cotton; *d*, soda lime; *e*, mercury; *f*, shielded cable; *g*, microburette; *h*, standard NaOH; *i*, hand screw for propelling mercury, covered with rubber.

The practice is to leave the digestion tubes in the oven overnight, (the tubes being covered with a sheet of tinfoil) after which they are placed in the digestion rack. The flames are kept low at first until the acid begins to reflux. Then the flames are gradually raised, but not at any time to such a height that the acid distils more than a few mm. above the constriction. The digestion is continued until the solution is colourless. The tube is cooled, 1 drop of the saturated persulphate solution added and the digestion con-

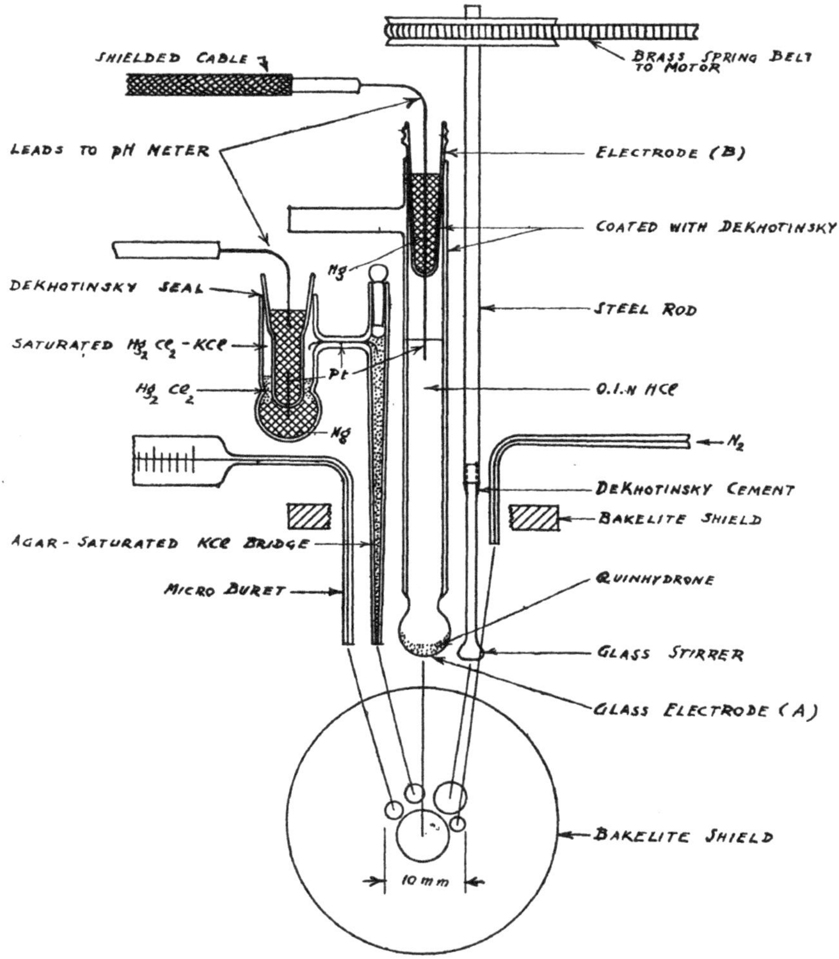

FIG. 33*b*. Electrode assembly details of titration apparatus used by Borsook and Dubnoff (as in previous Fig.).

tinued for 15 minutes after the digestion mixture is water-clear.

This digestion may be carried out without a hood. There is practically no loss of acid and hence the digestion may be carried on for hours.

When complete, the tubes are cooled and then water is added

to the mark around the constriction. The water is added in two portions, with mixing by rotation and cooling between additions. After being made up to the mark the contents are mixed by vigórous shaking and inversion.

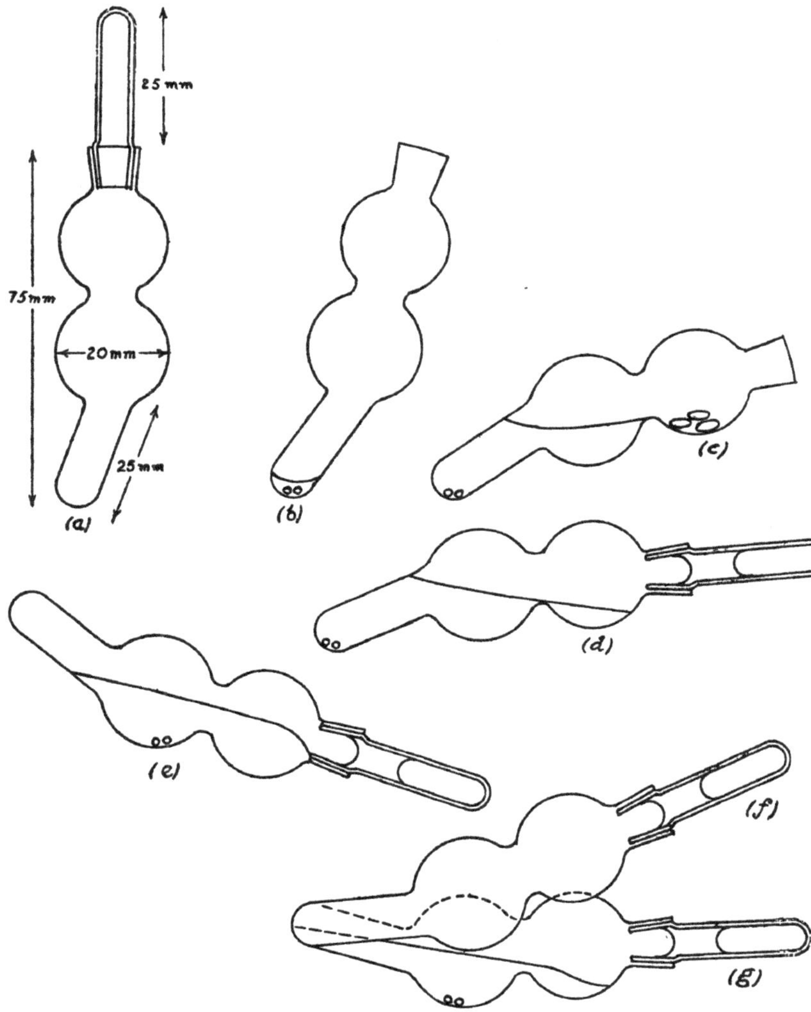

Fig. 34. Apparatus used by Needham and Boell for combined digestion and microdiffusion. Positions of apparatus described in text (from the *Biochemical J.*, **33**, 149, 1939).

Reagents. Digest mixture.—1 : 1 dilution of H_2SO_4 plus water containing 1 g. SeO_2 and 1 g. $CuSO_4$, $5H_2O$ per 100 ml. of diluted acid.

Saturated K *persulphate solution.* The salt used is described as ' Merck's Reagent Variety for D.D. Van Slyke's micro-Kjeldahl determination of nitrogen.'

(2) *Needham and Boell's procedure.*

The vessel used is shown in Fig. 34*a*. It consists of a digestion vessel with two bulbs blownin it (pyrex glass), some 7·5 cm. long. At the top there is an opening ground to receive a small tube of about the same dimensions as those used by Linderstrøm-Lang and Holter (1933) in their NH_3 estimations. Like their tubes also, it is coated on the inside with high melting paraffin wax. It is referred to as the receptor tube.

The digestion is carried out as follows. The source of nitrogen is placed or pipetted into the bottom of the digestion vessel (Fig. 34*b*) and 50–60 μl of the digestion mixture added. The

FIG. 35. Apparatus used by Tompkins and Kirk for combined digestion and microdiffusion (from the *Journal of Biol. Chem.*, **112**, p. 478, 1942).

A.—Simple digestion-diffusion vessel.

B.—Glass-stoppered vacuum digestion-diffusion vessel. Length of vessels—about 5 cm.

tubes are then placed in an air oven at about 120° C. for *ca.* 3 hrs. to drive off traces of water. At the end of that time a good deal of charring will be visible, and the digestion proper is carried out over pin flames constructed from old hypodermic needles. Two glass beads are usually added, and a mobile flame used. Digestion is complete in 3–5 minutes. When the liquid has become straw coloured, the heating is continued for ½–1 hr. over a pin point flame about 2 mm. high.

Digest mixture. 3 g. pure $CuSO_4$, 1 g. pure K_2SO_4 and 0·1 g. SeO_2 to 300 ml. of pure conc. H_2SO_4.

The next two methods like Needham and Boell's technique incorporate the digestion and diffusion processes in the one piece of apparatus (*vide* Fig. 34)—though this was practically arrived at in Kirk's original technique (Bentley and Kirk, 1936).

(3) *Tompkins and Kirk's procedure.*

The sample, containing from 0·5 to 20 μg of N, is measured and transferred with rinsing into the bottom small bulb of the special vessel (Fig. 37). To this is added about 0·1 ml. of digest mixture. The vessel is placed in a sand bath so that the temperature at the bottom is about 300°.

At first the vessel is placed at an angle of about 10–20° from the horizontal with the small bulb barely under the surface of the sand. After all the water is evaporated from the acidic mixture the vessel is placed upright and pushed deeply into the sand, where it remains until the digestion is complete. It is then removed and allowed to cool.

Digest mixture. 18 *N* sulphuric acid (N-free) containing 0·1 g. of SeO_2 and 0·1 g. copper sulphate per 100 ml.

(4) *Hawes and Skavinski's procedure.*

Samples are pipetted into ordinary 125 × 15 mm. Pyrex test tubes taking care to keep the upper wall as clean as possible,

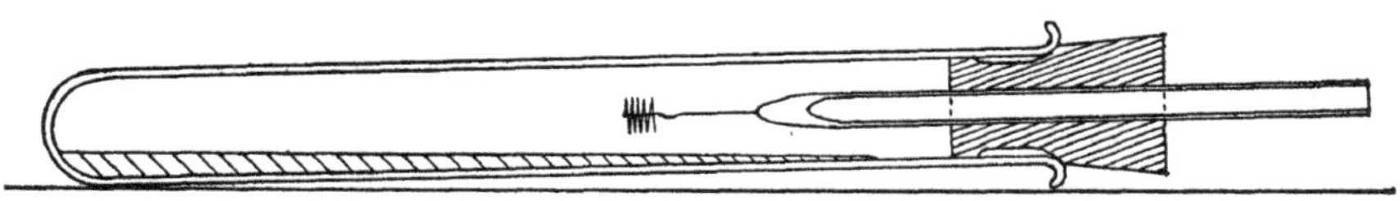

FIG. 36. Sectional view of arrangement as used by Hawes and Skavinski (from *Industrial and Engineering Chemistry*, Anal. Ed., **14**, 917, 1942).

0·2 ml. of acid digestion solution is added to each, and excess water is removed overnight in an oven at 100° C. If the sample volume is less than 0·5 ml., the preliminary drying may be omitted if the first heating is done cautiously.

Digestion is carried out in a sand bath heated by a Tirill burner so adjusted as to cause refluxing to about 4 cm. from the bottom of the tubes. Ebullition and bumping do not occur. The tubes are covered with glass caps cut from the bottoms of soft-glass vials 21 mm. in diameter, to prevent concentration of the digests and to permit operation outside the fume hood.

No attention is required during digestion as a rule, but if the digests are not nearly colourless in about 4 hrs. the flame is removed for 10 minutes, and then a small uniform drop of 30% hydrogen peroxide is placed in each tube, including blanks. Refluxing is continued for at least 30 minutes after the straw colour is gone.

When digestion is complete the tubes are placed in a rack and allowed to cool, then 0·6 ml. of water is added to each and all are stirred by shaking the rack.

Microdiffusion procedures.

(1) *Borsook and Dubnoff* (1939).

They used a small microdiffusion unit constructed of lucite, the dimensions of which are but little different from the Conway ' unit ' (No. 2) (Fig. 33). 0·1 or 0·2 ml. aliquot of the made up acidic mixture is pipetted into the outer chamber.

Since the alkalinising mixture used by them may be much improved for the purpose intended the following description is substituted. Into the central chamber is pipetted 0·1 or 0·2 ml. standard HCl, the strength depending on the level of expected ammonia.

The lid smeared with fixative is placed in position, and 0·5 ml. saturated potassium metaborate run in rapidly, the lid being immediately replaced in position as in the usual technique. The contents are mixed, and diffusion of the ammonia will be complete at room temperature within the hour when 0·1 ml. is used, and in 80 mins. with 0·2 ml.

They titrate the central contents electrometrically at the end, using a glass electrode with the portable Beckmann apparatus, and a long mercury screw burette (Rehberg type) (Figs. 33*a* and 33*b*). Titration with the horizontal burette previously described is simple and more convenient and in the writer's opinion as accurate (*vide* discussion at end of this chapter). If necessary, finer bored graduated tubes can be used so that the final mixture after titration need not much exceed 0·1 ml.

(2) *Needham and Boell's procedure.*

The vessels are now removed from their holders and the digest diluted by the addition of 0·7 ml. NH_3–free distilled water. This is done with a pipette ending at an angle of 45° so as to facilitate the access of the water to all the interior surface of the vessel. The vessel is now tilted as shown in Fig. 34*c* and three pellets of NaOH (corresponding to 300 mg. in all and giving 22% NaOH) are placed in the upper bulb.

Meanwhile the receptor tubes have been prepared. 70 c. mm. of distilled water are placed as a hanging film at the top of the waxed interior and 7 c. mm. of 0·294 *N* HCl delivered into the film from a constriction pipette (*vide* Linderstrøm-Lang and Holter, 1931). The acid contains bromocresol purple in double the strength used in the Linderstrøm-Lang NH_3 method. It is important to put enough water in the receptor tube, as owing to the osmotic pressure difference, the film loses water as it gains in ammonia, eventually falling to perhaps 30–40 c. mm. The receptor tube is well sealed into the digestion vessel with a grease composed of one part high melting-point paraffin wax to two parts white vaseline. The NaOH and the diluted digest are carefully mixed, care being taken to see the heat evolved does not blow out the receptor tube (*vide* Fig. 34*d*). The whole is then cooled in water.

If traces of strong acid are left between the alkaline digest and the film of acid in the receptor, gross errors will obviously result. To eliminate this source of error the vessel is now held as in Fig. 34*e* and rotated so that the alkaline digest runs right round the opening of the receptor tube. It is essential to cool the vessel well before carrying out this part of the technique.

To facilitate the diffusion and absorption of the ammonia the vessels are placed in a horizontal position in a rocker at 37° C. over-night and the titration carried out the following day. The rocker was easily made from an electric gramophone motor connected to an eccentric so as to give the motion seen in Fig. 34 *f*, *g*.

At the end of the period of diffusion the receptor tubes are removed, the ends carefully wiped, and the films titrated against 0·102 *N* NaOH from a Rehberg-Heatley micro-burette of 50 μl capacity (Heatley, 1939). Agitation of the film is carried out magnetically as described by Linderstrøm-Lang and Holter (1931).

In the above diffusion procedure, 95·2% of the ammonia is absorbed in 10 hrs., hence 99·5% may be presumed to have diffused in 18 hrs. The slowness of the diffusion is due, as pointed out by Needham and Boell, to the comparatively very small receptor surface.

(3) *Tompkins and Kirk's procedure* (1942).

The digest is diluted with about 0·5 ml. of water and, after cooling, 0·3 ml. of half saturated sodium hydroxide solution is added as a layer under the digest. This is accomplished by use of a long tipped capillary pipette having a little vaseline wiped over the tip which is inserted to the bottom of the digest. The liquid does not adhere and no nitrogen is lost. The absorption cup is greased around the rim with a little vaseline. To it is added, with rinsing, a measured amount of about 50 c.mm. of absorption acid containing methyl red. The volume of acid and rinsings should not exceed 0·1 ml. or the material might spill when the cell is rotated later. While the vessel is held by the small end, the large bulb is brushed several times with the flame of micro burner. The stopper carrying the acid cup is then inserted and the vessel allowed to cool, leaving in it a partial vacuum. After cooling, the digest and base are mixed in the small bulb and the vessel carefully turned to the horizontal position so as to wet completely the inside walls with the alkaline mixture, neutralising any acid which might have spattered or condensed on the walls. The diffusion is allowed to take place at room temperature, or at 37° in an incubator, for the necessary length of time. Finally, the absorption cup is withdrawn and the acid titrated back in the cup with standard NaOH by means of a capillary burette of a modified Rehberg type. Blanks are run for each series of determinations.

From considerations in Chapter III of the present volume, and as commented on by Hawes and Skavinski (1942), the time for equal percentage absorption of ammonia is independent of the quantity of ammonia examined. This has been likewise demonstrated in numerous analyses. Tompkins and Kirk (1942) have suggested a relation of timing to the quantity being analysed. In their published figures, however, it is clear that they gave adequate absorption times throughout.

(4) *Hawes and Skavinski procedure.*

When digestion is complete the tubes are placed in a rack and allowed to cool, then 0·6 ml. of water is added to each, and all are stirred by shaking the rack. After further cooling each in turn is treated as follows :

Three-tenths of a ml. of 50% NaOH is run down the side of the tube without agitation and so as to avoid wetting the rim and

1 cm. of the upper wall of the tube. Although the solutions become warm, the large difference in specific gravity prevents complete mixing which would cause loss of ammonia. On warm days it may be necessary to cool the tube in a beaker of ice water. It is left undisturbed while the receiver is prepared (see below). The latter is removed from the $1{\cdot}59 \times 12{\cdot}7$ cm. ($\frac{5}{8} \times 5$ inch) test tube in which it is kept, and rinsed with distilled water a few times. The water is shaken off the helix and wiped off the stopper with cleansing tissue. The helix is dipped in 1·0 *M* sodium dihydrogen phosphate solution, deeply enough to cover it, but not so deeply that the platinum-glass junction is wet with the solution. The stopper is lubricated with a drop of glycerol and then inserted in the tube, taking care that the helix does not touch the wall (Fig. 36). If it does it must be rinsed and recharged. Contact may be avoided by placing the rim of the tube against a notch in the bench edge, holding the receiver in the other hand in a position coaxial with the tube, and sliding the tube up over the receiver, which is held motionless while the tube is twisted to seat the stopper.

The digest is now overneutralised by shaking the tube while cooling with a stream of tap water. The entire wall of the tube is coated with a film of alkali by inverting and rotating and the assembly is left overnight on its side on a paper towel or blotter placed on a level surface.

When ready to titrate, the receiver is carefully removed by reversing the insertion procedure. The stopper is wiped free of alkali and the helix is immersed in 1 ml. of water containing either the indicator or a slight excess of quinhydrone, depending on whether the end-point is to be detected colorimetrically or electrometrically. The burette used is a modification of the Conway horizontal burette described in a previous chapter.

Preparation of the helix and stopper.

Five turns of 22-gauge platinum wire are close-wound on a rod 2·4 mm. in diameter (13-gauge), leaving a handle of straight wire 1·5 cm. long. The end of the latter is sealed into a short piece of 6 mm. soft-glass tubing inserted through the hole in rubber stopper. Before making up the receivers these latter are boiled 15 minutes in 1 *N* NaOH, then boiled and rinsed in distilled water. (Using a No. 0 pure gum rubber stopper obtained from the Chemical Rubber Company, Cleveland, they found in this way that less than 0·3 μg. N is contributed by the rubber stopper.)

To prevent the stopper from forcing itself out of the tube or moving so as to bring the helix in contact with the tube wall, a groove is cut about 1 mm. deep and 6 mm. wide and starts about 4 mm. from the small end of the stopper (see Fig. 36).

The helix is cleaned either by immersing for a few minutes in hot cleaning solution or by heating to incandescence for a second in a flame. It is then adjusted by pulling with fine forceps or spreading the centre turns with a knife blade, until the volume of the drop it carries is within 10% of 0·02 ml. (determined by titration). Uniformity is of more importance than the absolute size of the drop. The helices are recleaned at frequent intervals to maintain a uniform drop size.

Review of the foregoing methods.

While the use of a single apparatus for the digestion and microdiffusion would appear the more elegant procedure yet (with the exception of the Hawes and Skavinski method) specially blown apparatus is required which is not useful for other purposes. As will be seen also from the description of the Borsook and Dubnoff method (which the author has somewhat modified at the end) there can be very appreciable saving of time at the microdiffusion stage when the process is conducted by transference of an aliquot of the digested mixture made up to a special small volume. One is dealing with lesser amounts of ammonia than when the whole digested material is used but it need not be less than 0·4 times this quantity, and the microdiffusion technique with the standard 'unit' No. 2 can deal with 1–2 μg. N with no significant effect of the variable chemical error (see Chapter LVIII). This would be achieved with an end-point volume on titration of 0·1–0·2 ml., the Tashiro indicator being used with barium hydroxide titration. In other words, with such quantities the error of estimate will depend only on a total glass error at or about 0·2% (which may be considered rather rigorous) and not on the error incidental to the recognition of the end-point.

In Hawes and Skavinski's method also much longer times must be allowed for the microdiffusion, than when the process is taken as in the Borsook and Dubnoff procedure, the final titration being conducted in 1 ml. fluid, and electrometrically for the avoidance or minimising of the variable chemical error.

The use of electrometric titration in these procedures. The introduction of the electrometric titrations at the end may be said to have its rationale in the diminution or practical avoidance of the variable chemical error.

By keeping the titrated volume suitably low, the variable chemical error can be always reduced to one-half the variable glass error (further volume reduction being of little value). The latter is made as small as desirable by attendance to the principles outlined in Part III. For titrations of unbuffered absorbents of ammonia, and using the Tashiro indicator, the order of the volume limitation is given by the following simple relation (derived from equation 75, Chapter LIX).

$$V = 0{\cdot}5v_g \times 10^4 \times q$$

where V is the volume in c.c., and v_g the total glass error as a coefficient of variation. If we take a working glass error of 0·2% (about as delicate as advisable, considering the over-all error) this would mean that $V = 0{\cdot}1$ when $q = 10^{-4}$, which expressed in terms of N = 1·4 mg.

Keeping the volume of titration at or near to 0·1 c.c., therefore, makes the introduction of electrometric titration an unnecessary refinement.

Where buffered absorbents are used, with subsequent acidic titration, the position is somewhat different. Even under the best theoretical conditions (Conway and O'Malley, 1942) the variable chemical error—depending on the buffering at the end-point—is increased over that using the unbuffered absorption and alkaline titration. Yet this would only mean a further reduction of V above (to avoid the chemical error), or alternatively a corresponding increase in the ammonia level for analysis. Since the end-point buffering under suitable conditions would be increased at most 2 to 3 times over the unbuffered absorptions, it would mean a proportionate diminution of V or alternatively an increase of N analysed to about 3 mg. or upwards.

In the Hawes and Skavinski conditions the receptor volume compared with the donator volume is very small—0·020 ml. compared with about 1·0 ml. or only one-fiftieth. From equation (27) and taking there the *s* value as approximately 0·5, the *p*H for 99% absorption by the drop in the helix would work out at approximately 6·0. At the highest level of their range (100 μg N) they reach 6·5 *p*H, giving a 97% absorption.

To maintain these *p*H levels in the droplet requires the high buffering of $M/1$ NaH_2PO_4. This would correspond to approximately $M/50$ NaH_2PO_4 in 1 ml. fluid at the end. The ideal buffering *p*H (the *p*H at which least error will occur) is here approximately 4·4. At this *p*H the buffering is about 160×10^{-6}, giving $0{\cdot}04 \times 160 \times 10^{-6}$ as a standard deviation for the variable chemical error. If 10 μg are being estimated this results in a coefficient of variation of $(6{\cdot}4 \times 10^{-6}/0{\cdot}7 \times 10^{-3}) \times 100$ or of 0·9% and, similarly, to 0·09% when 100 μg N is being determined. Thus there will be a large region over which the chemical error will be similar to or less than the glass error, if we consider this to be about 0·2% over that region electrometric titration will not appreciably increase the accuracy, but at the lower levels it will be an advantage since the end-point can be more precisely determined and the chemical error reduced below the glass error.

(In connection with the above paragraph the reader is referred to Chapter LVIII.)

Suggested procedures using the standard microdiffusion units.

To the digest as carried out in the Hawes and Skavinski procedure, 1·0 c.c. water is added after cooling, etc. Since 0·2 c.c. digest mixture was originally added containing approximately 0·1 c.c. sulphuric acid, the total volume will be 1·05–1·06 c.c. provided no acid is lost in the digestion process, and in fact it would require 20% differences in the acid loss between one tube and another to give a difference of about 1%. If some care were given to uniformity of heating and consideration of the vaporised acid only a fraction of 1% difference may be presumed to occur.

The microdiffusion procedure may then be conducted with 1 c.c. of the mixed fluid in the outer chamber of the standard 'unit' (No. 1), 1 c.c. boric acid reagent in the centre and 1 c.c. of a solution containing a 20% KOH and 70% potassium metaborate as the alkalinising reagent. Absorption is complete after $1\frac{1}{2}$ hrs. at room temperature, the contents of the central chamber being titrated with $N/100$ acid. In this way the range of 10 to 100 μg N can be accurately covered.

CHAPTER XVI (*Aa* 8)

TOTAL NITROGEN (1–0·1 μg N)

As already dealt with in a previous chapter, Linderstrøm-Lang and Holter (1933) developed a micro-method for determining ammonia which is described as having an accuracy of 0·0055 μg N. From work carried out in the Carlsberg Laboratory these workers with Brüel and Rositz (1946) have elaborated a procedure for total nitrogen determination adapting their ammonia method without loss of sensitivity. In its present form (1946) the method they describe permits the determination of 0·1–1 μg N with the above accuracy.

It will be seen from Chapter VII that, using the microdiffusion principle as there described, 0·07 μg N may be determined with little apparent difficulty to an accuracy of $\pm$ 0·0007 μg N working on the drop-scale level, and it is obvious that such results do not set a limit to what is attainable. A micro-method for total nitrogen based on this method has not been worked out but it would necessarily imply the incineration and subsequent alkalinisation with whatever transfers etc., that would be found necessary. Such stages have been developed with remarkable elegance and thoroughness by Linderstrøm-Lang and collaborators. The method they publish has in addition the advantage of being long tested in practice.

They write: 'We consider the development of a suitable destruction technique as the main result of the present investigation and leave it an open question in what way the ammonia formed should be determined.' Readers specially interested in the procedure are referred to the original paper (Brüel, Holter, Linderstrøm-Lang and Holter, 1946) from which the following account has been taken.

General procedure.

The material for investigation is introduced into a narrow tube (40 mm., 1·8 mm. bore) 4μl digest mixture added, and the water removed in a vacuum dessicator over P_2O_5 (48 hours duration).

One μl concentrated sulphuric acid is added and the incineration carried out for about 6 hours in a small sulphuric acid bath at 295° C. After cooling the digestion mixture is transferred with

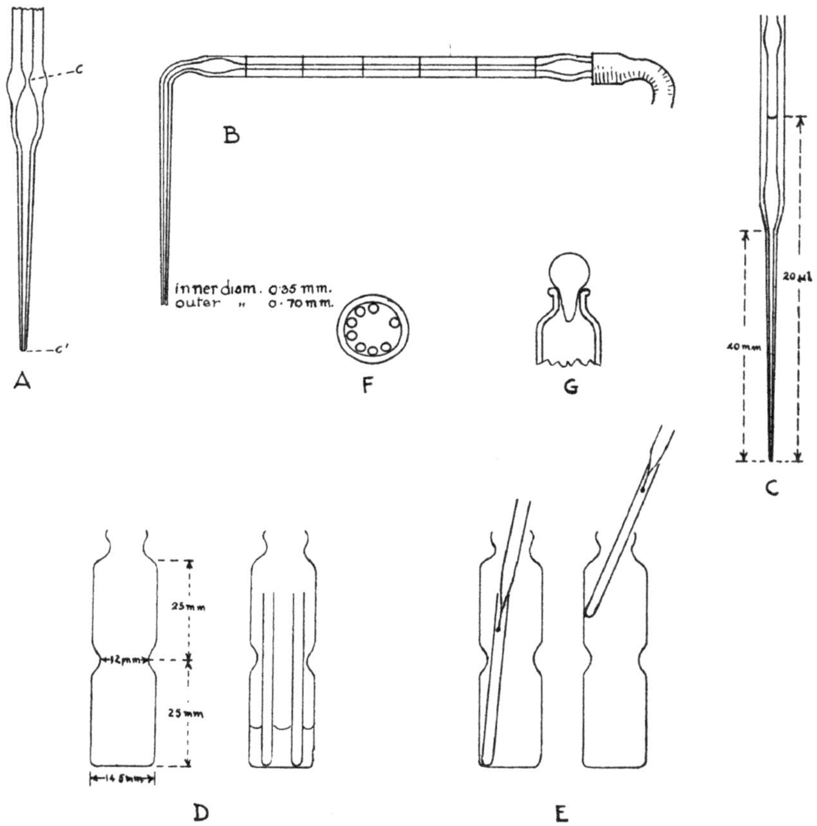

FIG. 37. Apparatus used in the technique of Linderstrøm-Lang and associates (from Brüel *et al*, 1946).
A, *B*, and *C*—various pipettes referred to in text.
D and *E*—digestion tubes and holders and manner of withdrawing tubes.
G—glass stopper for holder.

water to the bottom of a larger paraffin-coated distillation tube of the type previously described in Chapter VII and the ammonia determined as there described with some small changes in the new procedure.

The Incineration Procedure.

The incineration tubes. Tubes 40 mm. in length with internal diameter of 1·8 mm. and outer diameter of 2·4 mm. are used. They should not vary too much from these dimensions, especially with regard to the inner diameter. They may be easily made by sawing them from a wide glass tube of suitable wall thickness, and the end should be well rounded off.

The tubes are cleaned with chromic acid and sulphuric acid rinsed out with cold distilled water, which is led in a rapid stream through them, avoiding direct contact of the rim with fingers.

They are dried in vacuo at 100° for 10 min. and stored out of direct contact with the atmosphere.

Filling of tubes.

The organic material is introduced either by fine micro-pipettes or by means of a fine glass needle if solid. Holter (1943) describes a method for introducing living organisms. After the material $4\mu l$ of a ' digest mixture ' is introduced.

The pipette used for introducing the digest mixture is the Levy semi-automatic pipette (Fig. 37A). This is connected with an air pump and a pressure regulator, adjusted in such a way that the capillary forces at c and c' just counterbalance the pressure exerted when c', the tip of the pipette, is free, so that the pipette is discharged automatically as soon as c' touches or penetrates a liquid surface.

Care is taken to deliver the solution at the bottom of the tubes and that it is not drawn up between the pipette and the inner wall. The pipette is rinsed outside and inside before a series of pipettings.

The tubes are then placed in a vacuum desiccator and the water evaporated at room temperature. The main quantity of water is evaporated at 150 mm. Hg, which generally takes about 24 hours. Then the pressure is lowered to 0·1 mm. Hg and the drying continued for another 24 hours.

After this $1\mu l$ concentrated sulphuric acid containing selenium is introduced by means of a pipette of type shown in Fig. 37*b*. It is filled and blown out with the mouth. $1\mu l$ corresponds to a length of about 10 mm. (The scale is simply a strip of squared paper glued to the capillary.)

Care should again be exercised against seals forming in the upper part of the tubes, which however need not be discarded unless the acid is drawn up as high as the rim. The pipette before use should be rinsed well outside and inside.

Digest mixture. This is referred to as 'blue mixture' and contains—

1 g. of $CuSO_4, 5H_2O$ }
10 g. of K_2SO_4 } diluted to 100 ml.
0·2 g. of sucrose 5 ml. conc. H_2SO_4 }

Sulphuric acid with selenium. This is prepared by boiling 1 ml. H_2SO_4 with 10 mg. selenium, until the solution is clear.

The incineration. This is carried out in a small flask as shown in Fig. 37D. Such were made from insulin ampullae by pulling a constriction approximately in the middle. The bottom of the flask is covered with 1 ml. concentrated sulphuric acid containing 0·4 g. K_2SO_4 (about the same composition as the mixture in tubes. Before using the first time the solution is boiled in a flask to drive off water and gases.) When not in use the flask is closed with a small glass ball. Preliminary to the incineration the flask is placed in a small electrically-heated copper block with cylindrical holes fitting the flasks tightly and being about 14 mm. deep. The temperature of the block is kept constant at 295° at which temperature the sulphuric acid (containing K_2SO_4) is clear, fumes readily, but does not boil.

When inserting the tubes the flask is removed from the block permitted to cool for half a minute in the air ; the tubes are picked up one by one by means of a forceps and placed in one layer following the wall of the constriction (see Fig. 37F). During the first few minutes after replacing the flask, it is advisable to observe the tubes, and if liquid seals form, pressed up by gases evolved to beyond the middle of the tube, the flask is removed from the copper block for a minute or two, when the seals move downwards and break. The occurrence is rare and after one cooling process it generally disappears.

The process then requires no further supervision, and is allowed to proceed for 5 to 6 hours.

After this the tubes are removed from the slightly cooled

digestion flask by means of a conical glass rod, as shown in Fig. 37E. They are rinsed on the outside with distilled water, dried with a clean towel and placed in a desiccator over P_2O_5 where they remain until the next process.

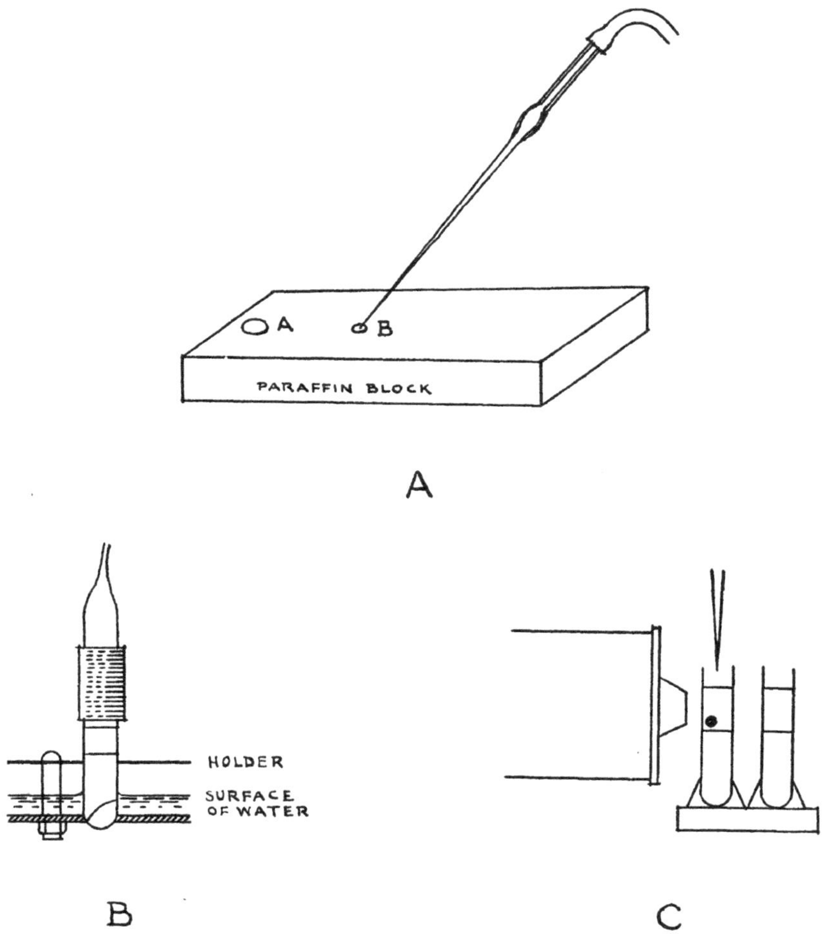

FIG. 38. Further apparatus connected with the technique of Linderstrøm-Lang and associates (from Brüel *et al*, 1946).
A—paraffin block and method of usage.
B—tube in position for ammonia distillation.
C—titration arrangement.

The Distillation stage (including microdiffusion).

(The reader is referred to the preface of this edition for the distinction drawn between these two processes.)

This procedure is carried out as described in Chapter VII with some changes mentioned below. Firstly the transference of the digested mixture to the distillation tubes will be described.

Transference of the acid digest. A drop of distilled water (volume about 20μl) is placed on a clean paraffin block by means of an ordinary hand pipette. About one-third of this drop is sucked up into another hand pipette (*P*) without constriction. This pipette is introduced by hand into the incineration tube (the end of this being previously heated slightly in a flame and its tip is led to the centre of the bottom of the tube. The water is blown out (by mouth) immediately sucked up again with the greater part of the acid digest, and finally blown out to form a new drop (*B*) on the paraffin block some distance from drop *A* (Fig. 38A).

The second third of drop *A* is now sucked up into *P* and introduced as before, but this time the entire inner wall of the tube is moistened with water by leading the stem of the pipette along the inner wall some 5–10 times around. The contents after sucking up are in turn transferred to drop *B* and the rinsing process is repeated with the last third.

With some experience it is easy to avoid sucking air through the tip of the pipette when the last trace of *A* leaves the paraffin.

Finally, drop *B* is sucked up completely in the pipette and transferred to the bottom of the distillation vessel. To avoid contact of the acid fluid with the paraffined wall where the receiving seal will be situated, the pipette is fastened in a vertical position and the distillation tube lifted towards the pipette by means of a pipetting stand with rack and pinion.

The pipette need not be rinsed with water to collect the negligible quantity of diluted digest adhering to the inner wall, and may be used repeatedly unless special conditions make a rinsing advisable.

The technique from this onwards is similar to that described in Chapter VII with some changes. To the transferred drop of water plus acid digest (21μl) 9μl of 18 *N* NaOH are added. The NaOH is introduced directly into the drop of diluted digest. Quickly afterwards a seal of water—45μl—is introduced into the neck of the vessel and 7μl of a 0·015 *N* H_2SO_4 are added to the seal using a semi-automatic pipette with pressure adjustment.

The vessel is covered with a rubber cap and placed halfway immersed in a water thermostat at 40°. The distillation (with microdiffusion) requires about 1·5 hours.

In titrating, excess solution of base with bromcresol green is added to the receiving seal and titrated back with 0·01 *N* HCl.

The base was added as $18\mu l$ of the following mixture—

0·5 ml. $M/15$ Na_2HPO_4

2 ml. bromcresol green (aqueous solution containing 0·4 mg. dye per ml., *p*H 4·6)

2·5 ml. boiled water.

Such solution is made up fresh every day and is kept in a well-closed vessel ; the highest accuracy is not required in making it. The acid used in titrating contains the same bromcresol green solution as that in the receiving seal. The end-point is marked by a green-blue colour corresponding to a *p*H of about 4·6. A colour standard is first made by titrating a blank. The position of the titration has been changed to the vertical (see Chapter VII) which simplifies the process and makes possible the use of ordinary titration stands.

The pipette used in the transfer. It is important that the pipette used in the transfer has suitable dimensions with regard to both length and thickness. The opening at the tip must not be too narrow because if so, small crystals, dust particles, or pieces of paraffin may block it up. Neither must it be too wide because it is very difficult to blow out the pipette by mouth without blowing air through the tip, which is to be carefully avoided because it gives rise to spattering on the paraffin block or in the distillation vessel. The stem of the pipette should not be too thin or it is liable to break when pulled along the wall of the tube, nor too thick on account of the drawing up of liquid between pipette and tube.

Though these conditions may look exacting, in practice it appears easy to make good pipettes of the required type (*vide* Holter, 1943, page 435).

Some Further Notes on the Method.

As the authors of the above method leave it an open question as to the way the ammonia formed should be determined, the following suggestions may be made. It will be seen from the determination of ammonia of the order of 0·07μg. in Chapter VII that at the stage where the diluted acid digest is collected in drop B this may be regarded as one of the drops in that description, the other being, say, 40% KOH. The drop receiving the ammonia could be suitably formed of borate with indicator as described in Chapter VII. In this way the process is very appreciably simplified. Thus, the transfer of drop *B*, the pipetting of the receiving seal with the exact amount of acid added, plus the addition of an exact amount of alkali mixture are all avoided as well as the heating process. The borate drop need not be accurately pipetted, provided it is at the right *p*H.

It would also appear that in this way the diffusion time could be much shortened.

CHAPTER XVII (Aa 9)

UREA (BLOOD AND URINE)

Determination of urea in 0·2 ml. blood. (Revised method.)

THE revised method using boric acid absorbent with indicator is suitable for blood urea determinations, as all ranges can be dealt with without altering the reagents.

Procedure.

Collection of blood (human subject). This is conveniently taken from the finger, using for the small incision a Franke spring lancet (see Fig. 39), which is very efficient and practically painless. The blood may be squeezed out into a small dry oxalated tube. This tube, of about 1 ml. capacity, is prepared by pouring into it some saturated potassium oxalate and then emptying out. The residual fluid on the walls forms a fine layer of the oxalate on drying either in the oven or quickly over a Bunsen flame. About 0·5 ml. blood may be collected as a rule, allowing duplicate 0·2 volumes for analyses. If the blood is not immediately analysed or is being sent by post, the tube is closed by a small rubber cork.

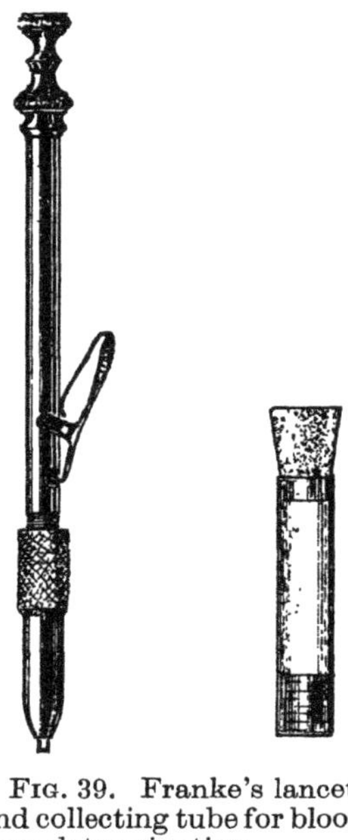

FIG. 39. Franke's lancet, and collecting tube for blood urea determination.

Urease action and ammonia determination. Into the central chambers of the 'units' (No. 1) is pipetted 1 ml. volumes of the boric acid reagent. Such deliveries need be only roughly measured. Into the outer chamber is pipetted 0·2 ml. of the blood from an Ostwald or straight tube pipette, with careful blow out at the end of the delivery. For the highest accuracy such pipettes may be calibrated for such blood delivery.

(This may be done by taking the mean weight of five deliveries into a weighing bottle, and calculating the volume as in Part III allowing, however, a sp.g. for blood as compared with water of 1·025.) 0·5 ml. of the urease-phosphate mixture is then dropped on the blood, and the lid—smeared with fixative—placed in position, the contents of the outer chamber being mixed by rotation as usual.

The ' unit ' is now left aside on the bench for 15 minutes, after which time 1 ml. (approximately) of saturated potassium carbonate is added in the usual manner. It is added without blowing and with the lid slightly displaced. After replacing the lid the ' units ' are left aside on the bench for 1½ to 2 hours. The lid is then detached and the central chamber titrated to a faint permanent red colour. A blank with the reagents is run at the same time, and an occasional control determination with 50 or 100 mg. urea/100 ml. is also advisable.

Calculation. With the standard solution used this is very simple—the large divisions on the burette corresponding to the absorbed ammonia are multiplied by 3 and the result corresponds to mg. urea/100 ml. blood. A correction for the pipette if this does not deliver 0·2 ml. exactly is unnecessary when the ratio of the blood and urine concentrations is being taken as in renal function tests.

Solutions, etc., required for the blood urea determination.

1. *N/50 hydrochloric acid.*
2. *Boric acid reagent with indicator.* This is made up as described on page 99.
3. *Saturated potassium carbonate or saturated potassium metaborate.* 112 g. potassium carbonate or 70 g. potassium metaborate may be allowed per 100 g. water.
4. *Fixative.* The best quality vaseline may be used, or alternatively one of the water-soluble fixatives described on page 97.
5. *Urease solution (glycerine extract of jack bean).* A preparation found very satisfactory in this laboratory is a glycerine extract of jack bean prepared in the manner described by Schmidt (1928). 22 g. of finely powdered permutite are washed with 2% acetic acid, which is then decanted off, the permutite being subsequently washed twice with distilled water. The permutite is mixed with 45 g. of finely powdered jack bean meal and 75 ml. of distilled

water. 225 ml. glycerol is then added and mixed. The whole is filtered and the filtrate collected.

It filters very slowly, but if a large folded filter paper is used, the whole may be poured on at once and allowed to filter overnight, when a considerable fraction of clear filtrate will have come through, the remainder being allowed to collect over some days. When the filtrate is collected it is stored in the refrigerator in a stoppered bottle. It will keep for several months at least. An advantage of this extract is that, besides being very active, it can be measured out volumetrically without the trouble of weighing very small quantities as with the solid preparation; at the same time any active preparation of urease will be found suitable. Besides the jack bean, the soya bean may also be used for the extract but is not so potent, and apparently also the seeds of the water melon, as reported by Damodaran and Sivarakrishnan.

6. *Phosphate solution.* 3 g. of anhydrous sodium phosphate (Na_2HPO_4) and 2 g. of anhydrous potassium acid phosphate (KH_2PO_4) are weighed out, dissolved in water and made up to 100 ml. (Both salts are stocked as Analar reagents by the British Drug Houses.)

7. *Mixed urease and phosphate solutions for immediate use.* 1 ml. of the phosphate and 1·0 ml. of the urease solution are diluted to 10 ml. and 0·5 ml. of the mixed fluid taken for each determination. Smaller or larger amounts of this mixture are made up according to requirements.

8. *Alternative preparation of urease-phosphate stock.* Kawerau (1941) found that incorporating the phosphate in the original preparation produced an extract which has very good keeping powers and filters much more quickly than the stock described above. The preparation is made as follows: To 10 g. of jack bean 20 ml. water is added and mixed well, then 100 ml. phosphate buffer at *p*H 7·4 (3 g. anhydrous Na_2HPO_4 and 2 g. anhydrous KH_2PO_4 in 100 ml. water) and 20 ml. glycerol are added. This mixture is shaken well, allowed to stand for a short time and the fluid decanted on to a large folded filter paper. (The mixture may also be largely cleared by centrifuging a short time before filtering.) After filtering the filtrates may be cleared of ammonia—if this is considered desirable—(though it could be allowed for afterwards in blank determinations)—by permutite. 5 g. finely powdered permutite are washed with 2% acetic acid and allowed to settle; the acid is decanted and the permutite washed twice with distilled water. To the permutite is added the urease-phosphate preparation, the mixture is well shaken for some time and allowed to settle, when the clear fluid is decanted off and stored in the refrigerator. A small volume of this is diluted three times when required for a series of determinations.

Note on the time required for the estimation.

The time of the above method could be reduced to one hour (including the period of ferment action), if care be taken to pipette the 0·2 ml. blood so that it is collected in one region and 0·2 ml. of a urease-phosphate mixture about 3 times stronger than as described, is carefully dropped on the blood, to ensure that it mixes adequately with it. Ten minutes is then allowed for the ferment action and 50 mins. after adding the alkali.

Alternatively, the estimate could be carried out without altering the volumes in about one hour if incubation at 38° C. were used throughout the fermentation action (allowing 10 mins.) and the microdiffusion (45 mins.). Other alternatives will occur to the reader.

Accuracy of the method with 0·2 ml. blood. The standard deviation of the single determination of urea in a standard solution of 50 mg./100 ml. was found to be 0·23 mg./100 ml., using the hydrochloric-barium-hydroxide procedure. This is a rather marked advance on the method as originally described, the standard deviation being then given as 0·6 mg./100 ml. The difference is apparently due to the introduction of the horizontal burette combined with the use of barium hydroxide for titration, the incorporation of the indicator in the acid, etc. This error practically corresponds to that resulting from the pipette (which was used to deliver without washing out) and titration errors alone, which under the conditions would be approximately 0·43%. It may be said, therefore, that the error of this determination depends only on the glass error, and this can be almost indefinitely lessened in accordance with the principles of Part III. *The method, in short, can be made to give almost any accuracy we wish.*

Using the boric acid-hydrochloric acid procedure as described here, from a series of 51 routine duplicate determinations the standard deviation of the individual estimate was calculated as 0·34 mg. urea/100 ml. It is thus scarcely appreciably less accurate in practice than the HCl-Ba$(OH)_2$ procedure.

Comparison of accuracy with that in the aeration method, using 2–3 ml. blood. The accuracy of the aeration method has been expressed by Addis and Watanabe (1916), and later by Kay and Sheehan (1934), as mean differences between duplicate analyses.

For the urea determination in blood as described above, with the first 150 duplicates in the protocols the mean value of the difference was found to be 0·3 mg./100 ml. Using 3 ml. blood as in the Van Slyke and Cullen (1914) method, Addis and Watanabe in 1916 carried out a large number of estimations and found with 90 duplicates an average difference of 1·5 mg./100 ml. Later they succeeded in reducing this to 0·5 mg./100 ml. Kay and Sheehan introduced into this aeration method certain improvements in technique designed for greater accuracy. Using 2 ml. blood, they found a mean difference for duplicate analyses of 0·4 mg./100 ml. Some of their experimental blood ureas were very high, but they state that this apparently did not affect the duplicate differences.

This statement of Kay and Sheehan may be taken as it stands for the normal blood urea concentration, for which the microdiffusion method is more accurate with only one-tenth the amount of blood used, as in ordinary routine working we have found a mean duplicate difference of only 0·3 mg./100 ml.

With regard to the very high blood ureas examined by them they have comparatively few observations with a recovery of only 96%, and for this reason, and also what we know of the errors of titration, etc., an independence of duplicate differences to height of blood urea may be regarded as very improbable.

The efficiency of urea recovery in the microdiffusion method. For the aeration method Kay and Sheehan found only 96% recovery, this being subsequently increased to 98–99% by Kay and Reid (1934) using much greater aeration volume. The recovery of urea added to whole blood has been shown to be quantitative in the microdiffusion method (1933). Gibbs and Kirk (1934) have confirmed these results, recovering the quantitative amount with as little as 0·00279 mg. urea N. Rappaport and Gutman (1935) also confirm. Lee and Widdowson (1937) have found a mean 99·6% recovery with urea added within the clinical range, and Abelin finds the recovery quantitative using the boric-sulphuric technique. Gibbs and Kirk, moreover, have demonstrated the agreement between urea whole blood analyses and corresponding protein-free filtrates, which they emphasise as a special advantage of the method.

Clinical use and applicability for renal function tests. The method is particularly suitable for renal function tests and also for general clinical use. The accuracy depends on that of delivering 0·2 ml. blood and on HCl titration of the contents of the inner chamber, using the horizontal burette (the error of delivering 1 ml. of the boric acid reagent not affecting the result). In Part III it is shown how any degree of accuracy desirable can be reached.

Criticism by Lee and Widdowson (1937) of the procedure previously used, namely, that the range of urea concentration determined was restricted, no longer applies to the method as described above.

It is to be noted that increased ammonia has never been observed from the action of the arginine-arginase system in the method as carried out here.

The use of the urinary and blood urea concentrations in determining renal function is given in the Appendix.

The *p*H and salt effects on the urease action.

The pH *effect.* Van Slyke and Zacharias (1914) very fully investigated the general *p*H and salt effects on the urease action. Kay and Reid (1934) have examined the *p*H effect on total recoveries. They worked, however, on comparatively large amounts of urea compared with the ‘ unit ’ range, and it was thought desirable to investigate again the question in relation to the present method of determining urea.

In the investigation the outer chamber in a series of ‘ units ’ contained 0·2 ml. of a 75 mg. urea/100 ml. solution and 0·5 of a 1 in 10 glycerine extract of urease as previously described under the blood urea determination. The diluted glycerine extract also contained phosphate in $M/20$ concentration of varying *p*H. The action was allowed to take place at room temperature (18° C.) and interrupted with 1 ml. saturated carbonate after a certain time period.

As may be seen from Fig. 40 *the optimum* pH *is somewhat to the alkaline side of neutrality,* but there is very little difference between the action at a *p*H of 7·0 and 8·0. There is a rapid fall in activity to the acid side of 7·0 *p*H, but only a slow fall as the *p*H rises beyond 8·0.

The salt effect of the phosphate. In this investigation the same

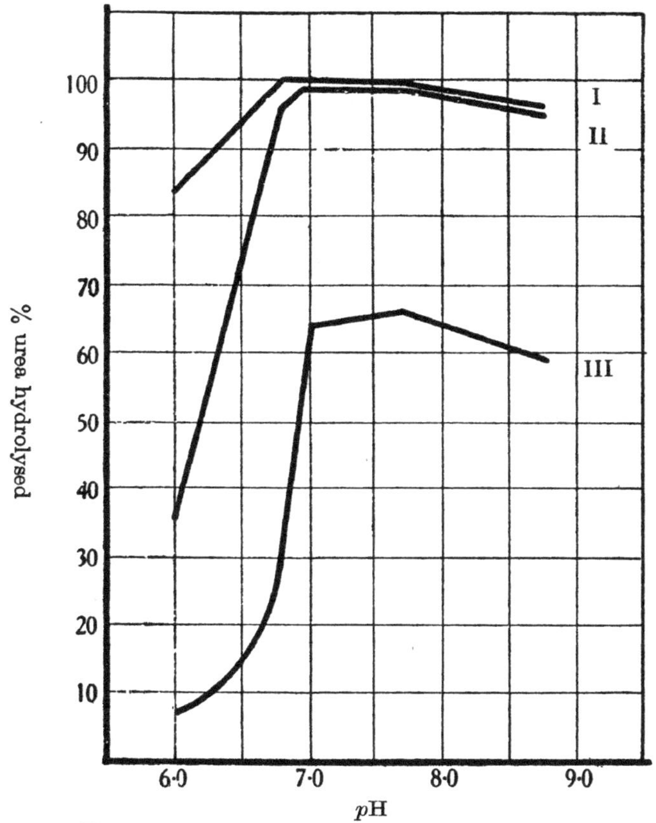

FIG. 40. pH effect on the urea hydrolysis. 0·2 ml. of 75 mg. urea/100 ml. solution plus 0·5 ml. urease solution containing $M/20$ phosphate at varying pH.

I. 10 minutes action.
II. 5 minutes action.
III. 1 minute action.

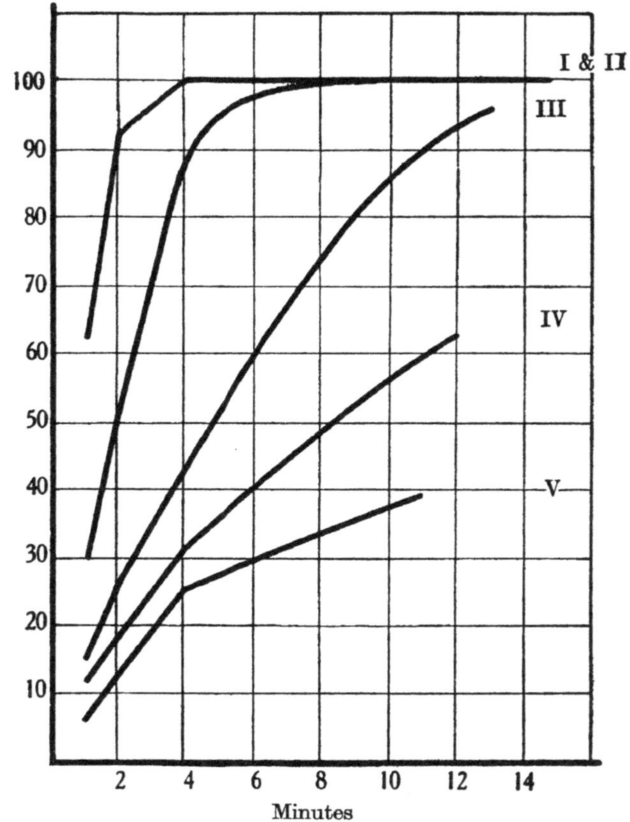

FIG. 41. Time course of urea hydrolysis with different phosphate concentrations at constant pH of 6·8, the mixture being otherwise the same as in Fig. 40.

I. No phosphate added.
II. $M/20$ phosphate in the urease solution.
III. $M/10$ phosphate in urease solution.
IV. $M/4$ phosphate in urease solution.
V. $M/1$ phosphate in urease solution.

procedure was adopted, except that the strength of the phosphate was varied at a constant *p*H of 6·8.

It will be seen (Fig. 41) that *the salt action of the phosphate is very marked, and to economise the time of action it is inadvisable to have a stronger phosphate solution than M/20 in the* 0·5 *ml. urease-phosphate mixture added.* The question of entirely omitting the phosphate was in fact considered since the rate appears to go as rapidly without it for the small amounts of urea hydrolysed. It was decided to include the buffer in its present composition (as described under the blood urea method) since it ensures against any wide change in the *p*H value.

Distribution of the normal blood urea in man. Where blood ureas are being determined it is often of interest to compare such with the normal distribution. Fig. 42 (*C*) shows the distribution of the normal blood urea figures for the human subject. The distribution is for 169 determinations in the literature from the figures of Addis and Watanabe (1916), Walker and Rowe (1927), Austin *et al.* (1921), McLean (1915), Conway, O'Connor and O'Donovan (1937) and some unpublished data from the author's laboratory. A maximum of five observations was allowed for any single subject, the first five listed being taken. It is to be understood that the subjects were not only normal in renal functioning, but normal in general health, hospital cases being excluded. The mean value of the 169 observations is 31·0 mg./100 ml. with a standard deviation of 7·5 mg./100 ml. It will be seen from the figure that the distribution does not correspond to a 'normal' distribution, but tends to show groupings of results in about three regions. This effect, if confirmed by more extended study, may be largely due to the rise following meals or protein intake, but may also be associated with other metabolic causes. The distributions (*A*, *B*) for single subjects investigated in the author's laboratory, for whom there is a large number of determinations, are interesting to compare with the general histogram. It has happened accidentally that the two subjects investigated had mean blood urea values in the upper and lower regions of the general histogram, the mean for one being 40·2/100 ml. with a standard deviation of 4·7 mg./100 ml. (122 observations) and for the other 26·8 mg./100 ml. with a standard deviation of 3·4 mg./100 ml. (64 observations).

Fasting values for D. K. O'D. (before breakfast and 3–5 hours after meals) gave an average of 38·4 with a standard deviation of 4·3 and the fasting values of E. J. C. gave 25·9 with a standard deviation of 2·5 mg./100 ml.

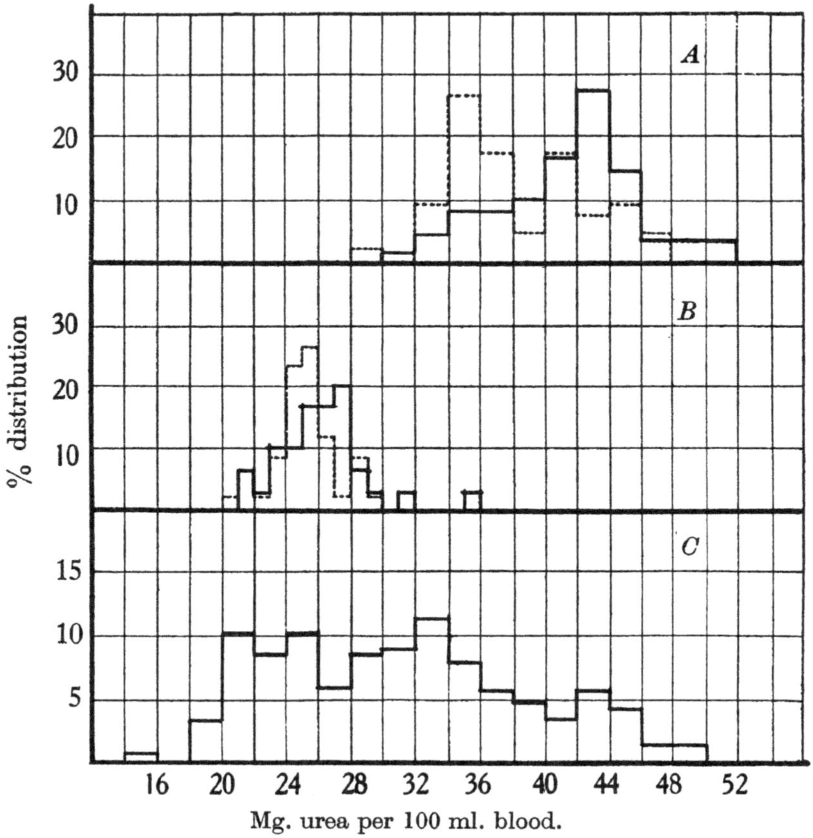

FIG. 42. Distribution of normal human blood urea.

A— 122 obs. on subject D. O'D. Dotted line—obs. before breakfast.

B—64 obs. on subject E. C. 34 before and 30 after light lunch. (Dotted line—before meal.)

C—160 obs. from general literature.

The histograms (*A*, *B*) show very clearly the influence of meals, which for D. K. O'D. shifts the mode or peak 8 mg./100 ml. and for a light meal with E. J. C. raises it 2 mg./100 ml.

Determination of the urinary urea. This determination is usually carried out in conjunction with that of the blood urea, and is done most conveniently after suitable dilution, so that the same solutions, pipettes, etc., may be used.

Seeing that the concentration in the urine is related to the rate at which urine is excreted, this rate must be considered in accurate working, the dilution being so made that a considerable fraction of the absorbing acid is always neutralised by the formed ammonia. Suitable dilutions may be made from the data in Table XXXVII, giving average concentration ratios of urine/blood in the varying urine volumes.

After diluting, the urine is analysed in the same manner as the blood, 0·2 ml. being introduced into the outer chamber, 0·5 ml. of urease phosphate solution added, etc.

One difference between the blood and urine determinations consists in the fact that even the diluted urine contains a small concentration of preformed ammonia which will amount to about 7% of the ammonia from the urea at 25 ml. per hour, 5% at 50 ml. and about 3% at or greater than 150 ml. In contrast with the urine the blood contains no ammonia on shedding (1935), and with human blood no appreciable ammonia compared with the urea content up to 2 hours after shedding. (After 24 hours, sterile human blood will contain preformed ammonia corresponding to about 1 mg. urea/100 ml. blood and formed chiefly by the breakdown of adenylpyrophosphoric acid.) Consequently it is necessary in urinary urea determinations to set up another ‘ unit ’ with 0·2 ml. of the diluted urine in the outer chamber, 1 ml. of saturated potassium carbonate being added as usual and the ammonia found subtracted from the total.

(At the same time it may be noted that it is now customary in determining the urea content of urine for renal function tests to ignore the relatively small free-ammonia content, including same as part of the urea ammonia formed after urease action.)

Calculation.

EXAMPLE. Urine rate - - - - - - 53 ml. per hour.

(From Table XXXVII the mean urinary concentration at this rate is 56·8 times that of the blood. A dilution of 56·8 times may then be made by adding 55·8 ml. distilled water from a burette to 1 ml. urine pipetted into a conical flask, and mixed. The urinary urea concentration will then on the average correspond to that of the blood. If considered more convenient a dilution of 50 times may be made by making 2 ml. urine up to 100 ml. in a volumetric flask.

It may be supposed that this latter is carried out with the following results :

Analytical ' unit ' -	- - -	13·30 large divs.
Blank ' unit ' (urine)	- - -	1·25 ,, ,,
Blank ' unit ' (urease)	- - -	1·10 ,, ,,

Then
$$(13{\cdot}30 - 1{\cdot}25 - 1{\cdot}10) \times 3{\cdot}0 \times 50$$
$$= 1642 \text{ mg. or } 1{\cdot}64 \text{ g./100 ml.}$$

(It is to be understood that the boric acid reagent has the end-point colour on first pipetting into the ' units '.)

Any pipette correction is made as usual, but if the ratio of the blood and urine concentrations are being measured—as in renal function tests—it is unnecessary when the same 0·2 ml. delivery pipette is used.

Distribution of urinary urea concentrations in the normal human subject pursuing an ordinary living routine. Unlike the urea concentration in the blood, that in the urine is very dependent on the rate at which the urine is being excreted. The distribution of the urinary urea for a number of normal subjects is given in Fig. 43 (*C*), and similar distributions for the two normal subjects considered under the blood urea. What is at first peculiar about the three distributions is their rather marked similarity, seeing that the blood distribution is very dissimilar, particularly for the two single subjects, and also that the urine rate has a very marked effect on urinary concentration. The main bulk of the distribution lies in each case between 1·8 and 2·8 g./100 ml., with the chief mode at 2·1, 2·3 and 2·3 g./100 ml. respectively. In each case there is an appearance of a higher mode at about 2·6 g./100 ml., this latter being due very likely to the influence of meals as shown by dotted line in distribution *B* for one subject (E. J. C.). The appearance of a third mode at a lower concentration level is obviously due to diuresis, and from that mode downwards we have scattered observations due to high diuresis.

Distribution of the normal urine volume. It is of interest in connection with the other distributions and in relation to the question of urine dilution for urea analysis, to consider also the distribution of normal urine rates, under apparently normal conditions. As shown in Fig. 44 the distribution is similar to that of the urine urea, but inversely related, high rates being associated with low concentrations and vice versa. From the distributions in Fig. 44 (corresponding to previous Figs. 42 and 43, *A* and *B*

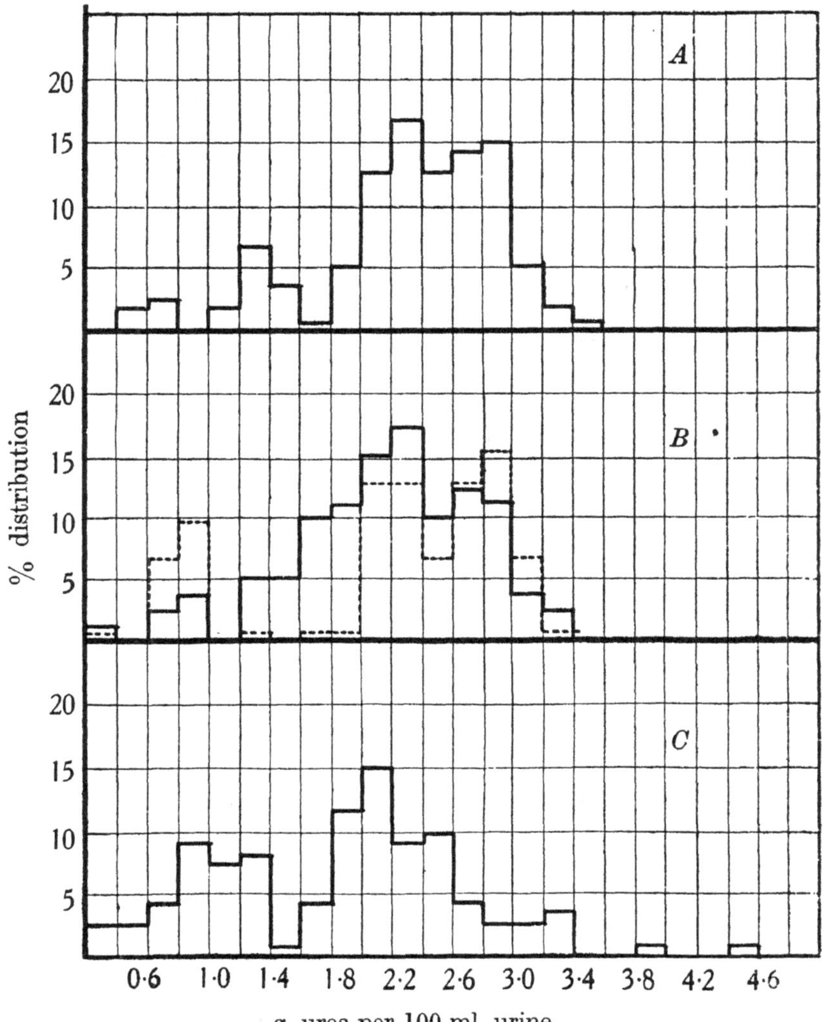

FIG. 43. Distribution of urine urea concentrations in the normal human subject. Water-drinking to produce diuresis excluded.

A—120 obs. on subject D. O'D.

B—81 obs. on subject E. C. Dotted lines gives obs. after light lunch.

C—111 obs. from general literature.

being for the same two subjects and *C* the general distribution) data after water drinking with the intention of producing diuresis are excluded. The main mode of the three histograms for the general group and two subjects lies around 40 to 50 ml. per hour, corresponding approximately to 1 litre per day. It is around this

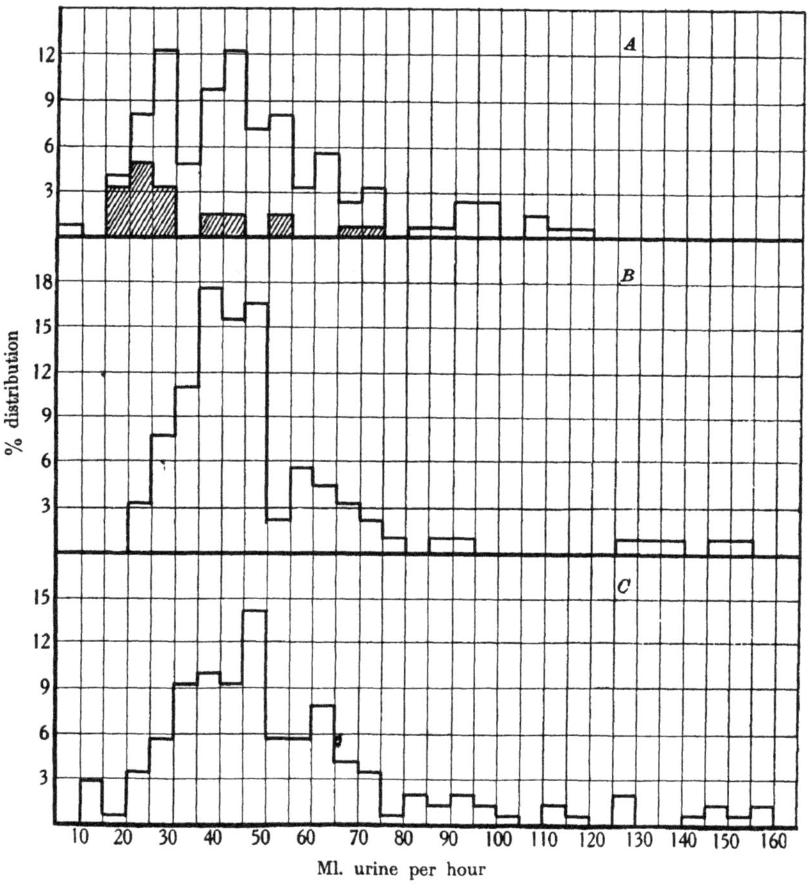

FIG. 44. Distribution of urine volume in normal human subject. Water-drinking to produce diuresis excluded.

A—obs. on subject D. O'D. as in Fig. 43. Shaded section—obs. before breakfast.

B—obs. on subject E. C. as in Fig. 43.

C—obs. from general literature as in Fig. 43.

point therefore that we can expect to have most observations in the normal subject. There is a tendency to a subsidiary mode at about 62 ml. per hour, corresponding to about 1,600 ml. per day.

CHAPTER XVIII (*Aa* 9)

UREA IN TISSUES

Principle. The tissue is ground with water and a little pure quartz sand, largely cleared by centrifuging and the urea hydrolysed in an aliquot part by the action of urease as with the blood and urine determinations, the remaining procedure being also similar, but, owing to the small amounts, the HCl-$Ba(OH)_2$ technique is used.

Procedure. The tissue section for analysis (100 to 200 mg. are usually suitable, though the amount may be varied greatly) is weighed in a covered watch glass and then transferred to a small mortar, 3 ml. of distilled water added from a standard pipette and a little pure quartz sand (Merck's). The tissue is ground to a fine suspension, the mixture poured into a small centrifuge tube and spun for a minute or two.

The following ' units ' are then set up, using the solutions of Range III (Table VI, p. 92) :

Analytical ' unit ' (urease plus extract) -	outer chamber—0·5 ml. extr. plus 0·5 ml. urease phosphate mixt. (as for blood urea).
Blank ' unit ' (extract) - - - -	outer ch.—0·5 ml. extr.
Blank ' unit ' (urease) - - - -	outer ch.—0·5 ml. urease-phosphate.

The inner chambers of the ' units ' are filled with 1 ml. of the absorbing acid—made up as described for ammonia determinations.

After 15 minutes on the bench 1 ml. of saturated potassium carbonate is added to each ' unit ' and the inner chamber titrated after two hours' absorption. (If it is thought very desirable to reduce the time required for this, twice the amount of tissue may be used in the grinding, and 0·2 ml. extract with 0·2 ml. urease-phosphate used instead of the above proportions, then 1 ml.

carbonate. Only 1 hour need then be given for the absorption at room temperature.)

It is advisable here as always to work with duplicates for each procedure. Such procedure as outlined above will no doubt require suitable modification in accordance with the experimental conditions, which may involve the injection of urea.

Calculation.

Urea in 1 *ml. extract.* (0·5 ml. used.) When the titrating fluid in the burette is made up as 16·7 ml. $N/10$ to the litre, each large division (0·01 ml.) is equivalent to 0·5 γ urea. Since 0·5 ml. of the extract is used, it follows that the number of large divisions on the burette corresponding to the urea ammonia gives the γ urea per ml. extract.

This is illustrated by the following example :

Analytical ' unit ' (urease plus extract)	- 43·7	large divs.
Blank ' unit ' (extract) - - -	- 59·4	,, ,,
Blank ' unit ' (urease) - - -	- 59·5	,, ,,
1 fill acid - - - - - - -	- 60·0	,, ,,
Urea ammonia in extract, 16·3 – 0·6 – 0·5	= 15·2	,, ,,

or 15·2 γ urea/ml. extract (0·0152 mg./ml.)

Urea in 100 *g. tissue.* (0·150 g. used, for example.) It will be assumed that the tissue contains 80% water, though small deviations from this will not appreciably affect the result. The total amount of fluid in the mortar is therefore 3·0 plus 0·8w, where w is the weight of the tissue in g.

Each g. tissue corresponds to $\frac{3\cdot0 + 0\cdot8w}{w}$ ml. extract and 100 g. tissue to $\frac{3\cdot0+0\cdot8w}{w} \times 100$ ml. extract.

Hence if U is mg. urea in 1 ml. extract, the amount of urea in 100 g. tissue is given by

$$U \times (3\cdot0/w + 0\cdot8) \times 100.$$

When U is 0·0152 as above and w is 0·150 g. we get the result as—31·6 mg./100 g. tissue.

Note on kidney tissue. Where the kidney tissue is being analysed it is necessary to boil the mixture and cool prior to taking a sample for analysis (in order to kill the ferment which forms extra ammonia). Usually it will be found most convenient to pour the mixture—washing out residue—into a small flask and heat to boiling, cool and make up to a special volume such as 10, 25, etc., and then take 0·5 or 1 ml.

The original weight of the tissue may be varied to suit Range II analysis. Alternatively the mixture may be brought momentarily to boiling in a small pyrex tube, and subsequently immersed in boiling water for a few minutes, the volumes if necessary being made up to the original—the level having been marked with a file. The amount of urea in the tissue water in general corresponds to that in the blood plasma, notable exceptions being brain and cord tissue, urea being present in the former to only 52% of the plasma concentration and in the latter to only 5–10% (Conway and Fitzgerald, 1942).

CHAPTER XIX (*Aa* 10)

ADENOSINETRIPHOSPHORIC ACID, ADENYLIC ACID, ADENOSINE, ETC.

SUBSEQUENT to the recognition of adenylic acid in blood by Bass (1914), it was discovered by Embden in muscle (1927). Lohmann showed, however, that it did not exist as the free adenylic acid in muscle, but as a compound with pyrophosphoric acid (1929), termed adenylpyrophosphoric acid or briefly A.T.P. In blood the adenylic acid also exists similarly bound. Adenylpyrophosphoric acid is intimately connected with the phosphorylation of hexose in its conversion to lactic acid, and as is well known, a similar but not identical substance exists in yeast (*vide* Schmidt, 1928).

Adenylpyrophosphoric acid breaks down rapidly on the death of the animal, and if it is to be determined in muscle this must be excised immediately after death, being ground in ice-cold trichloroacetic acid. It is also unstable in alkaline and in strongly acid media. In muscle it exists to the extent of about 4–6 mg. of amino N per 100 g., and in human blood, as determined by the author (1938) to the extent of 1·0 mg. amino N/100 ml., being in similar concentration in rabbit blood, and agreeing with the nucleotide data of Buell (1935).

Principles of determination. In the ensuing description the method described by Parnas *et al.* (1934, 1935) is followed, adapted for microdiffusion determinations, with HCl absorption and $Ba(OH)_2$ titration, using the solutions of Range II (p. 92).

The adenylpyrophosphoric acid in the muscle passes to a constant proportion of 90% into the trichloroacetic filtrate when the described proportions are used, the remaining 10% being present in the precipitate. On bringing the filtrate to a *p*H of 8·5 and adding excess of barium acetate the A.T.P. is precipitated as the insoluble barium salt, any adenylic acid, if present, being left in the fluid. In fact, however, adenylic acid as such will not be found in the fresh muscle.

The insoluble barium salt is subsequently dissolved in acid, the barium removed by sodium sulphate and centrifuging, and after neutralising, etc., is acted upon by some fresh muscle extract and the ammonia formed is determined.

In this way two special qualities of the A.T.P. are both availed of to ensure that this substance only is being determined. These are the insolubility of its barium salt at a particular *p*H and its specific deamination by the extract.

Procedure. The freshly excised muscle is quickly weighed and introduced into an ice-cold mortar with pestle. From a graduated 5 ml. pipette 10% ice-cold trichloroacetic acid is added, 1 ml. for every g. muscle and the muscle ground quickly with the aid of a little pure quartz sand (Merck's) to a fine suspension.

It may be noted here that all solutions used down to the solution of the barium salt in acid should be ice-cold like the trichloroacetic acid. Water is now added in the same volume as the acid, and the mixture is left for at least fifteen minutes in the refrigerator with frequent stirring. It is then poured into a centrifuge tube and quickly spun. 4 ml. of the clear fluid (corresponding to 1·43 g. tissue) are taken in a centrifuge tube and a drop of phenolphthalein solution added and the mixture carefully brought to a faint pink (about *p*H of 8·5) by the addition of 0·1 *N* NaOH, the tube being immersed in ice. 1 ml. of 25% barium acetate, which has been carefully neutralised, is now added and the mixture left aside in the refrigerator for 30 minutes, then centrifuged and washed twice with 1% neutral barium acetate.

The precipitate is then dissolved in 2·5 ml. of 0·1 *N* HCl, the barium removed by adding 1 ml. of 15% Na_2SO_4 and centrifuged. The clear fluid is poured into a 10 ml. measuring cylinder, the residue being mixed with 1 ml. water and a small drop of the sodium sulphate solution, centrifuged and the fluid poured into the cylinder.

The fluid in the cylinder is neutralised carefully with 0·1 *N* NaOH (using bromthymol blue), of which upwards of 2 ml. will be required. 0·1 ml. of *M*/3 phosphate (*p*H 7·0) is added and the mixture brought to the 10 ml. mark with water.

A complete reagent blank is carried out using 10 % T.C.A. instead of muscle extract.

Two glass tubes of 5 ml. capacity provided with stoppers are now prepared and into these are introduced the following:

Tube I: 2 ml. of fluid (from cylinder) plus 0·2 ml. enzyme.

Tube II: 2 ml. of reagent bland plus 0·2 ml. enzyme.

The two tubes are incubated at 38° for one hour. Subsequently 1·0 ml. volumes are pipetted into the outer chamber of prepared ' units ' containing 1 ml. acid of Range II, made up as on page 89, and 1 ml. of saturated potassium carbonate added, the ' units ' being titrated after 1½ hours at room temperature (or after 1–1½ hours at 38°). The amount of adenylpyrophosphoric acid amino-nitrogen corresponding to 130 mg. of the original muscle is finally analysed in this manner.

Calculation. If the ammonia—as large divisions on the burette—in samples from the two tubes be written I and II then the A.T.P. ammonia corresponds to

$$(\text{I} - \text{II}) \times 0{\cdot}3\ \gamma \text{ ammonia N.}$$

This corresponds to the A.T.P. amino N in 130 mg. muscle, but under the conditions—retention of 10% A.T.P. in the trichloroacetic precipitate—the result must be multiplied by 1·11. Milligrams A.T.P. amino N in 100 g. muscle is then given by

$$(\text{I} - \text{II}) \times 0{\cdot}256.$$

Solutions required.

10% trichloroacetic acid.

25% barium acetate and 1% barium acetate (both brought to 8·5 *p*H or faint pink with phenolphthalein).

0·1 *N* HCl.

0·1 *N* NaOH.

15% Na_2SO_4.

Phosphate buffer, *M*/3 *p*H 7·0 (3 volumes of *M*/3 alkaline to 2 volumes of *M*/3 acid phosphate).

Muscle extract 1 g. in 10 ml. of *M*/3 phosphate (*p*H 7·0).

The muscle is ground to a fine suspension with quartz, then the mixture is largely cleared by centrifuging.

Acid and alkaline solutions of Range II (pages 91 and 92).

Boric acid procedure using 0·1% of boric acid could also be used.

The method can obviously be scaled down to determinations in lower amounts and single frog's gastrocnemii could be used.

Adenylpyrophosphate in blood.

Principle and Procedure. This determination is carried out in practically the same way as for muscle. The blood may be received into a graduated cylinder containing about one-tenth of the expected volume of 1·3% potassium oxalate. After collecting and mixing the exact volume is read. The blood is then laked in the same volume of water. To 10 ml. of the laked blood is added 5 ml. 10% ice-cold trichloroacetic acid and the mixture centrifuged. 10 ml. of clear fluid are pipetted into a 25 ml. centrifuge tube a drop of indicator added, and the mixture brought to a faint pink, proceeding firstly with normal NaOH then with 0·1 *N*. 4 ml. of 25% neutral barium acetate are added, and from this onwards the method is identical with that of muscle, not only in general procedure but in volumes and quantities used.

Calculation. Similar to the previous method. The A.T.P. amino-nitrogen analysed corresponds here to 0·3 ml. approximately of original oxalated blood. If 1·3% potassium oxalate has been used to prevent clotting, this must also be allowed for (as also the fact that 1 vol. blood contains 0·80 vol. water).

Solutions required. Same as for muscle, with addition of 1·3% potassium oxalate (ammonia-free).

Determination of adenylic acid in the presence of A.T.P. in muscle and blood. If free adenylic acid is present it is left in solution after the precipitation out of the barium salt of the A.T.P. Suitable methods similar to that of the A.T.P. determination may therefore be devised, the barium being removed by sulphuric acid with subsequent neutralisation, etc.

A short method for determining the total A.T.P. and adenylic acid in blood.

Since there is no appreciable amount of free adenylic acid in freshly shed blood, this may be also taken as a short method for determining the A.T.P. of blood.

Principle. The blood is acted on directly by muscle extract, the preformed ammonia in the blood itself and extract under similar conditions being also determined.

Procedure. The following solutions are introduced into small stoppered Erlenmeyer flasks.

Flask I. 5 ml. laked blood plus 1 ml. buffer (maleic acid) plus 4 ml. muscle extract (1/10 with 0·600 NaCl).

Flask II. 5 ml. laked blood plus 1 ml. buffer (maleic acid) plus 4 ml. water.

Flask III. 5 ml. water plus 1 ml. buffer (maleic acid) plus 4 ml. muscle extract.

Flask II is analysed for ammonia at once after setting up, and flasks I and III after 1 hour at room temperature.

In the analysis 1 ml. fluid is pipetted into the outer chamber of a prepared ' unit ', 1 ml. saturated carbonate added and the ' unit ' left for $1\frac{1}{2}$ hours at room temperature. (Solutions of Range II used.) (0·5 ml. volumes may also be used with solutions of Range III (p. 92), 1 ml. being placed in the central chamber instead of 0·7 ml. as usual with this range.)

The total adenyl amino N is given as mg./100 ml. original oxalated blood by

$$(\text{I} - \text{II} - \text{III}) \times 4 \times 0{\cdot}0003 \times 100,$$
$$(\text{I} - \text{II} - \text{III}) \times 0{\cdot}12,$$

where I, II and III represent the ammonia from flasks I, II and III in large divisions on the burette, using alkali of Range II (Table VI). It is supposed in this calculation that the blood, as before, has been laked by an equal volume of distilled water. If the 1·3% oxalate solution is used to prevent clotting, allowance must be made for this in the final calculation.

In the above determination about 10 large burette divisions may be expected to correspond to the adenyl amino N in the blood. Much smaller volumes of blood may be used with similar accuracy.

The results with this method are similar to the Buell and Perkins determinations of total adenine nucleotide.

Solutions required for short method.

Maleic acid buffer, 20 c.c. 5·8% maleic acid plus 17·3 ml. *N*/1 NaOH.

1 in 10 muscle extract with 0·6% NaCl. The muscle is ground with some quartz sand and saline, then cleared by centrifuging.

Determination of other amino purine derivatives by the present technique. The possibility of these determinations may at present be only indicated.

Adenosine. Whole or laked blood deaminates adenosine with great rapidity and specificity, the process being complete in about 15 minutes at a *p*H of 7·0 and a 1 in 5 dilution of rabbit blood (room temp.) (Conway and Cooke, 1937, 1938). Provided adenylic acid (adenosine 5-phosphate) or adenylpyrophosphate is present in less than 0·1%, with $N/10$ phosphate, it does not appreciably interfere. No ammonia is formed from guanosine, cytidine, adenine, guanine, cytosine, glycine, aspartic acid, tyrosine, tryptophane, leucine, histidine, cystine, histamine, choline, *iso*-amylamine, butylamine, acetamide, propionamide, butyramide, glucosamine, urea, etc. (1938).

Adenine, guanine, etc. Ferments specific for such substances have been already investigated by many observers (e.g. Amberg and Jones, 1911, Schmidt, 1932), and it is found here that a very rich source of a number of such deaminases is the nucleated corpuscles of the fowl. They contain, for example, a high concentration of adenase, which is otherwise rare; also guanase, and ferments deaminating cytidine, cytosine, etc. (Conway and Cooke, 1938).

CHAPTER XX (*Aa* 11)

NITRATE, NITRITE AND AMIDE NITROGEN

Nitrate and Nitrite.

TOTAL nitrate and nitrite are estimated by reducing to ammonia by Devarda's alloy acting in the presence of strong alkali in the outer chamber of a microdiffusion 'unit', the ammonia being absorbed as usual in the central chamber. For more than 10 μg nitrate N, the absorbent may be boric acid reagent, and for less than 10 μg $NO_3 - N$, the $Ba(OH)_2 - HCl$ technique may be used.

Procedure. Into the outer chamber of the standard 'unit' is introduced about 0·1–0·2 g. powdered Devarda's alloy, and into the central chamber 1 ml. of the boric acid reagent (as described on page 98). 0·5 ml. of the nitrate solution is carefully pipetted into the outer chamber and the lid of the 'unit' placed in position. 1 ml. of the alkali mixture is then introduced into the outer chamber of the 'unit', and the contents mixed and allowed to stand on the bench for 50 minutes, the central chamber being then titrated with standard HCl from the Conway burette.

Solutions, etc. In addition to those already described for ammonia determinations :

Alkali mixture - - -	(Saturated K_2CO_3, 40% KOH 3 : 1).
Devarda's alloy - - -	Finely powdered.

Note on the determination. It is obvious that the above procedure gives the total of preformed ammonia and that of nitrate and any nitrite present. A determination of the preformed ammonia may be made simultaneously, and allowed for in the calculation. If nitrite is present in appreciable amount and it is required to estimate it separately from nitrate, it can be driven off by acidifying the 0·5 ml. in the outer chamber (not adding the Devarda's alloy until later) to about pH 1 by a drop of acid, and exposing on the bench for one hour, then proceeding with the determination as above.

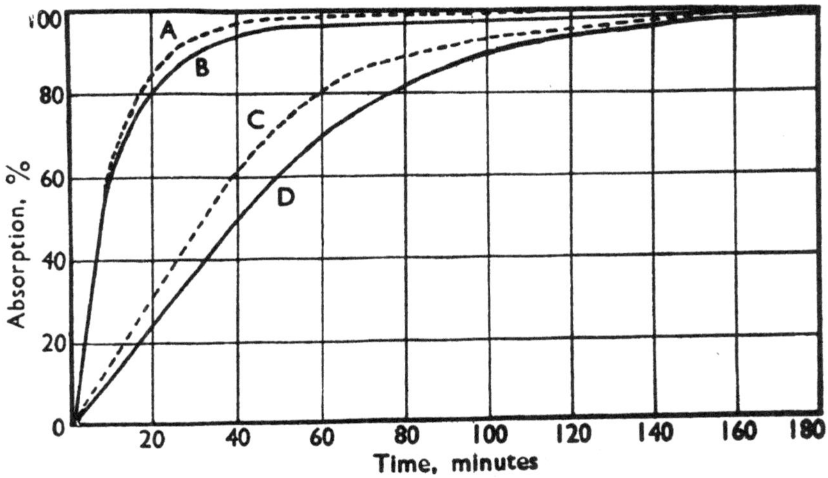

FIG. 45. Ammonia absorption curves at 31° C. from 1 ml. of 0·02*M* aqueous solution of: *A*. formamide, with oscillation; *B*. formamide, without oscillation; *C*. acetamide, with oscillation; *D*. acetamide, without oscillation (Hallam, *Analyst*, **80** 552, 1955).

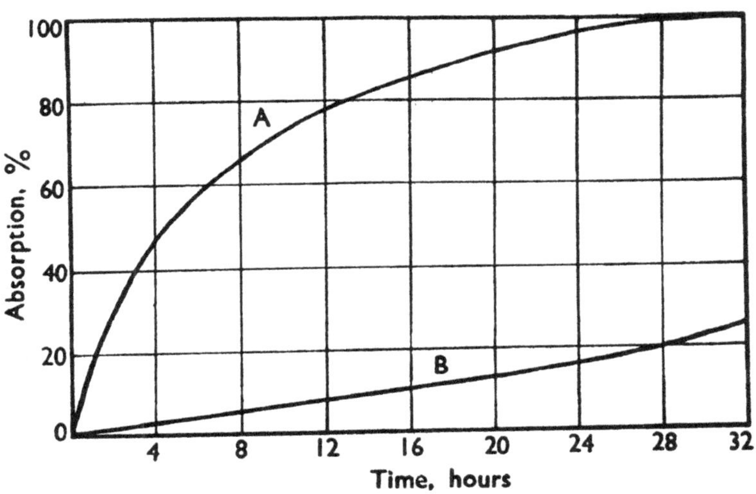

FIG. 46. *A*. Absorption curve for methylamine from 1 ml. of a 0·02*M* aqueous solution of N-methylacetamide; temperature 30° C. *B*. Absorption curve for ammonia from 1 ml. of a 0·02M aqueous solution of urethane; temperature 32° C. (Hallam, *Analyst*, **80**, 552, 1955).

Amide nitrogen in biological fluids.

Borsook and Dubnoff's method (1939). This was described for 0·2 ml. volume of dilute biological fluids. The amide N is split off as ammonia by heating with acid, the ammonia being then determined.

Procedure. To 1 vol. of the fluid is added 0·5 vol. of 3*N* H_2SO_4 in a small Kjeldahl digestion tube, the tube covered with a cap of tin-foil and set away in a boiling water bath for 3 hours. After 3 hours, the tube is cooled and 0·5 ml. of 2·9 *N* KOH added. The mixture is then made up to a definite volume. In the Borsook and Dubnoff procedure (as described on page 129) the small Kjeldahl tubes contain a constriction and suitable mark, and an aliquot is taken for the microdiffusion procedure in the usual way, using saturated metaborate for the alkali mixture as described.

Simultaneous determinations of the preformed ammonia are carried out.

The above procedure may be suitably modified as follows : If, for example, 0·5 c.c. of the fluid is taken and 0·25 c.c. of 3*N* H_2SO_4 added, then after digestion the remaining volume is carefully and as fully as possible pipetted into the outer chamber of a standard ' unit '. The digest tube is washed twice with 0·2 vol. of distilled water, these being transferred into the outer chamber. Into the central chamber is pipetted the absorbing reagent and the ' unit ' closed after which 1 c.c. saturated metaborate containing 10% KOH is introduced and the ammonia determination carried through in the usual way.

Solutions, etc. (*other than the usual reagents*) :

3 *N* H_2SO_4.

2·9 *N* KOH.

Saturated potassium metaborate : containing 10% KOH.

CHAPTER XXI (Aa 12)

AMIDES (continued). GLUTAMINE

Principle

This method is essentially the same as that of Krebs (1948), but adapted for microdiffusion. The method of Krebs in turn combines the principles used by Gale (1945, 1947) and Archibald (1944). Washed suspensions of Clostridium welchii (strain S.R. 12) were shown by Gale to decarboxylate L. glutamic acid and L. glutamine specifically. Krebs (1948) showed that when glutamine is decarboxylated an equivalent amount of ammonia is formed, and the glutaminase responsible is highly specific, more so than Archibald's (1944) glutaminase preparation from kidney. It is also more convenient to make.

Procedure for Glutamine determination

The fluids to be examined if very acid or very alkaline are brought to a *p*H of about 4·9 by addition of measured volumes of HCl or NaOH. One ml. volumes are then introduced into the outer chambers of No. 1 units. 0·1 ml. volumes of 3 *M* acetate (*p*H = 4·9) are added, and then 0·5 ml. of the Clostridium welchii suspension and the whole mixed by rotating.

The units without lids are incubated at 37° C. for 30 minutes. They are then removed and 1 ml. of boric acid reagent introduced into the central chamber. The lids, suitably treated with fixative, are placed in position and 1 ml. of saturated K_2CO_3 run into the outer chamber and mixed.

Blank determinations are carried out with 1 ml. volumes of water replacing the fluid for examination. The units are left on the bench for approximately two and a half hours and then the fluid in the central chamber is titrated with standard HCl. (*N*/200, or other suitable strength, from the Conway burette.)

The method should be occasionally tested with suitable L.-glutamine standards.

Reagents

Cl. Welchii suspension. The strain S.R. 12 of Cl. welchii (as recommended by Gale, 1945) is used. The organisms are grown in the medium suggested by Gale, with the addition of 50 ml. of yeast extract (prepared by autoclaving baker's yeast with an equal volume of water and filtering) per litre as in the procedure of Krebs (1948).

The washed cells from about 1 litre of medium are suspended in about 15 ml. of 0·45% (*w*/*v*) NaCl. (The dry weight of the bacteria in this suspension as determined from a 0·2 ml. sample, was found by Krebs to represent 1·1 to 1·5 g. dry bacteria per litre of medium.) This suspension does not appreciably deteriorate within a month when kept in a refrigerator.

L. glutamine. The commercial preparation may be purified by recrystallisation. After repeated recrystallisation of a sample, Archibald (1945) obtained 93% purity, arginine being among the impurities to the extent of 1·3%.

Boric acid reagent. 0·5% boric acid solution containing bromocresol green and methyl red as described in Chapter X.

Cetyltrimethylammonium bromide ('*cetavlon*') (see Note 3 below). Commercial cetavlon in water (2%) serves as a stock solution.

Notes on method

1. Glutamic acid may also be determined simultaneously with glutamine, since it is decarboxylated by the enzyme preparation, and one mole of CO_2 is produced from each mole of glutamic acid and of glutamine. The CO_2 evolved during the incubation could be received into 1 ml. of standard $Ba(OH)_2$ and after titration the central chamber washed out and 1 ml. of the boric acid reagent introduced, the glutamine being determined as above.

2. Special precautions have to be taken when glutamic acid and glutamine are to be determined in tissues as the concentration of these two substances can undergo rapid changes. The following procedure is in accordance with that suggested by Krebs (1948).

The tissue cut into 5 to 10g. pieces is introduced quickly into liquid air, and allowed to remain there for at least 15 minutes.

It is then taken and ground in a cooled mortar and the powder transferred to a tared beaker and weighed and for each g. of tissue 2 ml. of cooled 0·5 *N* HCl are added and well mixed. For immediate determination 5 ml. of the mixture are then pipetted into a small tube, and 0·25 ml. of aqueous 0·1% methyl orange, then sufficient *N* NaOH introduced from a graduated 1 ml. pipette to change the indicator to yellow (usually about 1·5 ml.). One ml. volumes of this mixture are introduced into the outer chambers of units and the procedure conducted as described.

3. Krebs showed (1948) that the detergent cetyltrimethylammonium bromide (' cetavlon ') considerably increases the rate of decarboxylation of glutamine, depending on the *p*H value. Similar results were produced by serum and by aqueous tissue extracts. At a *p*H of 4·9 the maximum effects of cetavlon and serum were about equal. The maximum effect of cetavlon at a *p*H of 4·9 is produced by a final concentration of approximately 0·024%.

The addition of cetavlon is recommended if the material to be examined does not contain protein.

4. In pure solutions various indicators (0·01%) such as phenolphthalein, bromocresol purple and anionic detergents such as dodecyl sulphate (0·05%) were found to inhibit the decarboxylation of glutamine. These inhibitions were largely abolished by the addition of cetavlon (0·05%) or serum (1 ml. to 3–4 ml. solutions).

5. Human serum contains about 8 mg. of glutamine and 2 mg. of glutamic acid per 100 ml.

CHAPTER XXII (*Aa* 13)

MONOAMINE OXIDASE AND HISTAMINASE IN TISSUES

Monoamine oxidase in tissues. (*Cotzias and Dole,* 1951)

It is pointed out by Cotzias and Dole that the determination of the oxidase by oxygen measurement has the disadvantage of relatively concentrated preparations, and in order to obtain an accuracy within 2 per cent with conventional manometric equipment, under optimal conditions, the oxygen consumption should be at least 10 μM per hour. Micro respirometers can deal with much smaller quantities but the volume must be similarly restricted so that they offer no advantage for the measurement of samples of low concentration.

The microdiffusion procedure allows of the use of solutions one-twentieth as dilute as the manometric methods.

Procedure of Cotzias and Dole

Preparation of tissue and dialysis. The tissue of liver or kidney has the grosser strands of connective tissue removed by forcing through the screen of a tissue press in the cold room. A weighed sample of the pulp is then homogenized with a suitable amount of the phosphate buffer. During dialysis the outer solution is stirred by bubbling with air washed free of ammonia. (The cellophane tube used by Cotzias and Dole was sausage casing 27/32, Visking Corporation.) After dialysis, the contents are washed into a calibrated cylinder and diluted with buffer to a final concentration of from 25 to 150 mg. of original wet tissue per ml.

Preparation of units. Standard units (No. 1) are then prepared so that the outer chamber is divided into two regions which will keep separate (by the use of vaseline ridges) the substrate and enzyme. Accurately measured 1 ml. samples of the mixed suspension of tissue are transferred into one side of the divided outer rings of four units and 1 ml. aliquots of 0·05 *M* tyramine solution into the other of two of the units, and similarly water into the other two, serving as blanks. Approximately 1 ml. volumes of the boric acid indicator solution are next introduced into the central chamber.

Enzymatic action. The units are covered with vaseline or silicone-greased glass plates and are immersed in a water bath ($20° \pm 0{\cdot}01°$). They rest on a horizontal screen about 10 cm. below the surface. After about 20 minutes have been allowed for temperature equilibration, the two solutions in each outer ring are mixed by suitable tilting, the exact time being noted.

Microdiffusion and titration. After one hour the units are taken out, surplus water removed and the enzymatic action ended by introducing 1 ml. of saturated metaborate into the outer chamber, in the usual manner, and a diffusion period of about 3 hours allowed. (Cotzias and Dole use a shaking platform.)

Titration with $N/50$ HCl is then carried out as previously described.

Calculation. The unit of monoamine oxidase activity is defined as the quantity capable of catalysing the production of NH_3 at the rate of 1 μM per hour under the conditions.

If V_s and V_k are the volumes in ml. of $M/50$ HCl required for titration of the sample and the blank, with t the time of digestion in hours, the number of units of enzyme (U) contained in the sample analysed is given by $U = 20\ (V_s - V_b)/t$.

Reagents

Tyramine hydrochloride (0·05 M, 0·87 g./100 ml.). This should be kept not longer than two days. The commercial preparation is normally heavily contaminated and should be freed of volatile amines. This is carried out as follows.

A nearly saturated solution of tyramine hydrochloride is adjusted to pH 10·5 with saturated $Ba(OH)_2$ and aerated overnight under reduced pressure. In this way the preformed volatile amines are swept out. Barium is removed as the sulphate after addition of 5 N H_2SO_4 in sufficient quantity to bring the pH to 4. The filtrate is evaporated to dryness in vacuo with protection of the capillary air intake by an acid trap, and then redissolved in a small volume of warm absolute alcohol. From this solution tyramine hydrochloride is crystallized at 4° C. by the gradual addition of ethyl ether, the crystals are removed by filtration with suction, washed several times with ether, and stored in a dark bottle in a desiccator.

Phosphate buffer. *p*H 7·3, 2·59 g. of KH_2PO_4 and 10·50 g. of K_2HPO_4 are brought to one litre volume with ammonia free water. The other reagents have been previously described.

Notes on method

1. The temperature coefficient of the reaction velocity is about 9% per degree.
2. Under 20° C. the deamination rate is independent of the substrate concentration.
3. Dialysis serves to free the tissue of about 90% of ammonia formed therein which could introduce a relatively large and inaccurate correction.

Histaminase

Determination of histaminase in tissues (*Cotzias and Dole*, 1952).

For precise determination, Cotzias and Dole (1952) considered a preliminary dialysis and reduction of the temperature of the reaction to be necessary.

Procedure of Cotzias and Dole

Dialysis of tissue pulp. After passage through a tissue press, the pulp is weighed in a tared homogenizer tube, homogenized with 5 volumes of phosphate buffer (*p*H 7·3, $\mu = 0{\cdot}2$) and then washed into cellophane casing for dialysis at 0° C., against six changes of the same buffer with a few drops of octyl alcohol added as preservative. After 48 hours the tissue is washed into a graduated cylinder and diluted with buffer to a final concentration of 80 to 200 mg. of wet tissue per ml.

Action of enzyme in outer chamber (No. ·1 Unit). Units are prepared with 1 ml. boric acid reagent with indicator in the central chambers. A division is made in the outer chamber with the aid of vaseline and 1 ml. of the vigorously stirred suspension of tissue particles is pipetted into one side, and either 1 ml. of 0·03 *M* histamine dihydrochloride in the buffer or one ml. of the buffer alone is added to the other side to serve as a blank. The unit is closed with a glass plate that has been greased with silicone or other suitable fixative and placed in a water bath at $10 \pm 0{\cdot}01°$ in a similar way to the procedure with monoamine oxidase. After 20 minutes of equilibration it is tilted to mix the solutions and initiate the reaction. After 24 hours (or 48 hours for samples of very low activity) the reaction is terminated by the addition of 1 ml. saturated metaborate.

Termination of enzyme activity and titration. When the diffusion is complete (3 hrs. given at room temperature) titration is carried out with 0·02 *N* HCl.

Calculation of activity. Cotzias and Dole define 1 *unit of histaminase activity* as the quantity capable of catalysing the release of NH_3 at the rate of 1 μM in 24 hrs. at 10°. If V_s and V_b are the volumes in ml. of 0·020 *N* HCl required for the titration of the sample and the blank, respectively, and t the time measurement in periods of 24 hrs., the number of enzyme units is given as $U = 20\,(V_s - V_b)/t$. Ammonia contributed by the purified enzyme itself should be negligible.

Reagents

Histamine. The commercial product is purified as for tyramine in the method for monoamine oxidase.

Phosphate buffer (pH 7·3 ; $\mu = 0{\cdot}2$) is prepared as for the monoamine oxidase determination.

Other reagents are prepared as described for the monoamine oxidase determination, and for ammonia determination by the boric acid procedure (Chapter X).

Notes on method

(1) The activity in the method described above is strictly linear with the time, and has been followed up to 54 hrs. by Cotzias and Dole.

(2) If the homogenate is not dialysed there may be a delay of many hours before ammonia is produced and then the slope may be much less than for the dialysed homogenate.

(3) The logarithm of the reaction velocity has a linear relation with time up to 12° C. As the effect of temperature up to 12° C.

may be described as an increase of 12 per cent activity per degree, it is necessary to control the bath to ±0·01° C. to obtain the highest precision obtainable under the conditions.

(4) The effect of *p*H is slight varying about 10 per cent between *p*H 6 and 8, with a maximum near 7·3.

(5) The activity is independent of the substrate concentration (range 0·003 to 0·06 *N* histamine) up to 12° C., beyond which the the higher substrate concentrations depress the activity.

Distribution of histaminase in tissues of rabbit, guinea pig, rat and mouse.

Histaminase was shown by Zeller *et al.* (1940) to consist of a protein and a flavin-containing prosthetic group, which was subsequently characterized by Kapeller-Adler (1949) as a flávin adenine dinucleotide. Its distribution in animal tissues is illustrated by Table XI*a* (from Cotzias and Dole, 1952).

Intracellular distribution

Cotzias and Dole conclude from a study of the enzyme in liver that the mitochondria must be regarded as a major site of accumulation.

TABLE XI*a*

	Histaminase, units per gm. wet tissue						
	Liver	*Spleen*	*Intestine (mucosa)*	*Lung*	*Heart*	*Kidney cortex*	*Brain*
Rabbit	6·53	1·34	3·35	21·2	2·52	4·80	0
Guinea pig	1·88	0·44	1·54	0·60	1·30	0·73	0
Rat	0·30	0	3·00	8·25	2·52	0·93	0·72
Mouse	0·82	0·74	3·06*	1·17	0·73	1·30	0·18
Pig	2·30					5·00	
Calf	0·97					0·62	

* Whole small intestine analysed.

CHAPTER XXIII (Ab)

AMINE GROUP

DETERMINATION OF VOLATILE AMINES

THE lower aliphatic amines are volatile substances resembling ammonia itself in many properties. Trimethylamine is important biologically as associated with bacterial decomposition. As a test for such it has been used in the Atlantic and Pacific Experimental Fisheries Stations. A method for determining trimethylamine—using the standard 'unit'—has been worked out by Beatty and Gibbons of the Atlantic Station (1937) and is described below.

The test was also used by Brockelsby and Riddell of the Pacific Station (Prince Rupert, B.C.) in studies on iced halibut (1937).

A paper by Richter (1937) indicates the metabolic importance of these lower aliphatic amines (methylamine, dimethylamine, ethylamine). It was shown that in the oxidation of adrenaline by the enzyme described by Blaschko, Richter and Schlossmann (1937), in liver, intestine, etc., of mammalian tissues methylamine was formed; from epinine and *l-p*-sympatol methylamine was also formed; from *dl*-alkalamine ethylamine and from hordenine dimethylamine.

In the identification and quantitative determination of these amines and ammonia, Richter used the standard 'unit' (No. 1).

Richter's method for the identification of the volatile amines formed in enzymic mixtures (from the *Biochemical Journal*, **31**, 2022):

Ammonia is liberated from the reaction products by treatment with an equal volume of saturated K_2CO_3 in a standard 'unit'. It is then converted into (*a*) the picrate, and (*b*) the 2:4-dinitronaphthol derivative by introducing about 0·05 mg. of the reagent suspended in a drop of water on a cover-slip attached by vaseline to the lid of the standard 'unit'. For identifying the volatile amines traces of ammonia coming from the enzyme preparation were first removed by keeping the solution with an equal volume of saturated K_2CO_3 for 45 minutes at 37° C. in a ' unit ' containing dilute sulphuric acid in central chamber.

In this time 99·5% of the ammonia, but only a much smaller fraction of the amines, passes over into the acid. The residual amine is then converted into the derivatives (*a*) and (*b*) as with ammonia, and the derivatives are identified by examining the optical properties of the crystals with the polarizing microscope in plane and polarised light (Klein and Steiner, 1928). The derivatives may be recrystallised on the cover-slips in order to obtain well-defined crystals.

It is possible to identify small quantities of amines (of the order of 10γ) very rapidly and with a high degree of certainty. Richter lists the following properties for identification, some of which were previously given by Klein and Steiner (1928).

TABLE XII

	Reagent.	*Ammonia.*	*Methyl-amine.*	*Dimethyl-amine.*	*Ethylamine.*
			Picrates.		
Habit	Prisms	Plates and Prisms	Plates	Plates	Plates and Prisms
Pleochroism	Weak	Distinct	Strong	Weak	Strong
Extinction	Parallel	Parallel	Parallel	42°	Parallel
Interference	Medium	Strong	Strong	Strong	Very strong
Crystal edge angles	α = 137°	α = 117°	α = 129° β = 141° γ = 102° δ = 78°	α = 96° β = 132°	α = 123°
			2 : 4-Dinitronaphthol derivatives.		
Habit	Prisms	Fine-pointed needles	Blunt-ended prisms, often curved	Short prisms, square-ended	Plates
Pleochroism	Weak	Weak	Yellow-brown	Marked yellow-brown	Weak
Extinction	0–10°	Angular	Parallel	20°	Angular

Quantitative estimation of volatile amines (Richter, 1937).

The absorptions of several volatile amines compared with ammonia are given in Fig. 47 (from the *Biochemical Journal*). Each 'unit' contained 1 ml. of 0·015M amine hydrochloride and 1 ml. saturated K_2CO_3 in the outer chamber and 2 ml. of $N/100$ sulphuric acid in the central compartments. Richter comments on the fact that *iso*-butylamine and *iso*-amylamine, although much less volatile than methylamine, were absorbed quickly, and considers the probable explanation as due to the lesser solubility of

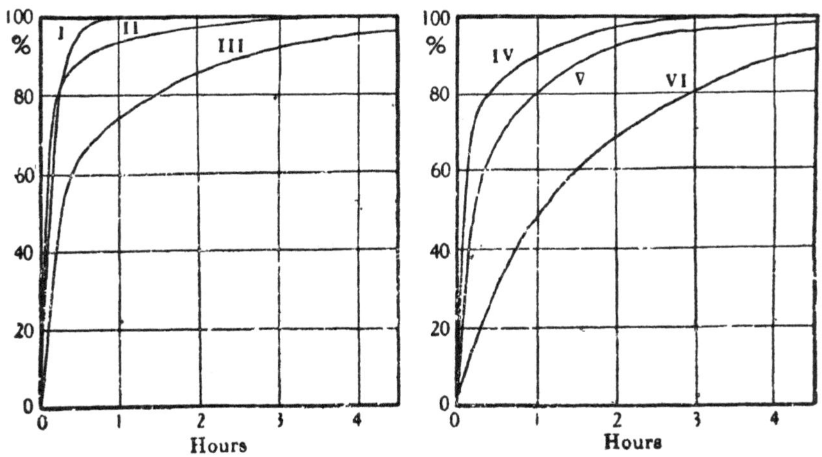

FIG. 47. Rates of absorption of amines. I, ammonia. II, *iso*-butylamine. III, methylamine. IV, *iso*-amylamine. V, ethylamine. VI, benzylamine. At 37° in standard 'unit' (from Richter's diagrams in the *Biochemical Journal*, **31**, p. 2020).

the first two. That this explanation is correct will appear from the considerations in Chapter IV.

Prior to Richter's paper some preliminary observations had been carried out here with regard to the amine absorption rates. These may be summarised provisionally in the following table.

These absorption rates into the central chamber of the 'unit' were investigated on $N/250$ solutions of the corresponding chlorides, the absorption taking place into 1 ml. of $N/150$ acid (solutions of Range I used). Into the outer chamber was run 1 ml. of the $N/250$ solution of the amine hydrochloride and then 1 ml. of 0·1N NaOH or 1 ml. of saturated potassium carbonate. The latter solution has a very marked effect—as with ammonia—in increasing the tension of the dissolved amine.

The amines listed in Table XII*a* would be completely absorbed in 3½ hours at room temperature, when about 10 times the half

TABLE XII*a*

Amine hydrochloride (1 ml. N/250 solution in outer ch.).	*50% absorption period (mins.).* 1 *ml.* 0·1*N* NaOH *added to* 1 *ml. in outer chamber.*	1 *ml. sat. pot. carb. added to* 1 *ml. in outer chamber.*	*p*H *at which the amine is 50% dissociated* (1926).
Ammonia	30	10	9·4
Methylamine	99	19	10·6
Dimethylamine	80	22	10·8
Trimethylamine	22	5	9·8
Ethylamine	106	16	10·6
Diethylamine	86	9	11·0
Triethylamine	16	4	10·7

Room temperature 20–23° C.

absorption period with saturated potassium carbonate is allowed. From Richter's figure (Fig. 47), however, the absorption of the last 10% of the methylamine is unexpectedly slow. Trimethylamine and triethylamine go very fast, and 1 hour would be sufficient for their complete absorption. The difference between the *p*H value for half dissociation of the amines gives a means if necessary of an analytical separation. Thus if a dilute borate buffer solution were used sufficient to maintain the *p*H approximately at 9·4 and the same volumes otherwise used, then after 260 minutes at room temperature (20° C. approximately) the following are the probable order of the amounts absorbed :

Ammonia	95%
Trimethylamine	90%
Triethylamine	73%
Methylamine	10%
Ethylamine	9%
Dimethylamine	8%
Diethylamine	5%

For the present the figures of this list must be taken only as very approximate and merely practical guides, since they have not been actually tested, but derived from calculations on the presumed course of the absorption and the data of the previous table.

Trimethylamine method of Beatty and Gibbons (1937).

Principle. The method was devised for determining trimethylamine in press juice obtained from the muscle of fishes. The reaction between formaldehyde and ammonia is utilised to hold back the ammonia while the trimethylamine diffuses over into acid in the central chamber, the amount being subsequently determined by titration.

Method. In obtaining the press juice a press cup of brass is taken which is perforated with holes 1/16 inches (*ca.* 1·6 mm.) in diameter. The cup is lined with Whatman No. 1 filter paper and filled with the muscle under examination. A circular sheet of filter paper is placed over the sample, a brass plunger inserted and pressure applied. In the inner chamber of the standard ' unit ' is placed a measured excess of standard acid, usually about one-seventieth normal. In the outer chamber is placed 1 ml. press juice and 0·5 ml. formaldehyde solution. Usually commercial formalin, shaken with an excess of magnesium carbonate and filtered or decanted, gives zero, or an extremely low, blank. After the addition of the formaldehyde the lid with fixative is placed in position and the dish rotated to ensure thorough mixing of press juice and formaldehyde. 1 ml. of saturated potassium carbonate is added in the customary manner, the lid being slightly displaced and subsequently replaced.

The dish is rotated to mix the contents and is placed in an incubator at 36° C. for 2 hours, after which the excess standard acid is titrated.

It is obvious that the standard acid may here be made up in the usual way of incorporating the indicator in the standard acid (page 94).

Suggested modifications. Instead of using press juice an extract formed by grinding with quartz sand and clearing by centrifuging would apparently work as well as the press juice—using weaker standard acid. Also if 0·5 ml. of this and 0·2 ml. formaldehyde be used the ' unit ' may be left on the bench for 2 hours instead of being incubated. In this way the solutions of Range I or II (Table VI, Chapter IX) will be found suitable.

To illustrate the kind of figures obtained—using the previous method—the following example may be given from the account by Brocklesby and Riddell. The halibut caught 8 hours previously (and presumably stored in the fish-hold, which is approximately at 1·5° C.) were dressed, iced and boxed, then stored on ice in a cool room. The trimethylamine content in the back muscle just behind the head, which was found to be 0·2 part per 100,000 immediately on arrival, remained very steady up to the ninth day, when it was 0·4 part. On about the twelfth day a rapid change set in, the trimethylamine content being then 1·3 parts, and on the 15th day 12·8 parts per 100,000. With halibut muscle juice (1937) stored at 35° the change set in rapidly on the sixth day.

Such changes in the trimethylamine content at 35° F. correspond in time to marked changes observed (1937, and unpublished results) in the permeability of the frog's sartorius muscle in Ringer at 2 to 3° C., with respect to sodium, potassium, magnesium and chloride. These permeability changes set in rather suddenly about the 4th to the 6th day.

In the above trimethylamine examination on iced halibut it was found by Brocklesby and Riddell that if the fish were given a five-minute dip in a 2% formaldehyde solution before icing, after 16 days the trimethylamine content was below 0·8 part per 100,000. No formaldehyde had penetrated the skin, which with the cut surface was also rendered impermeable to the further penetration of bacteria. The experiment illustrates the bacterial origin of the trimethylamine.

CHAPTER XXIV (*Ba* 1)

B. VOLATILE ACIDS

Ba. THE CARBON DIOXIDE GROUP

CARBONATES AND BICARBONATE

FOR quantities of carbon dioxide of the order of 1·0 mg. and bound as carbonate or bicarbonate, the method to be described, as worked out in conjunction with Dr. Isaccson (1936), gives an accuracy of about 0·5% as a coefficient of variation. With the standard 'unit' it is suitable for determinations down to 0·2 mg. or to 0·1 c.c. of the gas at N.T.P., 0·2 mg. carbon dioxide corresponding to 0·9 ml. of $N/100$ acid. Below this figure the carbon dioxide of the air begins to interfere very appreciably unless suitable modifications of the method are introduced.

With the No. 2 'units' the range can be stepped down to one-fifth.

The method is particularly suitable for the determination of the bicarbonate content of the blood ('alkali reserve' of Van Slyke, 1917), and also for the study of the oxidation rates of organic substances *in vitro*.

The carbon dioxide in the method is measured as ml. of a standard hydrochloric acid, and the titration figure can be converted at once into c.c. of carbon dioxide at N.T.P. (if required) by a single factor without any further corrections. In a method published in 1944, Lehmann independently arrived at a similar procedure for the determination of bicarbonate in blood using the standard 'unit'.

Principle of method. The total carbon dioxide is set free by acidification in the outer chamber and the free gas diffuses into barium hydroxide solution in the central chamber, where it precipitates an equivalent amount of barium carbonate.

The content of the central chamber is titrated after absorption subsequent to the addition of a drop of thymolphthalein solution, which can alternatively be incorporated into the alkali at the beginning. A mixture of thymolphthalein and phenolphthalein forms a very good indicator.

Procedure. The following procedure is applicable in general to bicarbonate and carbonate solutions with carbon dioxide range of about 1·0 to 0·2 mg. Into the outer chamber of a ' unit ' is run about 0·2 ml. of 1·0*N* sulphuric acid. This is delivered without blowing, which must be strictly avoided at every stage. It is better to have the acid at one part than spread over the whole surface. A lid is now smeared with vaseline and placed in a ready position when 1·0 ml. of the $Ba(OH)_2$ solution is run into the central chamber from a 2 ml. Bang burette with soda-lime guard (Fig. 11). It may also be delivered from a long tube pipette with 1 ml. mark. This is filled with the tip at the bottom of some barium hydroxide solution contained in a narrow test-tube, into which it is run from a stock bottle. The delivery into the ' unit ' should be quickly carried out and the lid placed immediately in position. The ' unit ' is now tilted by resting on a spare lid, the lid slightly displaced to allow the introduction of the tip of a pipette and 1, 2 or 3 ml. fluid for examination quickly run in, the lid being immediately replaced. These deliveries may be made very suitably from ordinary volumetric pipettes with the point of the tip broken off, and at the end of the delivery the tip of the pipette should touch the glass floor of the ' unit ' momentarily. The pipette is standardised for such delivery.

The contents of the chamber are now mixed by rotation and the ' unit ' left aside for 30, 45 or 60 minutes at room temperature. Fig. 48 shows the absorption rates in the No. 1 and 2 ' units '. The burette is then accurately adjusted, the lid of the ' unit ' removed and the fluid in the central chamber titrated as quickly as is compatible with accuracy. The end-point is a change from blue to colourless with thymolphthalein, or blue to red to colourless including phenolphthalein. (If B.D.H. Universal Indicator is incorporated, the end-point is a greenish tint.)

With regard to the accuracy of the rapid deliveries into the outer chamber, as described above, these can be carried out with coefficients of variation of about 0·3–0·4%, as already shown in Chapter V, when the delivery tip of the pipette touches the glass. If a higher degree of accuracy is required, the use of the

Krogh-Keys syringe pipette for introducing the barium hydroxide solution is recommended. This procedure was adopted by Lehmann (1944).

Calculation. The titration difference between the full 'unit' and the blank in large divisions on the burette multiplied by 0·02 gives the c.c. of carbon dioxide at N.T.P., which figure divided by the volume of fluid analysed and multiplied by 100

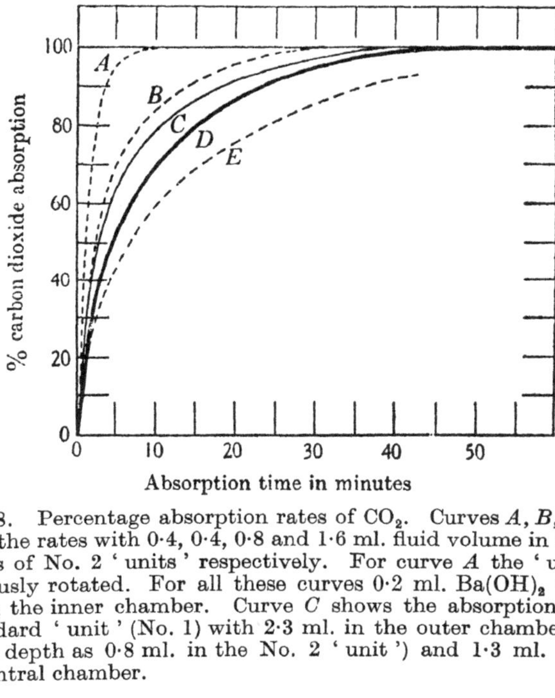

FIG. 48. Percentage absorption rates of CO_2. Curves *A*, *B*, *D* and *E* describe the rates with 0·4, 0·4, 0·8 and 1·6 ml. fluid volume in the outer chambers of No. 2 'units' respectively. For curve *A* the 'unit' was continuously rotated. For all these curves 0·2 ml. $Ba(OH)_2$ was contained in the inner chamber. Curve *C* shows the absorption rate for the standard 'unit' (No. 1) with 2·3 ml. in the outer chamber (giving the same depth as 0·8 ml. in the No. 2 'unit') and 1·3 ml. $Ba(OH)_2$ in the central chamber.

gives the vols. carbon dioxide per cent. The calculation is independent of temperature and barometric pressure.

Note on the effective sealing of the 'units'. For effective sealing of the 'units' against entry of the atmospheric carbon dioxide, the contact of lid and 'unit' must appear to form a circular transparent ring when looked at from above. Where the contact is not good this ring will be broken by an opaque patch. A small proportion of No. 1 'units' and lids as at present manufactured will not give an adequate contact for this determination and must be looked for on setting up the required number. If found they must be rejected. The No. 2 ' units ' can always be easily sealed.

Accuracy of method. Where care is taken to fill the ' unit ' and to titrate at the end in a uniform time, the standard deviation of the single determination need not exceed 0·6%. This accuracy was worked out using 2 ml. 1% lactic acid to liberate the CO_2, and may be considered to apply to the sulphuric acid procedure described. It is to be emphasised that the alkali surface must be exposed as little as possible to the air. It may also be mentioned that from the principles of Chapter IV practically the whole of the absorption time with volatile substances (such as carbon dioxide) is taken up by diffusion through liquid in the outer chamber.

Consequently, a considerable diminution in the surface of the internal chamber without appreciably extending the absorption time may be theoretically expected, and explains the similarity of curves *C* and *D* in Fig. 48. With No. 2 ' unit ' a higher accuracy may then be expected over certain ranges of CO_2. At the same time with the No. 1 ' unit ' the effect of surface exposure could be much lessened by removing 0·5 ml. and titrating in a stream of CO_2-free air.

Solutions required.

Hydrochloric acid. 0·179*N*. This is made by making 35·9 ml. of *N*/1 HCl to 200 ml. with distilled water. Each large division on the burette (0·01 ml.) is the equivalent of 0·02 c.c. carbon dioxide at N.T.P. or to 0·0395 mg. carbon dioxide. Once made up, this solution is permanent, and since it alone is referred to in the calculation the form of this latter is always fixed without further correction of temperature, barometric pressure, change of titre, etc. The slight alterations of fluid volume with temperature lie within the range of error.

Barium hydroxide with indicator. 44·9 ml. *N*/5 barium hydroxide plus 5·0 ml. 0·1% thymolphthalein* in alcohol (or alternatively 3 ml. B.D.H. Universal Indicator) are made up to 100 ml. with distilled water. Each ml. is the equivalent of 50 large divisions (0·50 ml.) on the burette with the standard acid.

* Or mixed thymolphthalein and phenolphthalein.

Standard sodium carbonate solution. $N/20$ sodium carbonate solution will be found convenient (0·265 g. anhydrous sodium carb./100 ml.). Each ml. is the equivalent of 27·8 large divisions on the burette. This solution may be used occasionally as a check or control.

Determination of the total carbon dioxide or of the bicarbonate content of blood ('alkali reserve'). (See below, method using 0·2 c.c. blood.) It is essential for any true picture of the conditions that the blood be collected anaerobically. With regard to the collection of blood, centrifuging and equilibrating the centrifuged plasma with alveolar air, Peters and Van Slyke write : "The procedure gave practical information of importance in the detection of the presence and extent of disturbance of the acid-base balance. However, it does not measure directly the concentration of either bicarbonate or total carbon dioxide in the circulating blood. For this reason it has been replaced by the simpler procedure of determining the CO_2 content of the true plasma or serum from blood which has been drawn and centrifuged with precautions to prevent loss of CO_2" (1932).

Collection of the blood anaerobically. The simplest way would appear to be a collection direct from the vein (or artery in the experimental animal) through a tube dipping under paraffin oil in a centrifuge tube lined with paraffin wax and containing a particle of heparin. The blood is then immediately sealed off with some molten paraffin wax (M.P. 45°), cooled on ice and centrifuged. A neat method of sealing due to Dr. Kane is used here as a routine and is of value when immediate determinations are not carried out. The liquid paraffin is sealed with a layer of molten paraffin (M.P. 45° or less), and in this is embedded a somewhat flattened wire spiral. The tube is cooled on ice and centrifuged, and in this way may be stored for days in the refrigerator. For analysis the tube is dipped momentarily in water at about 45°, then the seal is pulled out, using the spiral as a handle. The content of CO_2 in 1 ml. is determined as in the method described above.

Calculation. The number of large divisions on the burette corresponding to the CO_2 absorbed is multiplied by 2·0, giving the volumes CO_2/100 ml.

Variations of the blood CO_2 in health and disease. These are shown in the following table (Table XIII), from the data assembled by Peters and Van Slyke (1932).

TABLE XIII

	Plasma.		*Arterial blood* CO_2 *vols.* %.	*Venous blood* CO_2 *vols.* %.
	CO_2 *vols.* %.	*Bicarbonate content* (*as vols.* CO_2%).		
Normal resting subject	73–55	69–51	53–43	60–49
Mild acidosis, no pronounced symptoms	55–40	51–38	—	—
Moderate to severe acidosis symptoms may be apparent	40–30	38–27	—	—
Severe acidosis, acid intoxication	30	29	—	—

Note on the term 'alkali reserve'. The term 'alkali reserve' is somewhat of a misnomer, since the bicarbonate anion represents only a fraction of the total blood buffering, which in turn is only one-sixth the total buffering of the body at the normal or viable levels, and it would seem preferable to state the result as plasma bicarbonate.

Application of the foregoing method to the *p*H determination of circulating arterial blood. We may write the Henderson-Hasselbalch equation for very dilute solutions in the form

$$pH = pK'_1 + \log\frac{[HCO'_3]}{[H_2CO_3]} - 0{\cdot}5\sqrt{\mu}. \qquad (30)$$

In this, $[HCO'_3]$ is the bicarbonate ion concentration and $[H_2CO_3]$ may be taken as the concentration of free carbon dioxide in solution; $pK'_1 = -\log K'_1 = 6{\cdot}33$ at 38°, where K'_1 is the first apparent dissociation constant at infinite dilution. (The value 6·33 increases by 0·005 for every degree centigrade below 38°.)

μ is the 'ionic strength' of the solution and is obtained by multiplying the concentration of each ion by the square of that ion's valence number, summing all these products and dividing the result by two (Debye and Hückel, 1923).

For human blood plasma at 37° C. the equation may be written (taking account of the data of Dill, Daly and Forbes, 1937)

$$pH = 6{\cdot}10 + \log\frac{[HCO'_3]}{[H_2CO_3]}$$
$$= 6{\cdot}10 + \log\frac{[CO_2] - [H_2CO_3]}{[H_2CO_3]}, \quad (31)$$

where $[CO_2]$ is the total carbon dioxide, and $[H_2CO_3]$ the physically dissolved carbon dioxide. The value of $[H_2CO_3]$ is given as c.c./100 ml. or volume per cent. by $\frac{100\alpha p}{760}$, where α is the Bunsen absorption coefficient at body temperature (c.c. gas dissolved in 1 c.c. at 760 mm. pressure and reduced to volume at 0°). The exact value for α in serum as determined by Van Slyke, Sendroy, Hastings and Neill (1928) is 0·510, so that $[H_2CO_3]$ becomes $0{\cdot}0672p$, and we have

$$pH = 6{\cdot}10 + \log\frac{[CO_2] - 0{\cdot}0672p}{0{\cdot}0672p}. \quad (32)$$

In this equation by inserting the total carbon dioxide of the blood (vols. %) as determined in the preceding method for blood collected anaerobically, and for p the alveolar air carbon dioxide tension in millimetres of mercury, we get the required pH value from the equation. Thus, for example, if $[CO_2]$ is 55·0 vols. % and p is 38·0 mm., we obtain

$$pH = 6{\cdot}10 + \log\frac{55{\cdot}0 - 2{\cdot}56}{2{\cdot}56}$$
$$= 7{\cdot}42.$$

(This method was used by Kane and O'Connor (1937) in a study of the relation of the pH of the blood and the isodynamic law.)

Carbon dioxide combining capacity (similar to the older method of Van Slyke and Cullen as far as the carbon dioxide determination). This method (with the manometric analysis) is described by Peters and Van Slyke (*Quant. Clin. Chem.*, vol. ii, 1932) as still finding a definite place when provision for the anaerobic method is lacking. It is based on the determination of the carbon dioxide content of separated plasma equilibrated with alveolar CO_2 or with 5·5% carbon dioxide mixture in a 300 ml. separating funnel. The results are about 3 volumes per cent. too high by reason of the greater solubility of CO_2 in the cold (1·6 times as high at 20° C. as at 38° C.). " The relationship is quite constant

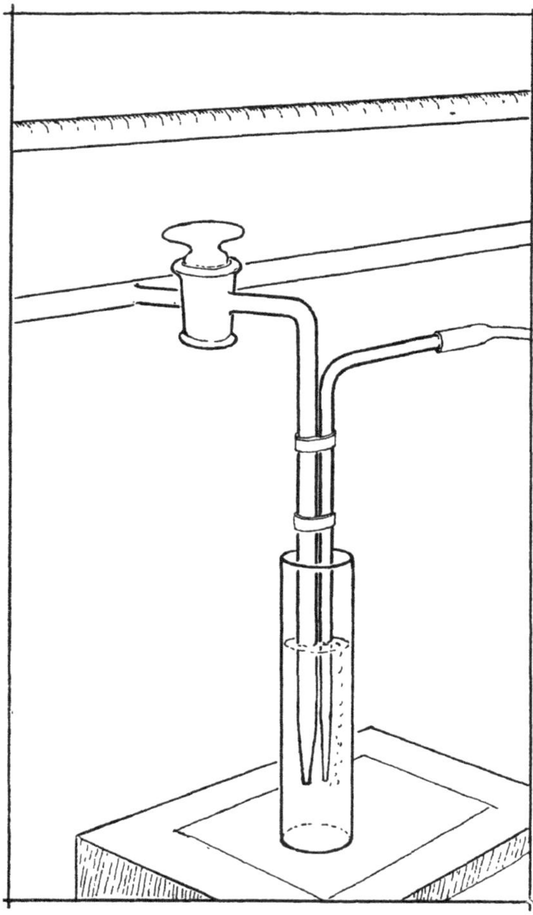

FIG. 49. Titration with the horizontal burette in a stream of carbon dioxide free air.

so that for the purpose of following pathological changes in the alkali reserve in the conditions above mentioned, the CO_2 capacity is entirely serviceable." It is also pointed out that in the majority of cases the physician wishes to ascertain if a diabetic or nephritic is threatened with acid intoxication or if vomiting or alkali therapy has produced a state of alkali excess in the blood. In these conditions the plasma bicarbonate changes are gross ones, and the results similar to the anaerobic method.

The conditions in which it differs much from this latter method are those associated with over or under ventilation of carbon dioxide.

Procedure.

Blood collection. The blood (5 c.c. or more) may be collected in a syringe and transferred to an oxalated centrifuge tube with the needle at the bottom of the tube, so that the blood does not fall through the air. The blood is agitated as little as possible—and only gently stirred for solution of the oxalate. It is then centrifuged and the plasma removed at the deepest level and transferred (somewhat over 2 ml.) to the separating funnel (Fig. 50).

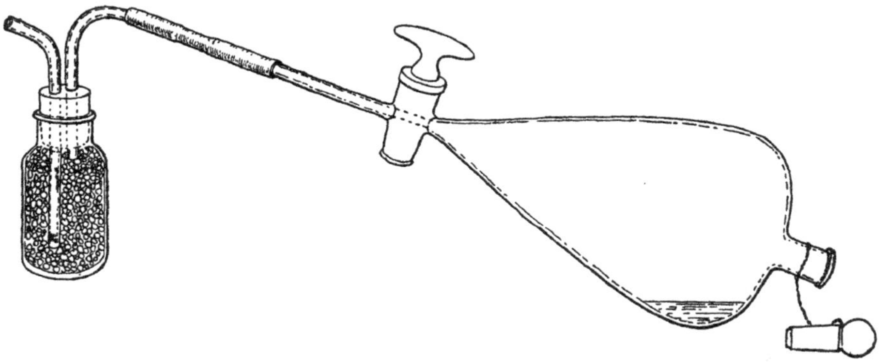

FIG. 50. Separating funnel and bottle with moistened glass beads for saturating plasma with alveolar air (from Van Slyke and Cullen, *J. Biol. Chem.*, 1917, **30**, 289).

The transference need not be immediate, it being possible to store the plasma throughout the day in a clean test tube or in a paraffined tube in the refrigerator for days.

Saturation with 5·5% carbon dioxide. Alveolar air may be used. Air is expired as quickly and as fully as possible after an ordinary expiration, and through the tube entering the bottle with moistened glass beads (Fig. 50). The air leaving the glass beads will be saturated with water vapour at the temperature of the room. Immediately after the expiration the stopper is inserted to prevent any back diffusion of air. The funnel is rotated for *two minutes* distributing the plasma in a thin layer over the interior. The stoppered funnel is then placed upright for a few minutes, so that the plasma drains to the lower end. One ml. plasma is removed in a long pyrex tube pipette with mark at the 1 ml. which has been standardised for delivery in 1 second or less with tip touching the glass bottom of the unit (this delivery gives an error of 0·4% as a coefficient of variation, as shown in Table IV). The introduction is made into a prepared ' unit ' which is tilted to one side, the lid opened momentarily just sufficient to introduce the tip of the pipette, which should touch the glass bottom of the ' unit ' at least momentarily after the delivery. The lid is instantly replaced and the ' unit ' rotated, the remaining procedure being as already described. A blank determination carried out at the same time on distilled water gives an allowance both for the dissolved CO_2 (not exactly, but sufficiently accurate for the purpose) and any deterioration in the $Ba(OH)_2$ solution.

Lehmann's microdiffusion procedure for plasma bicarbonate (1944)

This has been successfully used for many years by clinical technicians in Dr. Lehmann's Laboratory in Gothenburg. The mode of collection and transference to a centrifuge tube are apparent from Fig. 51. In this way a little blood is collected from the finger (which is made hyperaemic by warmth. By ' milking ' the finger 3 or 4 ml. blood can thus be obtained).

Procedure. Into one region of the outer chamber of a standard Conway unit, one ml. of 2 N H_2SO_4 is introduced, and 0·5 ml. of 20% $BaCl_2$ into the central chamber from a graduated 10 ml. pipette. The lids are smeared with *adeps lanæ* around the surface corresponding to their contact with the unit and placed in position.

One ml. of 0·1 N NaOH is then introduced in the manner shown in Fig. 52 and after that 1 ml. of 1·0 N NaOH is introduced into the central chamber in a similar manner. Then 1 ml. of plasma is placed in the outer chamber avoiding any contact of the plasma with the acid. (The use of 0·5 ml. acid instead of 1·0 ml. may be found more convenient for this purpose.) The lid is firmly placed in position and the unit rotated in the usual manner to mix the plasma and acid.

After one hour, 0·1 ml. thymol blue solution is introduced into the central chamber with an ordinary glass syringe, and then the contents titrated with 0·1 N HCl using a Manteaux syringe as shown in Fig. 52.

Calculation (ml. blank titration minus analytical) × 112 = Vol. % CO_2. Allowance would be necessary for any factors involved with the titrating acid and graduated syringes.

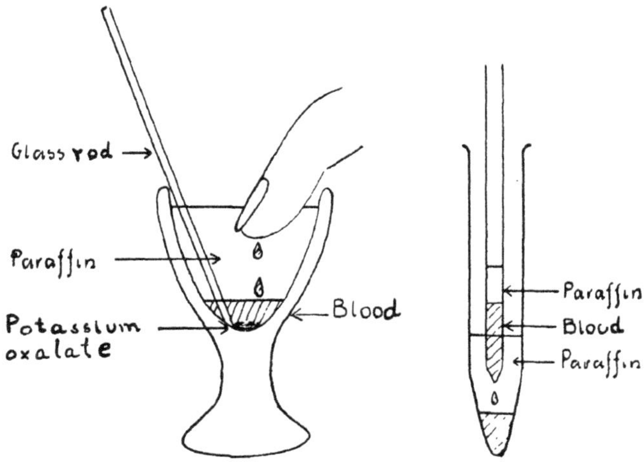

FIG. 51

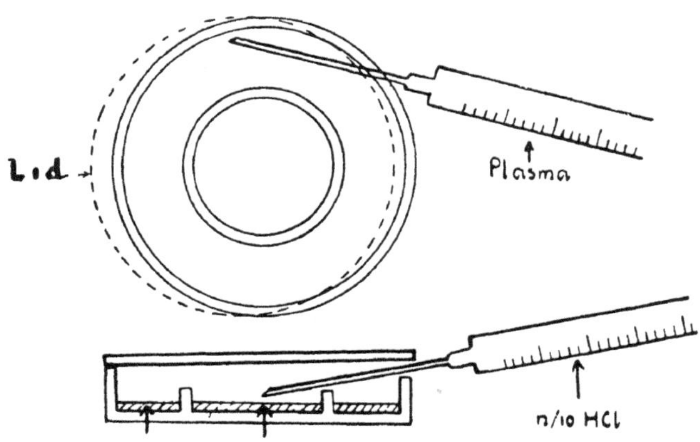

1 cc. $2nH_2SO_4$ 0·5 cc. 20% $BaCl_2$
1·0 cc. N/10 NaOH
0·1 cc. thymol blue

FIG. 52. Upper diagram shows the manner of introducing the plasma into the outer chamber and the lower, the procedure for titrating with 0·1 *N* HCl.

Notes on method

1. In Lehmann's clinical laboratory mixing is assisted by an oscillating magnet under the supporting platform of the unit, and a very small piece of iron wire sealed in glass in the central chamber.

2. With careful working the mean square error difference from results with Van Slyke's method is ± 1·1 vol. %. With routine analyses one may reckon on an error range of ± 3 to 5 vol. % the normal content of the plasma being 55 to 75 vol. %.

Total CO_2 in blood or plasma, using 0·1 ml. vols.

The microdiffusion method with 0·1 ml. blood using No. 2A ' units ' (or alternatively No. 2) is specially suitable. *M*/50 bicarbonate is used to control the determination.

Procedure. Into the outer chamber of No. 2A ' units ' is pipetted 0·5 ml. of *N*/20 HCl. The rims of the ' units ' are vaselined or smeared with liquid paraffin.

0·1 ml. volumes of *M*/12·5 $Ba(OH)_2$ with indicator are then introduced into the central chamber, using for convenience here a delivery between two marks, as, for example, in the pipettes supplied with the Warburg apparatus. The lids smeared with liquid paraffin (or tragacanth fixative as described below) are placed in position and the hole in the lid closed with the little glass cylinder, the base of which has been likewise smeared. The 0·1 ml. blood or plasma is then quickly introduced from a pipette inserted through the hole, removing the small cylinder momentarily for the purpose and using a similar pipette as for the barium hydroxide introduced into the central chamber. Immediately after withdrawing the pipette the hole is again closed by the little cylinder. The contents of the outer chamber are mixed and the ' units ' left aside for one hour on the bench.

Blank ' units ' are likewise set up. The contents of the central chamber are then titrated to a light green with *N*/50 HCl from the Conway burette.

Calculation. The number of large divisions on the burette corresponding to the CO_2 absorbed gives directly the millimols total CO_2 in the blood.

The procedure may be standardised with $M/50$ bicarbonate. The coefficient of variation was found to be about 0·8% in a series of determinations.

Solutions.

M/12·5 $Ba(OH)_2$. 40 ml. of $N/5$ $Ba(OH)_2$ are made up to 100 ml. with indicator (*vide* p. 192) incorporated. A little is poured into a narrow test tube and the 0·1 ml. vols. are removed by placing the tip of the pipette well below the surface.

M/50 $NaHCO_3$ or $KHCO_3$. Dissolve 168 mg. $NaHCO_3$ or 200 mg. $KHCO_3$ in distilled water and make to 100 ml.

Water soluble fixative (if desired). 6 g. tragacanth ground in a large mortar with slow addition of 80 ml. water ; 30 ml. glycerol and 4 ml. N HCl are then added and mixed.

Notes on the method. The method is very convenient and accurate for 0·1 ml. volumes of blood or plasma the titration results representing at once the mM. per litre. Instead of the No. 2A ' units ' the No. 2 may be used.

The blood must be introduced rapidly with as little loss of CO_2 as possible, and the delivery between two marks is here convenient. Such 0·1 ml. pipettes may be readily constructed.

The rate of absorption of CO_2 in the No. 2 ' units ' is shown in Fig. 48 compared with that from the larger standard apparatus and, as will be noticed, there is very little difference in the times required for the full absorption.

Measurement of fermenting power of yeast.

The CO_2 produced by the fermentation is absorbed in NaOH in the central chamber. After a short time (about 10 mins., but 30 may be allowed) the absorption becomes a linear function of the time and represents the CO_2 formed by fermentation.

By titrating some ' units ' at different times after setting up, the rate of fermentation may be determined.

Procedure. Into the outer chamber of No. 1 ' units ' is pipetted 1 ml. of the yeast suspension (1 in 20 of tap water). Into the central chamber is run 1 ml. of $N/10$ NaOH containing indicator. The lids are placed at once on the ' units ' and 1 ml. of 6% glucose then run into the outer chamber. After 30 minutes the central chamber is titrated, 0·5 c.c. of saturated $BaCl_2$ being first added and for ten or twenty minute periods after this 30 minute period (taken from the addition of the glucose) other ' units ' are titrated. In this way a line may be plotted giving the carbon dioxide output.

The method can be adapted readily for the No. 2 ' units '. 0·2 ml. $N/10$ NaOH is placed in the central chamber and 0·5 ml. of yeast suspension and 0·5 ml. 6% glucose added, etc. Before titrating a small drop of saturated $BaCl_2$ is added.

Charnley's application of the microdiffusion technique to measurement of fish spoilage (1945). In a previous chapter the determination of trimethylamine as a measure of fish spoilage was described (Beatty and Gibbons, 1937). Charnley (1945) has used the CO_2 value of the drained muscle tissue as a measure of the freshness of canned salmon. With his procedure 5 g. of drained muscle tissue from the interior of the sample are rapidly broken or minced and introduced into the outer chamber of a weighed standard ' unit '. The cover (with fixative) is replaced quickly and the whole weighed. 1·0 to 1·2 ml. of $N/10$ $Ba(OH)_2$ is quickly introduced into the central chamber, the cover replaced and the ' unit ' incubated at 50° C. After 30 minutes it is removed and the contents of the inner chamber titrated with $N/10$ HCl, using a drop of phenolphthalein as indicator.

(A suggested alternative procedure would be to use $N/10$ KOH with indicator in the central chamber, and to titrate with $N/5$ HCl from the horizontal burette, after adding 0·5 ml. saturated $Ba(Cl)_2$.)

Each ml. $N/10$ HCl = 2·2 mg. CO_2 (or each large division $N/5$ HCl from the Conway burette = 0·044 mg. CO_2). The CO_2 value of the sample is expressed as mg. CO_2 per g. This CO_2 value is found to be proportional to the mean value of the examiner's rating. A scale for grading freshness of canned salmon is given as follows.

Average CO_2 *value*	*Grade*
Less than 0·130	A
0·130 to 0·170	B
$>$ 0·170	Condemned

As fixative in the above method 50% paraffin with 50% beeswax was found serviceable ; also, 75% paraffin and 25% petrolatum.

CHAPTER XXV

BLOOD GLUCOSE AND FERMENTABLE SUGAR IN NORMAL URINE

BLOOD GLUCOSE

THE estimation of the blood glucose is an important clinical method, and the determination here given (O'Malley, Conway and FitzGerald, 1943) may be considered as generally applicable to minute quantities of glucose or other fermentable sugar.

A manometric method for determining the true sugar in blood was described by Holden (1937) and compared with the Somogyi (1927) modification of determining the true sugar by a subtraction of the reducing material present after, from that present before, fermentation. A method for the estimation of blood glucose is here described in which the CO_2 evolved by yeast action on blood is measured by a microdiffusion technique. The whole procedure of fermentation, diffusion and titration is carried out in the No. 2 or 2A 'units'. The results, for non-fasting subjects, are in agreement with those of the Fujita and Iwatake (1931) modification of the Hagedorn-Jensen method on the same specimens of blood. In this modification the blood is deproteinised with Cd instead of Zn, and the method is claimed to give a reliable measure of the true blood glucose. With blood from fasting subjects the microdiffusion method gives slightly lower values, which indicates an even closer approach to the actual value.

In the method as already published the *autofermentation* of the yeast in the presence of blood as compared with saline under similar conditions remained to be determined. It has now been shown (McCauley, Conway and Dillon, 1944) that no difference arises from the interchange. It would appear then that the method gives the true value of the fermentable material in blood which is almost altogether glucose.

While the general principle of the method is here very simple the disturbing factor of the autofermentation of the yeast must be borne in mind and carefully allowed for, also the yeast suspension should be practically as nearly CO_2 free as possible before introduction, and the preformed CO_2 of the blood removed. Another difficulty will appear from the following.

Yield of CO_2 from fermented glucose.

From the equation

$$C_6H_{12}O_6 = 2CO_2 + 2C_2H_5OH$$

1 mg. of glucose should yield 0·489 mg. CO_2 corresponding to 1·11 ml. $N/50$ acid. As is well known, less than this theoretical yield is obtained. We have found an average of 78% with baker's yeast and of 92% with brewer's yeast (Guinness). Van Slyke and Hawkins (1929) found 71–79% with baker's yeast, and Holden (1937) an average of 86% with a certain type of brewer's yeast (L-Anheuser-Busch).

The yield is also somewhat affected by the age of the yeast though from 2–19 days little variation was found with baker's yeast. It is also affected by the nature of the medium, being a little higher with saline than with blood. At the same time rather wide variations of *p*H have little effect; thus it is practically constant from 3–5 *p*H.

It will appear then that an internal control is necessary; by which is meant that the amount of CO_2 produced from a definite small quantity of glucose added to a measured sample of blood under the conditions of the determination is required to measure the percentage yield of CO_2 and so control the estimate.

Procedure. Firstly, in order to have the yeast suspension practically free of CO_2, 1·5 ml. volumes of the suspension are introduced into the outer chambers of the standard 'units' (No. 1). These are left exposed on the bench while the others are being made ready. Duplicate samples of 0·1 ml. blood are then introduced into No. 2 (or No. 2A) 'units', 4 'units' being prepared for one sample of blood. Two 'units' are also set up with 0·1 ml. water. To all of these except 2 of the 4 'units' with blood from the one specimen, 0·5 ml. of the acidifying mixture is added, this is mixed with the blood: 0·5 ml. of the acidifying mixture containing glucose is then introduced into the 2 remaining 'units'.

Approximately 10 min. from the introduction of the acid, 0·2 ml.

of the solution of $Ba(OH)_2$ plus indicator is introduced into the central chamber, and each ' unit ' closed immediately. Into all the small ' units ' 0·2 ml. of the yeast suspension is now introduced, and is taken from the large ' units ' after stirring. This introduction is also made quickly from a 0·2 ml. tube-pipette and after the ' unit ' has been closed the contents of the outer chamber are mixed by rotation. After 90 minutes at room temperature the contents of the central chamber are titrated with $N/50$ HCl from the Conway burette, to a *p*H of 7·5–8·0.

With regard to the fixative, vaseline or liquid paraffin may be used, or a water soluble fixative, such as the tragacanth mixture described below. The No. 2 ' units ' may be used as described in the last chapter, under the description of the estimate of total CO_2 in 0·2 ml. blood.

Calculation. The number of divisions on the burette corresponding to 1 mg. glucose/100 ml. blood is obtained by subtracting the mean titration for the 2 (of the 4) blood ' units ' which contain added glucose from the mean of the 2 without added glucose, this value being then divided by 200. For the remaining ' units ' the titration values are subtracted from the mean of those containing water, and divided by the result obtained above, giving the concentration as mg. glucose/100 ml. blood.

EXAMPLE.		
	(1) Mean titration of ' units ' with added glucose	11·0 (large divs.)
	(2) Mean titration of corresponding ' units ' without added glucose	28·0
	(3) Blank with water instead of blood	36·8

Concentration of glucose in the blood

$$=\frac{(36{\cdot}8-28{\cdot}0)}{(28{\cdot}0-11{\cdot}0)}\times 200=104 \text{ mg./100 ml.}$$

Other blood samples require the setting up only of duplicate ' units ' as in (2).

Range of method. The range for the blood sample with the internal control is 300 mg./100 ml., but for other blood samples done at the same time it is 500 mg./100 ml.

Solutions, etc.

N/20 $Ba(OH)_2$ *with indicator* (*vide* p. 204).

N/50 HCl.

Acidifying mixtures for outer chamber. (1) *N*/20 HCl containing 4% ethanol. (2) *N*/20 HCl containing 4% ethanol and 0·04% glucose which may be prepared by using a stock 2% glucose in 2% NaCl solution of which 2 ml./100 ml. are included.

Yeast suspension. Approximately 10 g. of yeast are made into a suspension with 100 ml. *tap* water, left for a day on the bench, and then placed in the refrigerator. It may be used after the first day for about 2 weeks. If the suspension is made in 0·5% KH_2PO_4 it will be usable for a somewhat longer period.

Fixative. Vaseline, or liquid paraffin or some acidic water soluble fixative, such as tragacanth may be used. In making the tragacanth fixative, 6 g. tragacanth are ground in a large mortar with the slow addition of 80 ml. water ; 30 ml. glycerol and 4 ml. *N* HCl are then added and mixed.

Further notes on the method.

Effect of glycolysis. The microdiffusion method like Holden's is independent of glycolysis up to 4 or 5 hr. exposure of blood samples to room temperature. If the samples are stored in the refrigerator much longer times may be allowed.

Anti-glycolytic substances such as fluoride must not be added. Oxalate or citrate used to prevent clotting have no effect on the determination.

Autofermentation of the yeast. Besides the CO_2 coming from the fermentation of blood sugar, yeast steadily produces an appreciable quantity by autofermentation. Such autofermentation decreases with the age of the yeast suspension, the fall being most rapid in the first 24 hours. It is advisable, therefore, to allow it to stand for 24 hrs. at room temperature and then to place it in the refrigerator. After the first day the autofermentation during the course of the determination and using baker's yeast will account for about 0·03 ml. (3 large divs.) of *N*/50 acid.

Ethanol in the acidifying mixture. Before using the special acidifying mixture described it was found that the CO_2 yield for small quantities of added glucose (1 mg./ml. blood) was proportionally less than for larger additions (2–4 mg./ml.). This effect was found to be suppressed by adding a little ethanol to the acidifying fluid introduced into the outer chamber.

Accuracy of the single determination. From a study of the duplicate results the coefficient of variation was found to be 1·7 %.

Fermentable sugar in normal urine.

For glucose in diabetic urine the ordinary methods turning on copper reduction will probably be found more convenient. The delicacy of the microdiffusion method lends itself, however, to accurate determination of the minute amounts of glucose or fermentable sugar in normal urine. The principle of the method is essentially similar to that of the blood glucose already described.

Procedure. In order to have the yeast suspension practically free of CO_2 when introduced into the ' units ' 1·5 ml. vols. may be introduced into one or more No. 1 ' units ' and left exposed on the bench.

Some of the urine is then fermented. For this purpose 10 c.c. may be taken, 1 g. yeast added and mixed and maintained at 30° C. for 15 minutes. It is then centrifuged at high speed and filtered free from yeast cells.

For each urine sample three 25 c.c. flasks with the following additions are prepared :

(*A*) 10 ml. fermented urine + 2 ml. acidifying mixture No. 1.

(*B*) 10 ml. urine + 2 ml. acidifying mixture No. 1.

(*C*) 10 ml. urine + 2 ml. of acidifying mixture No. 2.

These are brought to boiling and then cooled and a duplicate set of No. 2 or 2A ' units ' prepared for each. Into the outer chambers of these is put 1 c.c. of the urine-acid mixture and into the inner 0·2 c.c. of the $N/20$ $Ba(OH)_2$ with B.D.H. Indicator. After closing the ' units ' 0·2 ml. of 1/20 yeast suspension is added as described (see previous method or that for CO_2 in 0·2 c.c. blood). Each ' unit ' is rotated after the addition of the yeast and at the end of 3 hours the contents of the inner chamber are titrated with $N/50$ HCl from the Conway burette. (The end-point with Universal Indicator is taken when a light green appears.)

Calculation. This may be represented as :

$$\frac{a-b}{a-c} \times 50 = \text{mg. glucose/100 ml. urine.}$$

a, *b*, and *c* are the titration results for *A*, *B*, and *C* flasks as described above.

Solutions, etc.

N/20 $Ba(OH)_2$ with indicator (*vide* p. 192).

Acidifying Mixture No. 1. (*p*H = 4·5) This is prepared by dissolving 50 g. of citric acid in distilled water, adding 15 g. NaOH and 40 ml. absolute alcohol and making up to one litre. The solution is stored in the refrigerator.

Acidifying Mixture No. 2 (*with glucose*). This contains 250 mg. % glucose. 10 c.c. of stock glucose solution (2·5 g. % in 2% saline) are made up to 100 ml. with acidifying mixture No. 1.

N/50 HCl.

Fixative. Vaseline, or liquid paraffin or the soluble tragacanth mixture described on page 204.

Yeast suspension 1/20. 10 g. of yeast are made into a suspension with 200 ml. of 0·5% KH_2PO_4, in tap water. The mixture is allowed to stand on the bench overnight and then placed in the refrigerator. (If required the same day, it may be allowed to rest on the bench for 4 hrs.)

Notes on the Method.

(1) The results from the above method were compared with the West and Petersen (1932) modification of the Shaffer and Somogyi method. The following gives a comparison of the methods for 5 normal fasting urines.

	Microdiffusion method mg. %	*West and Petersen method mg.* %
(1)	1·6	0·0
(2)	0·0	0·0
(3)	2·2	5·5
(4)	22·3	24·3
(5)	0·0	0·0

(2) With the use of rapidly fermented urine to obtain the blank value in the above microdiffusion method, as well as exact recoveries of added glucose, the method would appear to give a true measure of fermentable sugar in urine.

(3) Reviews of early work on such determinations are given by Benedict, Osterberg and Neuwirth (1918) ; Folin and Berglund (1922) and Greenwald Gross and Samet (1924) ; more recently by Harding and Selby (1931).

CHAPTER XXVI

DETERMINATION OF CARBONIC ANHYDRASE

METHODS previously used for carbonic anhydrase determination include the *manometric* method of Meldrum and Roughton (1933) and the *colorimetric method* originally due to Brinkman (1933) and since modified by Philpot and Philpot (1936) and again by Keilin and Mann (1940). The microdiffusion procedure can be readily adapted for convenient measurements of the enzyme (Conway and O'Rourke, 1945).

In applying the principle 0·4 ml. of a mixture of the enzyme with the standard bicarbonate is exposed in the outer chamber of a No. 2 microdiffusion 'unit' for one minute, the 'unit' being rocked on the apparatus shown in Figs. 6–8. The remaining carbon dioxide is then determined.

The increase in rate of emission of the CO_2 from the outer chamber goes in linear relation with the enzyme concentration up to a certain emission rate.

Procedure for blood. 0·2 ml. of the buffer solution containing 1 in 4 of the enzyme solution, suitably diluted (1 in 500 or 1000) for blood, is run into the outer chamber of a No. 2 'unit' and spread in a layer over the floor of the chamber. Then at a definite time—best taken with a stop watch—0·2 ml. of the bicarbonate is added to the buffer and the solutions well mixed for 15 seconds and placed on the rocking platform. Five seconds from the end of the exposure time the 'unit' is removed and 0·2 ml. of the barium hydroxide solution is then run into the central chamber so that the end of the addition coincides as exactly as possible with the lapse of 1 minute from the beginning of the addition of the bicarbonate, and the 'unit' immediately closed, giving in all one minute exposure of the bicarbonate/buffer mixture. Subsequently, excess acid (0·2 ml. of N H_2SO_4) is added to the outer chamber and the 'unit' left aside for 1 hour before titration. This is carried out with $N/50$ HCl from the Conway burette.

A companion estimate is carried out with water used instead of the enzyme solution. For a single determination of carbonic anhydrase triplicate series should be set up.

Calculation. If E is the titration figure (in large divisions) for the 'unit' with enzyme and B that for 'unit' with water, $(E-B)$ represents the increased CO_2 emission due to the enzyme in the one minute exposure. The calculation for Roughton-Meldrum enzyme 'units' is :

$$(E-B) \times 97 \ldots \text{ (where the blood is diluted 1000 times).}$$

The calculation applies for 17·5° C. An exact correction for other temperatures has not yet been worked out, but a reduction of 10% per rise of one degree may be provisionally allowed.

Solutions.

Buffer solution. Na_2HPO_4 solution ($M/4$) 3·55%.
KH_2PO_4 solution ($M/4$) 3·40%.

These are mixed in equal volumes.

Buffer-enzyme solution. 3 parts of the buffer solution plus one part of the enzyme solution.

For blood, this enzyme solution should be 1 in 500 or 1 in 1000.

Bicarbonate solution. 0·40 g. $NaHCO_3$ is made up with 0·5 ml. N NaOH to 100 ml.

$Ba(OH)_2$ *solution.* $N/10$ containing indicator (*vide* p. 204).

$N/50$ HCl.

$N/1$ H_2SO_4.

Notes on the method.

Other fluids and tissues. Besides blood, other biological fluids and tissues may be used, the tissues being brought into suspension by grinding. With dilutions of 1 in 50 or over, no attention need be paid to the carbon dioxide already present in the tissue suspension. Under this dilution level the suspension may be slightly acidified and the CO_2 allowed to diffuse into the atmosphere.

Limit of enzyme concentration. If the number of burette divisions corresponding to the enzyme action is above 15, the solution should be further diluted to bring the reading to about 10. Fig. 53 shows the effect of different blood concentrations, and the fact that only below a certain level of enzyme concentrations is the relation to CO_2 emission a linear one.

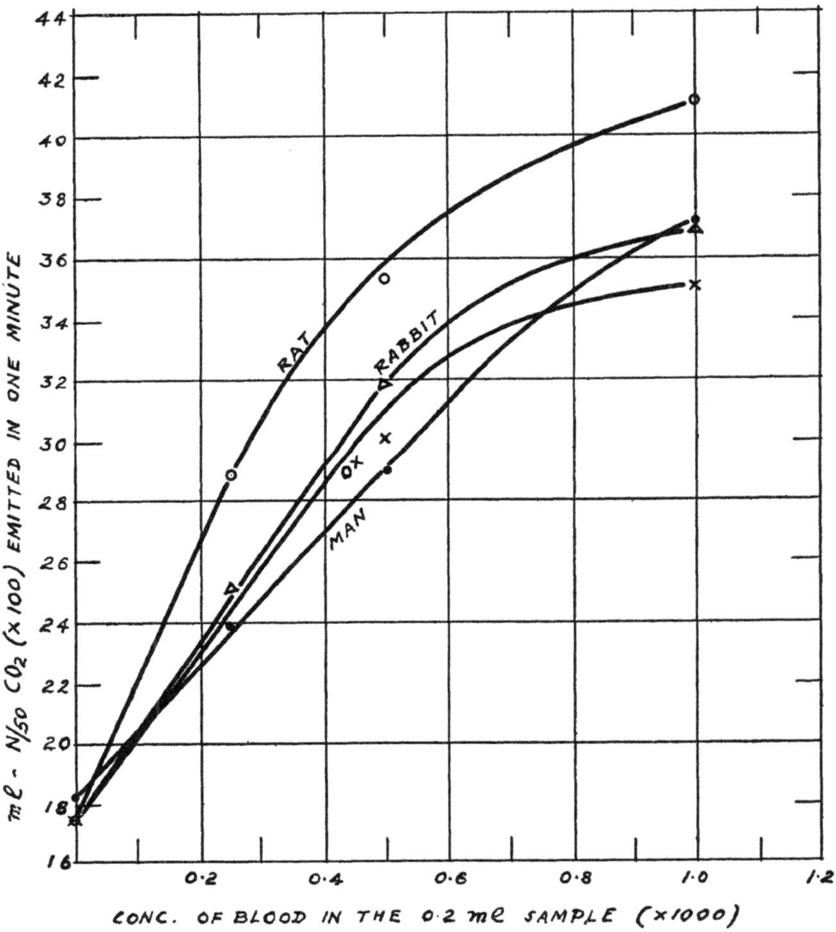

FIG. 52*a*. Increased losses of CO_2 with increasing concentration of carbonic anhydrase as described in text.

'*Units*' *of enzyme concentration.* Meldrum and Roughton (1933) defined the 'unit' of enzyme concentration as that which when present in 4 ml. at *p*H of 6·8 gave a $(R - R_0)/R_0$ of 1·0 at a temperature of 15 degrees. Here R and R_0 are the rates of CO_2 emission with and without enzyme. Applying this to the above method, the results in Table XIV have been obtained with the blood of different animals. The results are compared with those in Roughton's table.

TABLE XIV. (Carbonic anhydrase in blood.)

Blood.	*Enzyme concentration (Meldrum-Roughton 'units').*	*Meldrum and Roughton's figures.*
Human - - -	552 (3) (500–590)	550 (8) (370–680)
Rat - - - -	1260 (2) (790 ; 1720)	1350 (1)
Ox - - - -	700 (1)	1100 (6) (600–1800)
Rabbit - - -	820 (1)	1210 (2) (710 ; 1720)

The bracketed single figures after the means give the number of observations. The range is given below the means.

CHAPTER XXVII

OXIDATION RATES OF ORGANIC SUBSTANCES WITH A STANDARD OXIDANT WITH APPLICATION TO DETERMINATION OF MINUTE AMOUNTS OF CALCIUM AS OXALATE

THE following account of rates of oxidation of various organic substances of biochemical significance by a standard oxidant may be of use in various ways, such as the estimation of such material under certain conditions or the investigation of stages in the path of oxidation of particular kinds. If, for example, an oxidant mixture oxidises an organic substance producing carbon dioxide, it is reasonable to suppose that a substance in the main path of this oxidation will not be oxidised at a lesser rate than the original substance.

In the experiments as originally carried out (Conway and McVerry, 1937) the standard ' units ' (No. 1) were used, but the procedure could be easily modified to suit the No. 2 or No. 2A ' units '. The oxidant mixture mostly used was one made up as follows: One volume of $M/20$ $KMnO_4$, with one volume of $M/20$ KH_2PO_4 and 0·1 volume of 8% $MnSO_4$. Where a stronger oxidant was required the $M/20$ KH_2PO_4 was changed for $2N$ H_2SO_4 or $4N$ H_2SO_4. The mixture so made up is muddy but easily pipetted, and should be freshly prepared before using.

The mechanism of the ordinary permanganate oxidation as investigated by Skrabal (1904) consists in the formation of manganous ions, then the oxidation of these to manganic ions, which carry out the rapid stage of the oxidative breakdown of the organic substance. The presence of excess manganous ions in the mixture described above assists in abridging the first stage.

Factors determining oxidation rates. These are in a general way as follows:

(1) Redox potentials of the systems involved.

(2) Internal molecular stability.

If, for simplicity, we are dealing with reversible systems then the oxidation of one by the other (by which is meant the change of reductant in the first system to its oxidant form), is determined by the fact that its redox potential must be less. This states in other words that a necessary condition of the oxidation is a transfer of free energy from the system oxidised. When a reversible system is oxidised its free energy is decreased. In the case of an organic substance the oxidant stage may be very unstable and break down to smaller molecules with a lesser total of free energy, heat being liberated in the process, which is in effect irreversible.

For the simplest reversible systems and where the 'oxidising' mixture is in large excess a difference of approximately 60 millivolts ensures a 99% oxidation of the 'oxidised' system, and a difference of 120 millivolts 99·9% 'oxidation'. With about 100 millivolts or more difference the substance with the lesser potential may be said to be fully oxidised.

The question arises, is the *rate* of the oxidation influenced by a difference of 200, 300, or more millivolts? In general this would appear true using the same redox system throughout as the 'oxidiser', and the potential changed by acidification, but no definite theoretical relation can be stated. The greatest discrepancies in rates occur using different systems, or in other words a constant definite potential difference between the oxidising and oxidised systems does not imply anything like a constant rate of oxidation. Thus the standard oxidant mixture described above has a potential of 1·12 volts and a cerium mixture consisting of 25 c.c. of 20% cerium sulphate diluted to 100 c.c. by N H_2SO_4 has a potential of 1·48 (against the normal hydrogen electrode), but the standard oxidation mixture oxidises a number of organic substances at a far greater rate, such, for example, as glycerol, glucose, glyceric aldehyde. The reason for such differences may be sought in a consideration of the second factor—the internal molecular stability. Such stability may be lessened considerably by a catalyst, and the oxidising system may function in this way as well as by being a purely redox system, with a higher potential. The catalyst by combining with the organic substance lessens the internal stability and the product becomes more readily oxidised.

In the oxidations to be described, the measure of the oxidation rate is taken as the production of CO_2. It may be recalled that in biological oxidation the liberation of CO_2 is not regarded as in itself an oxidative stage, but is produced by carboxylases or allied enzymes with little or no change in the free energy of the organic substance. Yet the stages that lead to the production of terminal –COOH groups are oxidative stages, and thus the CO_2 liberation appears as a measure of the total oxidative rate, and its appearance here may be similarly regarded. The method used in the oxidative studies was as follows: 0·2 c.c. of $\frac{N}{4}$ solution of the substance for investigation is introduced into the outer chamber of the 'unit', and this is followed by the rapid introduction of 1 ml. $\frac{N}{20}$ (approx.) KOH from a Bang burette into the central chamber, and the 'unit' is immediately made airtight by means of a glass lid smeared with vaseline, 1·5 c.c. of the standard oxidation mixture is quickly introduced into the outer chamber, the 'unit' again sealed and rotated several times. The oxidant is allowed to act on the substance for a fixed length of time, after which period the oxidation is stopped by the introduction of 0·75 ml. $FeSO_4$ (saturated in $2N$ H_2SO_4) into the outer chamber. The 'unit' is allowed to remain for at least an hour, so that all the carbon dioxide evolved may be driven over, and absorbed by the KOH in the central chamber. The alkali is then quickly titrated with $\frac{N}{40}$ HCl using thymolphthalein indicator, having added a few drops of saturated barium chloride. Several 'units' are used in the manner described, allowing the period of action of the oxidising agent to vary in each, making sure that it acts sufficiently long in some 'units' to establish an accurate end-point. With the aid of blank estimations, the results of the titrations will show the percentage oxidation for each particular time, and a curve can be thus established showing the rate of oxidation.

For purely analytical purposes the time for complete or practically complete oxidation is no doubt the most serviceable; but since this time may be indefinite, and some expression of the rate of oxidation of value in deciding comparative speeds, the half-value period or the time for 50% oxidation is convenient. The relation that the rate of CO_2 formation is proportional to the remaining fraction of the substance or its products in the outer chamber is not sufficiently well obeyed to render it serviceable in the form

$$\frac{1}{t}\log a/(a-x)=k.$$

In Table XV is given a list of half-value periods for a number of

TABLE XV

1 *Substance.*	2 *Formula.*	3 *½ value period with standard oxidant (mins.).*	4 *½ value period with standard oxidant with H_2SO_4 (mins.).*
Methyl alcohol	$H.CH_2OH$	8000	
Formaldehyde	$H.CHO$	70	
Formic acid	$H.COOH$	0·12	
Ethyl alcohol	$CH_3.CH_2OH$	No CO_2	
Acetaldehyde	$CH_3.CHO$	No CO_2	
Acetic acid	$CH_3.COOH$	No CO_2	
Glycine	$CH_2NH_2.COOH$	No CO_2	
Ethylene glycol	$CH_2OH.CH_2OH$	240	
Glycollic aldehyde	$CH_2OH.CHO$	8	
Glyoxal	$CHO.CHO$	7	
Glyoxyllic acid	$CHO.COOH$	4	
Oxalic acid	$COOH.COOH$	1	
Alanine	$CH_3.CHNH_2.COOH$	No CO_2	
Glycerol	$CH_2OH.CHOH.CH_2OH$	285	
Lactic acid	$CH_3.CHOH.COOH$	84	1·2 (4*N*)
Glyceric aldehyde	$CH_2OH.CHOH.CHO$	13	20 (2*N*)
Pyruvic acid	$CH_3.CO.COOH$	7	0·08 (4*N*)
Butyric acid	$CH_3.CH_2CH_2COOH$	No CO_2	
Succinic acid	$HOOC.CH_2CH_2COOH$	No CO_2	
Erythrol	$CH_2OH.CHOH.CHOH.CH_2OH$	276	
Malic acid	$HOOC.CHOH.CH_2.COOH$	6	
Tartaric acid	$HOOC.CHOH.CHOH.COOH$	4	
Maleic acid	$HOOC.CH:CH.COOH$	1·6	
Adonite	$CH_2OH(CHOH)_3CH_2OH$	720	
Arabinose	$CH_2OH(CHOH)_3CHO$	420	50 (2*N*)
Mannitol	$CH_2OH(CHOH)_4CH_2OH$	1440	
Glucose	$CH_2OH(CHOH)_4CHO$	720	60 (2*N*)
Citric acid	$HOOC.CH_2.C(OH)COOH$ CH_2COOH	20	

substances with the standard oxidant and also with mixtures incorporating $2N$ and $4N$ H_2SO_4.

From the following table the conclusions with reference to the standard oxidant may be drawn :

(1) If a C atom is associated with three hydrogen atoms, it has a high stability to oxidation and no CO_2 is formed from this carbon atom or from an associated $-COOH$ group.

The same holds if one of these H groups is substituted by NH_2 or by a $CH_2 . COOH$ group.

(2) For an organic substance to be completely oxidisable to CO_2 at an appreciable rate it is necessary that each carbon atom be associated with one oxygen or OH group.

(3) Terminal $-CH_2OH$ groups as evidenced, say, by ethylene glycol require about 4 hours for the $\frac{1}{2}$ value period change of even one $-CH_2OH$ to $-CHO$, and this is likewise apparent from glycerol, erythrol and adonite studies, when allowance is made for the rate of oxidation of the corresponding aldehyde. With increasing number of carbon atoms (CHOH groups between the terminal radicles) the increasing size of the molecule would need to be allowed for and the allowance may be made by dividing the $\frac{1}{2}$ value period by the number of carbon atoms. Using this division two terminal $-CH_2OH$ groups cause a delay of 1 to 2 hours in the $\frac{1}{2}$ value period per carbon atom. Two aldehyde groups would require only a few minutes to reach their next stage, and a similar order of time would naturally apply to one $-CHO$ to reach its COOH stage of oxidation.

(4) When the H ion of the mixture is increased, thereby increasing the oxidation-reduction potential, the rate of oxidation markedly increases. There are some exceptions, notably, formic acid and glyceric aldehyde. With formic acid it would appear that the un-ionised compound is much more slowly attacked than the formate ion. Figs. 53 *A* and *B* show the $\frac{1}{2}$ value curves for formate and oxalate at different *p*H levels, using phosphate and arsenate buffering. Fig. 53 *C* and *D* show rates of oxidation of glycollic aldehyde and glyceric aldehyde with the standard oxidant and arabinose and glucose with the $2N$ H_2SO_4 mixture.

A study of such oxidation rates may be of analytical service in a variety of conditions.

The mechanism of the above oxidation processes. In the oxidation processes as above considered, there is a conversion of $-CH_2OH$ to $-CHO$ and to $-COOH$ groups, and of $-CHOH-$ to $-CO-$, besides which CO_2 is split off from $-CO.COOH$. This latter may be considered as a non-oxidative process akin

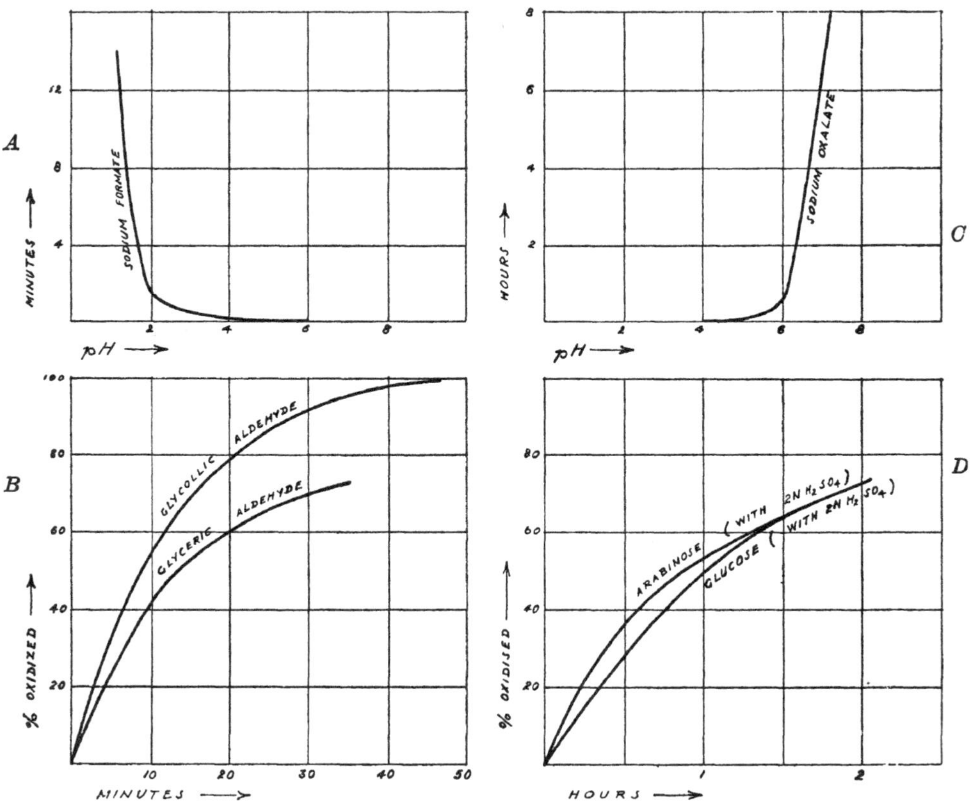

FIG. 53. Oxidation rates (as measured by CO_2 formation) of certain organic substances with a standard permanganate oxidant.

A.—Formic acid (as the sodium salt).
B.—Oxalic acid (as the sodium salt).
C.—Glycollic aldehyde and glyceric aldehyde.
D.—Arabinose and glucose (using the oxidant with $2NH_2SO_4$).

to the action of carboxylase. The oxidation of the alcohol and aldehyde groups on the other hand may be interpreted as a hydrogen transfer regarded, primarily at least, as an electron transfer (with release of hydrogen ion) to the manganic system, and in this way resembling biological oxidations.

Murnaghan's Procedure for determining Ca Content of Small Volumes of Serum (1949)

Principle. The Ca is precipitated as oxalate, which is then dissolved in H_2SO_4 solution, and an aliquot transferred to the outer chamber of a No. 2A unit. It is oxidised therein with acidified cerium sulphate, the CO_2 evolved being absorbed in barium hydroxide solution in the central chamber. The titration is then carried out as previously described.

Procedure

Precipitation and isolation as oxalate. Into a 3 ml. centrifuge tube with pointed end, pipette 1 ml. of distilled water and 0·5 ml. of the serum (or 1 ml. when a very low value is anticipated). Add 0·5 ml. of filtered saturated ammonium oxalate solution, which should not touch the sides of the tube. Mix with a glass rod, and rinse the rod with distilled water until the total volume is about 3 ml. Set aside for 30 minutes and then centrifuge.

Remove the supernatant fluid with the syphoning arrangement shown in Fig. 54, until the surface of the fluid is 1 to 2 mm. above the opening in the capillary tube. The rate of syphoning should not exceed 1 ml. per minute. Rinse the sides of the tube with 2 ml. of dilute ammonia solution (2% *v/v*) without disturbing the precipitate, centrifuge and syphon. Repeat the washing, centrifuging and syphoning once more and put the tube in an oven to dry at 105° C. When the tube is dry, pipette into it exactly 0·5 ml. of *N* sulphuric acid so that the acid washes the sides of the tube. Place in a hot water bath, stir with with a fine glass rod to dissolve the precipitate and allow to cool.

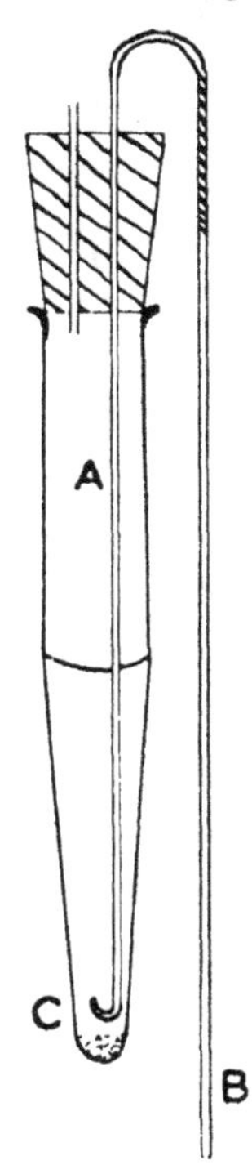

FIG. 54. Syphoning apparatus.
A. Capillary with up-turned end.
B. Constructed outlet.
C. Calcium oxalate ppt.

Determination of the oxalate. Pipette exactly 0·4 ml. of the solution into the outer chamber of the unit (No. 2A). Then pipette 0·2 ml. of the 0·05 *N* barium hydroxide into the central chamber and seal the unit as quickly as possible. (Seal each unit before pipetting barium solution into the next.) Displace laterally the cork over the hole in the lid, rapidly pipette 0·3 ml. of the oxidant solution through the hole into the outer chamber and rapidly

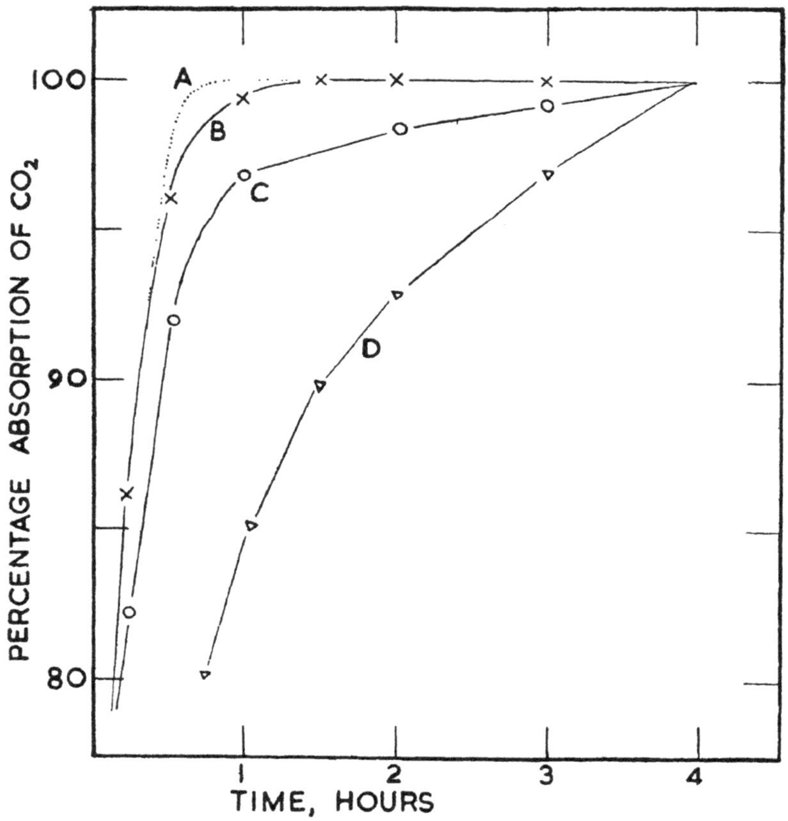

FIG. 55. Percentage absorption rates of CO_2; 0·7 ml. of fluid in outer chamber of Number 2A unit (Murnaghan, *Analyst*, **74**, 1949).
Curve *A*. Acidification of bicarbonate.
Curves *B*, *C* and *D*. 134 μg. of $(COONa)_2$ with :

B – $Ce(SO_4)_2$ (saturated) and $MnSO_4$ (trace) in *N* H_2SO_4.
C – $Ce(SO_4)_2$ (saturated) in *N* or 0·2 *N* H_2SO_4
D – $Ce(SO_4)_2$ (saturated) in 5 *N* H_2SO_4

replace the cork to form a firm seal. Rotate the units several times, to facilitate intimate mixing, and set them aside at room temperature for 90 minutes. Then remove the lids and titrate the excess of barium hydroxide with 0·02 *N* hydrochloric acid from the Conway burette until the pink colour of the phenolphthalein disappears. Carry out blank determinations with N sulphuric acid.

Calculation. If 0·5 ml. of serum was taken, the volume of 0·02 N hydrochloric acid used, in large divisions of the burette, divided by 2 = mg. of calcium per 100 ml. of the serum. (Each large division = 0·01 ml.)

Reagents

Oxidant solution. Saturated ceric sulphate in N H_2SO_4, containing a trace of manganous sulphate.

0·05 *N barium hydroxide solution.* 25 ml. of 0·2 N barium hydroxide, 20 ml. of absolute alcohol, 3 ml. of 1% phenolphthalein solution in alcohol and distilled water to 100 ml.

Notes on method

1. Using this method the Ca content in 0·5 ml. serum (about 50 μg.) was determined with a greater accuracy (coeff. of variation = 2%) than the determination by the Clark and Collip method (1925) in 2 ml. (coeff. of variation = 3%).
2. Manganese as manganous sulphate has a specific effect in hastening the oxidation of the oxalate (*vide* Fig. 55).
3. As described above 20 μg. Ca can be determined (in duplicate) with a coefficient of variation of 2·5%, but a greater refinement can be achieved if necessary.

CHAPTER XXVIII (*Bb* 1)

Bb VOLATILE FATTY ACIDS

ACETIC ACID AND OTHER LOWER FATTY ACIDS

ACETIC acid and the lower volatile fatty acids being volatile may be determined by microdiffusion from mixtures where they could not be directly determined by titration. Owing, however, to the very small tensions they produce in watery solutions the diffusion process is very slow even at raised temperatures. After trying several reagents which might be expected to increase the tension it was found that, among these, much the greatest effect was produced by powdered anhydrous sodium sulphate. With the use of this substance the tension may be increased to over ten times, and the time for full diffusion reduced to about four hours even at room temperature. The following procedure has been worked out (Conway and Downey, 1950).

2 ml. of sodium citrate solution, the strength of which depends on the acetic acid strength of the sample for analysis, are pipetted into the outer chamber. If the organic acid in the sample is about $N/20$ approximately $N/20$ citrate is used but a very wide latitude is permissible.

0·1, 0·2 or 0·5 ml. vols. of the mixture are then pipetted into the central chamber. If 0·1 or 0·2 vols. are added it is advisable to cover the floor of central chamber with a piece of filter paper, cut if necessary to ensure its fitting snugly. This is done to ensure full spreading of the sample. The 0·5 ml. sample may be spread around the surface without filter paper.

The solution may be first acidified adding 10 vols. of the solution to 2 vols. of 10% (by volume) of sulphuric acid or as described later for the determination of acetic acid in blood it may be acidified after pipetting into the unit.

The unit is then exposed, without putting on lid, for 15 minutes to free completely from carbon dioxide in the central chamber. During this time only a very small amount of acetic or other volatile fatty acid escapes (about 1·5% if 0·1 ml. is used and 0·3% if 0·5 ml. is used) but the carbon dioxide is fully lost beyond the small equilibrium value with the atmosphere. Alternatively, at the end of the diffusion period, the lids of the units are removed and the fluid left exposed for about 20 minutes when CO_2 will have diffused away from the outer chamber and without appreciable loss of volatile fatty acid.

The central chamber is now covered quickly with a thick layer of powdered anhydrous sodium sulphate (of Analar grade). This is conveniently done with a spoon spatula. Altogether about 2 to 3 g. sodium sulphate are used in this way—no exactness being required.

If a few grains of the powder escape into the outer chamber accidentally—this has no appreciable effect. The unit is now covered with a lid smeared with a paraffin fixative and left aside on the bench. The time taken for the diffusion depends in some measure on the amount of the sample used. If 0·1 ml. of the acidified sample is used it is complete (99%) in about 5 hours at room temperature. If 0·5 ml. is used it takes about 8 hours at room temperature, and the diffusion, may, with convenience, be allowed to take place overnight.

After the diffusion period the lid is removed and the outer chamber contents removed by a glass tube with rubber teat and introduced into a small tube for titration, the chamber being washed out about twice with 0·5 ml. distilled water, making roughly 3 ml. in all for titration. A little neutral phenolphthalein in alcohol is added and the titration carried out with barium hydroxide solution from a Conway burette, the strength depending on the order of acetic concentration expected.

Calculation. If, for example, 0·5 ml. of the fluid for analysis is used, and $N/50$ alkali is used for the titration the result may be expressed as

$$(X - B) \times \frac{1}{50} \times \frac{60}{0{\cdot}5} = \text{mg. acetic acid per cent}$$

where X and B are the large divisions (each 0·01 ml.) on the burette, corresponding to the analytical and blank values.

Formic, Propionic and Butyric Acids.

These acids with the above procedure diffuse over into the outer chamber at similar rates to acetic. Propionic diffuses at an almost identical rate, butyric at a definitely slower rate. Thus, using 0·1 ml. volumes butyric acid diffuses to 37% of the full value whereas acetic acid diffuses to about 53% at room temperature in 1 hour. Butyric acid takes about 8 hours to reach 98 % diffusion, and using 0·5 ml. about 16 hours may be assumed, and is suitable for leaving overnight. Formic acid diffuses a little faster than butyric but somewhat slower than acetic acid.

Members of the fatty series containing more than four carbon atoms have not been investigated.

If a mixture of volatile fatty acids is present in the sample for analysis, distilled water may be used in the outer chamber, with results almost identical with the use of citrate, but a pure solution of volatile fatty acids is obtained. Individual members may then be determined by special methods. The relative preponderance of fatty acids higher than propionic could be determined by observations of the diffusion after one hour, and after 24 hours compared with standard acetic acid under the same conditions.

Colorimetric Method for Acetic Acid

The method of Hutchens and Kaas (1949) may be applied to the pure solution of volatile fatty acids (without citrate) in which substances interfering with the colour development are either absent or present in greatly reduced quantity.

Determination of Acetic Acid in Blood Serum.

The method described above may be used for the determination of acetic acid in blood serum. The procedure adopted is as follows.

Procedure. 0·5 ml. of the serum is pipetted into the central chamber of a No. 1 microdiffusion unit. 0·1 ml. of 3*N* sulphuric is then pipetted on to the serum, and mixed with a fine glass rod. Such units are exposed then for 15 minutes, during which 2 ml. of *N*/500 sodium citrate are run into the outer chamber.

After 15 minutes, powdered anhydrous sodium sulphate is heaped on the fluid in the central chamber, and the lid fixed in position as before. A blank determination is set up in a similar way using 0·5 ml. water to replace serum.

The units are left overnight, when the contents of the outer chamber are pipetted with washing as above, and titrated, after adding a drop of phenolphthalein solution, with $N/200$ barium hydroxide from a Conway burette.

Calculation. To obtain the absolute value of the volatile fatty acid which has come over into the outer chamber, it is necessary to apply a correction for the small amount of acid, very likely HCl, which diffuses across, and is not acetic or closely related acids. This amount has been investigated by allowing the diffusion to proceed for 48 hours and observing the difference in titration between the 24 hour period. This difference may be expressed as 3·5 large division (or 0·035 ml.) of $N/200$ alkali. The calculation in terms of mg. acetic acid/100ml. is therefore:

$$(X - B - 3{\cdot}5) \times \left(\frac{60}{0{\cdot}5}\right) \times \left(\frac{1}{200}\right)$$

or

$$0{\cdot}6 \times (X - B - 3{\cdot}5)$$

where X and B are the serum and blank titrations in large burette divisions (each division = 0·01 ml.).

A mean value found for abattoir blood (4 samples) was 3·9 mg. acetic acid per cent.

Notes on the method.

(1) Owing to the very low tensions of acetic acid in dilute solution, it is essential that the units be perfectly dry, and no droplet of water contained in them ; also, that there is no trace of alkali on the inner surface of the units or the lid.

(2) The fixative should be a pure paraffin mixture (Page 93), or pure liquid paraffin, and cover the whole under-surface of lid.

(3) Recoveries of acetic acid added to blood plasma—to give an additional 15 mg. per cent—have been quantitative.

(4) Spilling of some grains of the anhydrous sulphate into the outer chamber which is better avoided, as far as possible, has no very appreciable influence on the titrations.

While this 5th edition was in press a highly specific enzyme method for determining acetate was described by Lundquist, Fugmann and Rasmussen (*Biochemical Journal*, Vol. 80, p. 393, 1961). The procedure involves microdiffusion in the No. 1 units and reaction with sulphanilamide by means of a pigeon liver enzyme preparation, stable at 20° for at least a year. The enzyme catalyses the acetylation of sulphanilamide through reactions involving the presence of CoA and ATP. (Soodak and Lipman, 1948; *Fed. Proc.* **7**, 190). The residual sulphanilamide is determined colorimetrically.

Procedure of Black for the Determination of Volatile Fatty Acid (1949)

The procedure of Black in which temperatures of 100–105° C. are maintained for 14 to 16 hours would seem obviated by the procedure described above, using anhydrous sodium sulphate.

He used as a microdiffusion unit (*vide* Fig. 3*a*I), a 50 ml. glass stoppered Erlenmeyer flask with small glass cup not fixed to the bottom of the flask and surrounding a detachable stainless steel inner cup. One ml. of the fluid to be examined, and then one ml. conc. H_2SO_4 are introduced into the flask, and then the glass cup with its inner steel cup containing K_2CO_3 solution as absorbent.

CHAPTER XXIX (Bb 2)

ASSAY OF ACETYLCHOLINESTERASE

Method of Serlin and Cotzias (1955)

Principle. The acetic acid formed by the action of the enzyme in a buffered mixture, on an acetylcholine substrate is transferred by microdiffusion from the outer into water in the central chamber. The enzyme action is stopped by citric acid, before transferring an aliquot to the outer chamber containing 4 to 7 g. of anhydrous sodium sulphate powder.

A special feature of the use here of anhydrous sodium sulphate is the observation by Serlin and Cotzias that if the units, sodium sulphate, and the 1 ml. aliquot transferred have been all maintained at 4° C., and the subsequent microdiffusion conducted at 22–26° C., this latter is completed in three hours, whereas if the whole process were carried out at 22 to 26°, it is comparatively very slow. (It may be noted that in the method described in the previous chapter 0·5 ml. serum was used.)

Procedure. The contents of an acetylcholine ampoule (containing 100 mg. acetylcholine ; Merck and Company) are transferred to a 50 ml. volumetric flask and made up to the mark with 0·0001 N H_2SO_4. (This gives approximately 11 μmol. acetylcholine per ml.) One ml. of a tissue homogenate in the phosphate buffer is introduced into a test tube (15 mm. × 120 mm.). After equilibration in a water bath at 37·5° ± 0·01° 1 ml. of the acetylcholine solution is added to the test tube, the contents mixed, and the time noted. (A blank determination is run at the same time with the homogenate which has been pre-heated in boiling water for 5 minutes.) At the end of 1 hour the test tube is immersed in ice cold water and the reaction stopped by adding 0·2 ml. of the citric acid solution. The tube is transferred to a cold room at 4° C. where a 1·0 ml. aliquot (from the test mixture or the blank) is spread in the outer chamber of a Conway unit (No. 1) previously prepared as follows : The floor of the outer ring is covered with

4 to 7 g. of Na_2SO_4 powder, and 1·5 ml. water are placed in the central chamber.

Spilling of some of the Na_2SO_4 powder into the water does not matter. After addition of the aliquot the lid suitably smeared with vaseline is placed in position and the unit brought to room temperature. Diffusion is allowed to proceed for 3 hours, at the end of which the water in the central chamber is titrated with standard barium hydroxide solution, ($N/50$ solution would be a suitable strength in the Conway burette) using a phenol red indicator (one drop of a saturated solution in 5% ethanol).

Serlin and Cotzias introduced the following device for facilitating the titration. The unit is resealed with a lid which has a 4 mm. hole drilled slightly off its centre. The tip of the burette as well as a fine capillary with a bent tip are introduced through the hole. A stream of nitrogen is passed through the capillary. The advantage gained by this arrangement is that the fluid is stirred automatically, and the sharp stable end-points obtained due to the exclusion of atmospheric CO_2.

Calculation. One unit of acetylcholinesterase activity is defined as the quantity per ml. capable of catalysing the release of acetic acid at the rate of 0·1 μmole per hour at 37·5° C. If V_s and V_b are the volumes in ml. of $N/50$ $Ba(OH)_2$ required for titration of the sample and the blank respectively, t the time in hours, and V_a the volume in ml. of the aliquot tested, the number of enzyme units (U) per ml. of the fluid analysed is given by

$$U = 220 \times (V_a - V_b)/(t \times V_a).$$

Reagents. The special reagents required are :

Potassium phosphate buffer. 1·29 g. of KH_2PO_4 and the 5·25 g. of K_2HPO_4 per litre ; $pH = 7{\cdot}2$.

Acetylcholine chloride. Ampoules containing 100 mg ; (Merck and Company).

Citric acid monohydrate. 0·4 M (8·4 g. per 100 ml.).

Notes on method

1. The coefficient of variance found in 10 recoveries of 0·963 μmole of the acetic acid was 0·8%.

2. For a net enzymatic production of 0·0506 μmole per hour the coefficient of variance was 4%.

CHAPTER XXX (*Bc* 1)

VOLATILE WEAK INORGANIC ACIDS AND PHENOL

CYANIDE, AZIDE, SULPHIDE, PHENOLS

Cyanide in Biological Fluids.

This method of Feldstein and Niel C. Klendshoj (1954) combines microdiffusion with the colorimetric procedure of Boxer and Richards (1952).

Procedure of Feldstein and Klendshoj

The microdiffusion procedure is carried out in well cleaned standard units (No. 1).

Into the central chamber is pipetted 2·0 ml. 0·1 *N* NaOH and into the outer 2·0 ml. (or other convenient volume) of blood plasma, urine, stomach contents or tissue homogenate (1 in 5 dilution homogenised by means of a Waring blender).

The ground glass lid is smeared with petroleum jelly and a final addition of 0·5 ml. 10% sulphuric acid is made to the outer chamber. The lid is then moved into position. A blank is prepared in a similar way using water instead of the blood or other fluid.

After 2 hours the lid is removed and 1 ml. of the alkaline solution in the central chamber is removed and introduced into a test tube and similarly with the blank unit. The test tubes are placed in an ice bath and to each tube 0·2 ml. of chloramine T-phosphate reagent is added. The contents of the tubes are mixed, and allowed to stand for two or three minutes in the ice bath. This is followed by the addition of 3·0 ml. of pyrazolone-pyridine reagent. The contents are mixed and allowed to stand at room temperature for 60 minutes. The optical density of the blue colour is then determined. A Beckmann Model B spectro-photometer at 630 mμ with the blank set at zero density is very useful for this purpose.

Calculation. Standard solutions of cyanide containing 0·1 to 2·0 mg. of cyanide per ml. of 0·1 *N* NaOH are prepared, and the colour developed as above. A calibration curve relating optical density and concentration is obtained and the cyanide content of the material analysed is read from the curve.

The calibration curve is linear and passes through the origin (or very close thereto). (An optical density of 0·600 is given by 2·0 μg cyanide/ml.)

The special reagents used

1. 0·25% solution of chloramine T (stored in refrigerator).

2. Saturated aqueous solution of 1-phenyl-3-methyl-5-pyrazolone, stored in brown bottle (Eastman Kodak Reagent, recrystallised from 95% alcohol).

3. One-tenth per cent solution in pyridine of bis-(1-phenyl-3-methyl-5-pyrazolone) prepared by boiling a mixture of 50 ml. ethylacetoacetate and 150 ml. phenylhydrazine under reflux until the solid bis-(1-phenyl-3-methyl-5-pyrazolone) separates. Filter and wash with hot alcohol. The pyridine solution is prepared fresh as needed.

The following reagents are prepared immediately prior to colour development :

1. Chloramine T-phosphate reagent. To 1 part of 0·25% chloramine T solution add 3 parts of 1 *M* NaH_2PO_4. Mix and keep in the ice bath.

2. Pyrazolone-pyridine reagent. To 5 parts of saturated aqueous solution of phenylmethylpyrazolone add 1 part of the bis-phenylmethylpyrazolone in pyridine.

Determination of Azide.

The method of Brady and O'Callaghan (1955) consists in the liberation of volatile hydrazoic acid from an azide solution by acidification with approximately 1 *N* strong acid in the outer chamber of a unit. The hydrazoic acid reacts with silver nitrate in the central chamber forming insoluble silver azide. Excess silver nitrate is then titrated with ammonium thiocyanate.

Procedure of Brady and O'Callaghan for fluids in the range 0·095 *M* to 0·0003 *N* azide.

The central chamber of No. 2 units are filled with 0·1 ml. 0·1

N $AgNO_3$ containing indicator and Lissapol N. If the azide concentration in the solution to be analysed is high 0·1 ml. is added to the outer chamber, followed after the units are sealed by 1 ml. N H_2SO_4. If the azide concentration is at the lower end of the scale 1 ml. is added followed by 0·1 ml. 10 N H_2SO_4.

Two and a half hours, are allowed for the diffusion and the central chamber is then titrated to the first tinge of yellow. The titration figure subtracted from a blank gives the amount due to azide. 1 division on the burette = 12·7 μg. N_3.

Solutions. 0·1 N $AgNO_3$ to which 3 ml. of a saturated solution of ferric nitrate are added as an indicator. This forms a stable solution which lasts indefinitely. Ferric alum cannot be used as it slowly forms a precipitate of silver sulphate. Two drops of Lissapol N are added in each 50 ml., and effectively prevent the formation of a skin of AgN_3 in the units.

1 N H_2SO_4 and 10 N H_2SO_4.

0·003 N NH_4CNS is used in the Conway burette.

Method for fluids in the range $9{\cdot}5 \times 10^{-3}N$ to $3 \times 10^{-5}N$ azide

At concentrations lower than 0·03 N thiocyanate the colour of the ferric thiocyanate was too faint and no colour change could be observed. Out of a larger number of adsorption indicators used for the determination of Ag, the only indicator giving a colour change when used in the microdiffusion procedure was Bromophenol blue. This gives a sharp end-point, the mixture changing from blue to purplish violet due to the sudden formation of a colloidal solution.

The procedure is the same as previously described, 2½ hours being allowed for diffusion. With azide concentrations at the lower end of the scale 1 ml. of solution plus 0·1 ml. 10 NH_2SO_4 is used. The colour in the central chamber of the 'blank' units is at first purplish violet, and on addition of 0·003 N NH_4CNS they change to blue with no evidence of violet. At the end-point they change to purplish violet. The colour in the analytical units is blue and changes at the end-point to purple violet.

1 division on the burette corresponds to 1·27 μg N_3.

Solutions

$0{\cdot}01N$ $AgNO_3$ *with indicator.* 16 ml. of a 0·1% Bromophenol blue solution in 50% alcohol is added to 5 ml. of $0{\cdot}2$ N $AgNO_3$. 40 ml. of $0{\cdot}1$ M Sodium acetate are added to stabilise the pH and 4 drops of Lissapol N. The solution is made up to 100 ml. with distilled water.

$1{\cdot}0N$ H_2SO_4 *and* $10N$ H_2SO_4.

$0{\cdot}003N$ NH_4CNS is used in the horizontal burette.

Notes on method

1. It is possible that the lower end of this range could be extended somewhat, but with solutions ten times more dilute no colour change was obtained.

2. A further possibility which is being investigated is the use of an acid-alkali titration. The reaction, as can be seen

$$HN_3 + AgNO_3 = AgN_3 + HNO_3$$

from the equation liberates nitric acid which could be titrated with a dilute alkali.

Sulphide

The following method is adopted from that of Feldstein and Klendshoj (1954) as described in their account of the determination of Volatile Poisons.

Procedure

Microdiffusion absorption is carried out as in the determination of cyanide by Feldstein and Klendshoj as described above.

After the two-hour period, one ml. of the absorbing solution ($0{\cdot}1$ N NaOH) is taken, one ml. of bismuth reagent added and the mixture diluted to 5 ml. with water. Nitrogen is bubbled through the solution to remove oxygen and the optical density is determined in a spectrophotometer at 350 millimicrons against a reagent blank set at zero density. The absence of any brown colour of bismuth sulphide is evidence of the absence of sulphide in the biological sample. The method will detect 5 micrograms of sulphide. Recovery of added sulphide is good. Standards are prepared from pure sodium sulphide treated as described above.

Reagents

Bismuth reagent. Dissolve 4·28 g. bismuth nitrate pentahydrate in 300 ml. glacial acetic acid and add 1·5 litres of water.

Phenols

These may be also determined as described by Feldstein and Klendshoj (1954), the absorption by microdiffusion into 0·1 *N* NaOH being carried out as with cyanide and sulphide.

After the 2 hours absorption period to one ml. of the absorbing solution in a test tube is added 2–3 drops of 10% NaOH and 0·5 ml. phenol reagent. The solution is diluted to 10 ml. and the optical density is determined in a spectrophotometer at 700 millimicrons against a reagent blank set at zero density. The absence of any blue colour is evidence of the absence of volatile phenol in the biological sample. Standards are prepared from phenol and treated as described above.

Reagents. The only special reagent required here is the phenol reagent of Folin and Ciocalteu (1929). This is prepared as follows: Into a 1500-ml. Florence flask introduce 100 g. of sodium tungstate, $Na_2WO_4 . 2H_2O$, 25 g. of sodium molybdate, $Na_2MoO_4 . 2H_2O$, 700 ml. of water, 50 ml. of 85% phosphoric acid, and 100 ml. of concentrated hydrochloric acid, and reflux gently for 10 hours. Add 150 g. of lithium sulphate, 50 ml. of water, and a few drops of bromine. Boil the mixture for 15 minutes without condenser, to remove excess bromine. Cool, dilute to 1 litre, and filter. The reagent should have no greenish tint. Protect from dust. Dilute a portion with an equal volume of water before use.

C. VOLATILE ALCOHOLS

Ca. METHANOL AND ISOPROPANOL GROUP

Methanol in Biological Fluids.

Method of Feldstein and Klendshoj (1954)

Principle. Methanol diffusing from the outer chamber of the microdiffusion unit is absorbed in a solution of sulphuric acid in the central chamber and is then determined quantitatively by oxidation to formaldehyde and subsequent reaction with chromotropic acid (Boos, 1948).

Procedure. Pipette 2·2 ml. of 10% sulphuric acid (*v/v*) solution into the central chamber of a No. 1 unit. Into the outer chamber place an aliquot of biological material.

The liberating agent used is 1 ml. saturated K_2CO_3 introduced in the usual manner and the diffusion time two hours at room temperature. At the end of this time place 1·0 ml. of the absorbing solution into each of two test tubes. To one, add 1 drop of 5% $KMnO_4$. Stand ten minutes. To each add 1 drop of saturated $NaHSO_3$ to decolourize the $KMnO_4$. To each add 0·2 ml. of 0·5% chromotropic acid solution.

Place in an ice bath and add 4 ml. concentrated sulphuric acid to each tube, while maintaining cooling. Place the tubes then in boiling water bath for 15 minutes, cool and dilute to 10 ml. Read in spectrophotometer at 580 millimicrons. Optical density of the unoxidised tube is subtracted from the density of the oxidised tube. Gross contamination with formaldehyde makes method inapplicable. Compare with standards in 10% sulphuric acid, 1·0 ml. volumes being treated as above.

Multiply results by 1·21 to correct for 85% diffusion in case of blood and urine. The factor 1·21 is used because the tension of the methanol is not reduced to zero in the absorbing fluid and the diffusion equilibrium of approximately 82·5% absorption is reached.

Reagents. The special reagent required is chromotropic acid 0·5% aqueous solution. It is prepared fresh each week and is stored in a refrigerator.

Notes on the method

1. Methanol by this method of Feldstein and Klendshoj can be determined in biological samples containing as little as 0·5 mg. per cent of methanol and in general to within a range of ±3·0%.

2. Tissue homogenates may also be used, and Feldstein and Klendshoj use 2·5 ml. of tissue slurry (equivalent to 0·5 g. tissue) in the outer chamber and then allow 3·5 hours at room temperature.

3. Methanol gives a positive reaction in all procedures for ethyl alcohol based on the reduction of dichromate-sulphuric acid. These workers point out that between 1948 and 1953 over 600 tests for methanol were performed on routine ethyl alcohol cases, and in 11 of these, or in 1·8%, sufficient methanol was found to interfere seriously with the determination of ethyl alcohol by dichromate reduction. 'In rúling out the presence of formaldehyde and methanol, the dichromate-sulphuric acid reduction methods for ethyl alcohol are rendered more specific.'

Isopropanol

The determination of isopropanol is carried out in a similar way to that of methanol. After the two hours at room temperature 1 ml. is removed from the central chamber and placed in a test tube graduated at 10 ml. to which is added 1 drop of 5 per cent potassium permanganate.

Reagents. The special reagent required here is salicylaldehyde, 0·1% in absolute alcohol (freshly prepared).

Notes on Method

1. Isopropanol can be recovered with an error of ±5% in the range of 10 to 100 mg. per cent.

2. As methanol and isopropanol are not absorbed to zero vapour pressure, but rather to a diffusion equilibrium depending on solubility differences, it is advisable for quantitative purposes to carry out determinations in separate microdiffusion units.

CHAPTER XXXII (*Cb* 1)

Cb. ETHANOL GROUP

ETHANOL

In methods for determining alcohol in tissue and body fluids for the most part the principle of the oxidation of the alcohol to acetic acid by dichromate, with subsequent titration of the dichromate, has been used. Such methods differ only in the manner of separating the alcohol and titrating the dichromate (*vide* Levine and Bodansky, 1939).

In Widmark's alcohol method (1922) the alcohol passes by diffusion from a small cup suspended from the bottom of a stopper inserted into a 50 ml. flask, into a solution of dichromate in sulphuric acid in the bottom of the flask. The excess dichromate after the oxidation of the alcohol is determined iodometrically.

Winnick (1942 *c*) has shown how the standard microdiffusion unit can be applied to the determination of alcohol in blood and urine. In the following description much of his procedure is followed. It has been modified (Conway and Nolan, 1948) in such a manner that full absorptions occur in 1½-2 hrs. at room temperature. The ease, accuracy and delicacy of the method now make it specially commendable.

Procedure for blood or other fluid containing over 20 mg. per 100 ml. alcohol (**Conway and Nolan, modified after Winnick**).

If the specimen or sample is expected to contain more than about 300 mg. per cent it should be suitably diluted to bring it under this value. If it is expected to contain less than 20 mg. per cent, the procedure described subsequently with the No. 2 unit is followed.

Pipette 1 c.c. of 0·03 *N* dichromate (in 10*N* sulphuric acid) into the inner chamber. Run approximately 0·5 c.c. of saturated potassium carbonate solution into the outer chamber. Place the lid in position using liquid paraffin as fixative. Slide back lid sufficiently to introduce a pipette and quickly but accurately introduce 0·1 c.c. of the sample into the outer chamber. Reseal the unit and mix 0·1 c.c. of sample and potassium carbonate by rotating several times in the usual manner. Leave aside on bench for 1½–2hrs.

At the end of this time remove the lid and add 0·1 c.c. of 20 per cent potassium iodide to the dichromate. The liberated iodine is titrated at once with 0·05 N thiosulphate from the Conway burette.

The thiosulphate is added rapidly with efficient stirring until nearly all the iodine is reduced, and then dropwise until the colour of the solution is a light yellow-green. A drop of 1 per cent starch solution is added, and the titration is completed to the point at which the deep blue starch-iodine colour is replaced by the light blue-green colour of the chromic ion. Owing to the small volume of the solution being titrated, this colour change is abrupt.

A blank determination is also carried out at the same time using 0·1 c.c. water instead of the sample for analysis. The result as milligrams per cent of alcohol in the sample is given by :

$$576 \times (B - A)$$

where B and A are the c.c. of 0·05N thiosulphate required for the blank and the analytical units.

Solutions.

Dichromate solution (0·03N in 10N sulphuric acid). This contains 1·47 g. of pure dry potassium dichromate in 1 litre of 10N sulphuric acid. Such 10N sulphuric acid contains 266 ml. of the pure acid per litre.

0·05 *thiosulphate solution.* This contains 12·42 g. of crystalline sodium thiosulphate per litre. To preserve the solution 50 mg. of sodium carbonate are added per litre.

20% *potassium iodide.*

1% *starch solution.*

Fixative. Liquid paraffin.

Method for fluids containing alcohol less than 20 mg. per cent.

Into the inner chamber of a No. 2 unit, 0·1 c.c. of 0·03N dichromate in 10N sulphuric acid is introduced. If more than 5 mg. per cent alcohol are anticipated in the sample, 0·5 c.c. of saturated potassium carbonate is introduced into the outer chamber and the lid placed in position with fixative. Then 0·1 c.c. of the sample is introduced in the usual way, and the lid firmly sealed. The contents of the outer chamber are mixed by rotation. The unit is now left aside for three hours when the fluid in the inner chamber is titrated with 0·01*N* thiosulphate after adding one drop of 20% potassium iodide. The titration is carried out slowly with efficient stirring.

If it is anticipated that the sample contains less than 5 mg. per cent alcohol, 0·5 g. of solid powdered potassium carbonate is used instead of 0·5 c.c. saturated potassium carbonate solution, and 0·5 c.c. of the sample is pipetted into the outer chamber instead of 0·1 c.c. as above. Otherwise the procedure is the same. Blank determinations with water are carried out at the same time.

Calculation.

Where 0·1 c.c. of the sample is used (above 5 mg.% anticipated) the alcohol as milligrams per cent is given by:

$$115 \times (B - A)$$

where B and A are the c.c. of 0·01*N* thiosulphate required for the blank and analytical titrations.

Where 0·5 c.c. is used (levels below 5 mg. % expected) the calculation is given by:

$$23 \times (B - A)$$

Notes on method.

Recoveries and accuracy. Winnick obtained recoveries of 97·5 to 103% of alcohol added to blood or urine by the method and his duplicate determinations agreed to within about 2%. A higher accuracy can be obtained by the modified method depending on the accuracy of introduction of the 0·1 ml. sample, (*vide* Part III of present volume).

In a series of analyses on a standard solution containing 100 mg. per cent alcohol, the standard deviation of the single determination was 0·6 mg. %

Collection of samples for analysis. Blood and freshly voided urine specimens are collected in tubes tightly fitted with rubber stoppers. The tubes for blood samples contain oxalate or citrate. Winnick finds, confirming Cavett's observation (1938) that the blood or urine can stand for 24 hours in the ice box without appreciable change in alcohol concentration.

Interfering substances. Acetone and chloroform do not interfere, and though ether (di-ethyl) is slowly oxidised by the dichromate, it is unlikely to interfere seriously, since alcohol is at a relatively much higher level during intoxication than is ether following anaesthesia.

The following substances which are of industrial importance and may give rise to intoxication, do not interfere—benzene, carbon tetrachloride, naphtha and trichloroethylene. On the other hand methyl alcohol, isopropyl, amyl (also isoamyl acetate) and octyl, as well as the aldehydes acetaldehyde and propionaldehyde do interfere.

Blood and urine levels of alcohol. Winnick found, using the above method, that the blood and urine of six individuals who had taken no alcoholic beverages for several days contained no measurable amounts of alcohol.

He also found the blood concentration in mild to moderate intoxication to be 130–220 mg.%, in moderate to severe—170 to 380 mg.% and in very severe intoxication—about 530 mg.% The urinary concentration was higher than that of the blood (upwards of 1·5 times), confirming the observation of Jetter (1941).

Rate of disappearance of alcohol from the blood. Confirming Newman, Lehmann and Cutting's observations for the dog (1937), Winnick showed that the alcohol in the blood of a human subject falls in a linear manner with the time. Over 12 hours the concentration fell from 307 mg.% (moderate to severe intoxication) to about half this amount in 7 hours and to 35 mg.% in 12 hours.

MacLeod's procedure (1949)

Principle. In this the alcohol diffuses into 1 ml. of alkaline permanganate in the central chamber and after addition of barium chloride to precipitate the manganate (the intense green colour of which would interfere with the titration) the mixture is titrated with standard thiourea.

Procedure. This is the same up to the titration as the method already described, except that 1 ml. of alkaline permanganate is the absorbent ; also waterglass (very slightly diluted with water) is used as the fixative. The units are ready for titration after 2 hours on the bench at room temperature.

To the central compartment 0·2 ml. of 10% barium chloride is then added, and the mixture gently stirred using a glass rod suitably drawn out. The titration is then carried out until the last trace of pink colour has disappeared. The end-point is marked by the precipitated manganate suddenly changing colour from steel blue to dark green.

Calculation. This may be made from the formula

$$(B-E)\times 1000\ C = \text{mg. per cent of alcohol in the blood.}$$

Where B is the blank value and E the analytical, C being a factor per unit volume. If a 2 ml. Bang burette is used B and E may be written as millilitres and C is the mg. alcohol equivalent per millilitre of titrating fluid ; or if the Conway burette is used B, E and C may be conveniently written as large divisions on the graduated tube.

Reagents. (Apart from those already described)

Alkaline permanganate solution. This is approximately 0·04 *M* potassium permanganate in 3 *N* potassium hydroxide. 0·63 g. of pure potassium permanganate is added to 100 ml. of a solution containing 17 g. of potassium hydroxide (analytical reagent used). The solution is stored in an amber glass bottle and shaken at intervals until the permanganate is totally dissolved. The stopper should remain detached as short a time as possible to minimise CO_2 absorption.

Barium chloride solution. 10% (*w*/*v*) barium chloride ($BaCl_2$, $2H_2O$) in water, kept in a well stoppered bottle.

Thiourea solution. A stock 0·04 *M* solution of thiourea in water (3·04 g./litre). If the 2 ml. Bang burette is used, this is conveniently diluted 25 times for immediate use, or if the Conway burette is used a 5 times dilution is convenient. This solution is best standardised by using known amounts of alcohol in the Conway units.

Standard alcohol solution. A convenient standard is to deliver accurately 1 ml. of pure ethanol into a 500 ml. flask and dilute to mark with distilled water. This solution contains 159 mg. per cent alcohol.

Notes on Method

1. Each ml. of the stock thiourea solution if (0·04 *M*) diluted 25 times was found equivalent to 0·177 mg. alcohol, or to 177 mg. alcohol per cent in the blood. If diluted 5 times, each large division on the Conway burette will be equivalent to 8·85 mg. alcohol per cent in the blood.

2. The acid dichromate absorbing solution has the advantage of not being sensitive to acetone, whereas the alkaline permanganate is reduced thereby. This, however, is of small importance owing to the rarity of free acetone in any considerable quantity in blood. The method could be adapted to determining alcohol in the presence of acetone by substituting a buffered solution of potassium sulphate for the saturated potassium carbonate, but it would seem preferable to use the acid-dichromate procedure.

Procedure of Sunshine and Nenad (1953)

These workers modified Winnick's method to give rapid results by raising the temperature and in this way obtained complete absorption at 90° C. The final quantitative measurement was carried out on the acid-dichromate from the central chamber, using a photoelectric photometer with a blue filter (450 mμ.).

Procedure. Introduce 3·0 ml. of the acid-dichromate solution into the central chamber. Add 1 ml. of 20% sodium carbonate to the outer compartment. Apply silicone stopcock grease to the surface of the ground glass plate, and proceed in the usual manner, introducing last 0·5 ml. of the blood or fluid containing ethanol.

Then incubate the units at 90° C. for 20 minutes. For quantitative measurement the colour of the fluid in the central compartment is compared, after removing, washing out twice and making up to 25 ml. and determining the extinction at 450 mμ. This is compared with a calibration curve obtained with standards. (A prepared blood sample containing 600 mg. ethanol per cent suitably diluted and processed as above will serve.)

Notes on this procedure

1. The use of saturated potassium carbonate instead of 20% sodium carbonate should considerably hasten the process and allow of a lower temperature.

2. The use of 0·1 ml. instead of 0·5 ml. would also hasten the diffusion but Sunshine and Nenad prefer 0·5 ml. on the ground that capillary blood is then frequently used but the 0·1 ml. could be taken from a blood sample like the 0·5 ml. as used by them.

3. The authors find that the precision of each measurement is within 3% and that it is very suitable for accurate multiple determinations carried out by the average clinical technician.

Reagents

Acid dichromate. 4·262 g. of analytical grade potassium dichromate are dissolved in about 100 ml. of distilled water; 500 ml. of reagent grade sulphuric acid is then added with suitable cooling and the solution made to one litre with distilled water. One ml. of this reagent is equivalent to 1 mg. of ethyl alcohol.

Dow Corning Silicone Stopcock Grease is used as the fixative.

The other reagents do not require description here.

Procedure of Feldstein and Klendshoj

This is similar to the procedures involving acid dichromate described above. Feldstein and Klendshoj conduct the determination at room temperature (2 hours) or at 37° (1 hour). They use 2·0 ml. of the acid dichromate in the central chamber, 0·8 ml. of the fluid for analysis in the outer chamber and 1·0 ml. of saturated K_2CO_3. At the end the contents of the central chamber are transferred with washing and made up to 10 ml. the extinction value being determined in a Beckman Model B spectrophotometer at 450 mμ., blank determinations being also carried out.

An important point in the procedure of Feldstein and Klendshof lies in the concentration of the dichromate solution. This contains 3·370 g. of $K_2Cr_2O_7$ (note they use 3·70 g. of $K_2Cr_2O_7$ in their article) and 280 ml. of concentrated H_2SO_4 made up to 650 ml. *and is claimed to be of critical composition with relation to the oxidation of acetone.* Acetone present to the extent of 400 mg. per cent contributes only 1 to 7 mg. per cent as apparent ethanol.

They state that in the methods of Winnick (1942), MacLeod (1949) and Sunshine and Nenad (1953) it can seriously interfere. They also note that if the dichromate solution remains a bright orange the alcohol content may be assumed to be negative and provides an excellent screening test when large numbers of samples are run at the same time.

They describe an excellent agreement between the results obtained and those of other workers using different methods (e.g. Dubowski and Witherow (1952))

CHAPTER XXXIII (*Ct* 2)

ETHANOL FROM URETHANE

The Determination of Urethane in Animal Tissues

(*Boyland and Rhoden*, 1949)

Principle. Urethane can be hydrolysed by caustic alkalis to yield ethanol as shown by Archer, Chapman, Rhoden and Warren (1948) who used the reaction for its determination. Boyland and Rhoden (1949) used the microdiffusion procedure to determine its distribution in tissues after administration.

Procedure of Boyland and Rhoden (1949)

Method of estimation. A standard volume of 1 ml. 0·04N $K_2Cr_2O_7$ in 10 N H_2SO_4 is placed in the central compartment of a Conway unit. The sample which should be less than 0·7 mg. urethane, contained in 1 ml. fluid, and 1 ml. 4 N KOH, are placed in the outer ring of the unit. A blank is set up with a 1 ml. sample of urethane and 1 ml. 2 M K_2CO_3. The alkalis used must not contain volatile reducing substances and are kept in glass-stoppered bottles. The units are sealed and incubated at 37° for 16 hours or more. At the end of the incubation KI is added to the acid $K_2Cr_2O_7$ and the liberated I_2 titrated with 0·02 N $Na_2S_2O_3$. The difference in titre between the two units is proportional to the amount of urethane present ; 1 ml. 0·02 N $Na_2S_2O_3$ is equivalent to 0·445 mg. urethane.

Alternatively, the titration can be carried out with a Conway burette using 0·1 N $Na_2S_2O_3$. With this modification 1 ml. of 0·25 N $K_2Cr_2O_7$ in 10 N H_2SO_4 is placed in the central compartment and the urethane sample (e.g. 0·1 ml. blood) is incubated with 2 ml. of M K_2CO_3 as the blank, and a further 0·1 ml. blood treated with 2 ml. 2 N KOH in the outer ring of the second unit. Thus by using a sensitive microburette the estimation can be carried out on 0·2 ml. blood. The amount of urethane and volatile reducing matter in the sample must not be more than the equivalent of 0·5 mg. urethane.

Duplicate estimations agreed within 5% and the method could be applied directly to blood, urine or tissue extracts deproteinised with tungstic acid or trichloroacetic acid. Determinations on normal human blood and on blood of normal rats and rabbits gave no indication of the presence of any substance reacting as urethane in this reaction.

Reagents

Acid-dichromate. 0·04 N $K_2Cr_2O_7$ in 10 N H_2SO_4.
Potassium hydroxide solution. 4 N KOH.
Potassium carbonate. 2 M K_2CO_3.
Sodium thiosulphate. 0·02 N $Na_2S_2O_3$.

Notes on the results

1. After injection, urethane becomes evenly distributed in all the tissues of the body.

2. Only a small amount is excreted in the urine, the greater part being metabolised.

3. The metabolism of urethane is slower in tumour-bearing rats than in normal animals—which appears to be due to deficiency of enzymes concerned in the breakdown of the urethane molecule.

CHAPTER XXXIV (*Da* 1)

D. VOLATILE ALDEHYDES
Da. FORMALDEHYDE GROUP

FORMALDEHYDE

Principle. The colour change to purple on heating with chromotropic acid (MacFadyen, 1945) has been used.

(Methods employing this reaction with microdiffusion have been used in determining formaldehydogenic steroids, as described later.)

Procedure of Hollander, DiMauro and Pearson (1951) (using No. 1 Units)

In this (after oxidation of the formaldehydogenic steroids with periodic acid), 0·4 ml. of the mixture containing the formaldehyde is pipetted into the outer chamber of a unit and 0·80 ml. chromotropic acid reagent is placed in the inner. The previously greased units are sealed by means of stiff stopcock grease and distillation allowed to proceed in the dark overnight or for about fifteen hours. The contents of the inner well are then transferred to a small glass-stoppered test tube and heated in a boiling water bath for thirty minutes in the absence of light. This transfer does not happen to be fully quantitative.

The tube is then cooled in ice, allowed to come to room temperature and read at 570 mμ. in the spectrophotometer. To prepare a blank, a solution containing 0·02 ml. glacial acetic acid, 0·38 ml. water and 0·80 ml. chromotropic acid is heated in the water bath along with the samples.

(In determining formaldehydogenic steroids in urine extracts as later described, it is necessary to do a reagent blank in which distilled water is substituted for urine and the entire extraction procedure and microdiffusion gone through. This blank value is subtracted from the values obtained for the standards and urine extracts.)

Calculation. A curve is prepared by diffusing pure formaldehyde 0·4 ml. fluid in the outer chamber into 0·8 ml. chromotropic reagent in the central well, taking the same time and developing the colour in the same way as above. The concentration in the unknown solution can be read off then from this curve.

Reagents.

Chromotropic Acid Reagent. 0·2% in 15 *M* sulphuric acid. The chromotropic acid is purified by treating an aqueous solution of Eastman Kodak Practical Grade Chromotropic Acid with charcoal and then re-crystallising from aqueous acetone. The final reagent is stored in the icebox.

Procedure of Edwards and Kellie (1954)

The procedure was adapted to the use of sodium bismuthate as the oxidant for the formaldehydogenic steroids as described later (Chapter XXXVI). The principle of the transfer of the formaldehyde and the colorimetric procedure is the same as that of Hollander *et al.* but a higher temperature was used and smaller apparatus (i.e. a special modification of the No. 2 Conway unit).

Apparatus

The modified Conway unit. The original No. 2 Conway unit is somewhat large for the small quantities measured, and the modified unit (Fig. 56) has a centre well of about 0·25 ml. capacity.

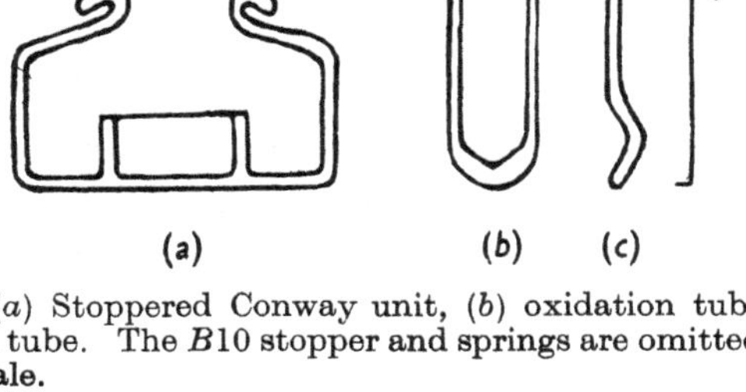

FIG. 56. (*a*) Stoppered Conway unit, (*b*) oxidation tube, (*c*) stirrer for oxidation tube. The *B*10 stopper and springs are omitted. Approximately to scale.

The units employed are blown from borosilicate glass and fitted with *B*10 standard socket and stopper. Lugs on both socket and stopper are fitted with springs to withstand the small internal pressures produced on heating.

Procedure. The modified Conway units are prepared by pipetting chromotropic acid reagent (400 μl.) into the outer compartment, using a 1 ml. graduated pipette, the last centimetre of which is bent 45° off axis. The supernatant liquid (100 μl.) from the oxidation mixture is pipetted into the centre compartment of the unit, and the stopper, lubricated with a trace of 9 *M* H_2SO_4 is fixed in position by the springs.

The units are placed on a platform out of bright light, in a simmering-water bath (95–100°) so that the top of the chamber is just immersed. After 30 min. the units are removed and stood to a similar depth in cold water (15°) for 5 min. Within 30 min. of cooling, the fluid from the outer compartment is transferred to a micro-colorimeter cell of a Spekker Absorptiometer (0·5 ml., optical depth 1 cm.) and the optical density read against 9 *M* H_2SO_4 as blank, using a yellow light source (Ilford Spectrum Yellow light filter No. 606 ; transmission 5600–6000).

When assaying extracts of biological material an ' unoxidised determination ' is carried out by the above procedure substituting an equal amount of kieselguhr for $NaBiO_3$. This inert substance also removes any cloudiness. The difference between the ' oxidised ' and the ' unoxidised ' determination is due to the formaldehyde produced by the oxidation of the steroid. A blank determination is also made by following this procedure but omitting the steroid. The optical density of this blank should lie between 0·02 and 0·05 ($\equiv$ 0·7–1·7 m-moles).

Besides the Spekker instrument the Beckman spectrophotometer may be also used as in the procedure of Hollander *et al.* (1951).

Calculation. A graph is prepared as in the procedure of Hollander *et al.* (1951). The extinction has been calculated as 0·0302/m-μ mols. When pure cortisone was used as the source of formaldehyde, a standard deviation of $\pm$1·5% was found in the mid-range of the calibration.

Chromotropic acid reagent. If necessary, chromotropic acid is purified by treatment with charcoal and by recrystallisation from aqueous acetone. Pure chromotropic acid (20 mg.) is dissolved in water (1·30 ml.) and, whilst cooling, H_2SO_4 (76% *w/w* 8·62 ml.) is added. The volume is made up to 10·0 ml. with water. The required concentration of H_2SO_4 may be prepared by running conc. H_2SO_4 (A.R., 10 ml.) into water (4·8 ml.). This chromotropic acid reagent will absorb one-quarter of its own volume of water to become 9 *M* with respect to H_2SO_4. The reagent is stored at 0°.

Formaldehyde standard. Formalin solution is redistilled and standardised as detailed in Analar Standards for Laboratory Chemicals (1951).

CHAPTER XXXV (Da 2)

FORMALDEHYDOGENIC STEROIDS

Periodic Acid as Oxidant

CORTICAL hormones may be determined in urine by oxidation of the ketone side chain with periodic acid (Lowenstein, Corcoran and Page, 1946) liberating formaldehyde.

For greater specificity the formaldehyde may be removed by distillation before colour development with chromotropic acid (Daughaday, Jaffe and Williams 1948; Corcoran and Page, 1948).

Bassil and Hain (1950) reported the use of diffusion for removing the formaldehyde but this required 3 or 4 days; Hollander DiMauro and Pearson (1951) describe a convenient procedure, using the Conway unit (No. 1) leaving the diffusion to proceed overnight and getting quantitative recoveries of added formaldehyde.

Procedure of Hollander, DiMauro and Pearson (1951) for urine

Preparation of extract

1. Usually a 4 to 5% aliquot of an exact 24-hour specimen (corresponding to 15–20 μg. formaldehydogenic material) is taken, and acidified with concentrated H_2SO_4 to a *p*H of 1·0 and let stand at room temperature for 24 hours.

2. This is now extracted twice with double its volume of chloroform. This is washed once with 10% of its volume of ice cold 0·1 *N* NaOH and twice with 10% of its volume of water. (Each washing is back extracted with 50% its volume of chloroform.)

3. The combined chloroform extract is evaporated to a small volume at 45° C. It is then taken up by alcohol (aldehyde-free) and transferred to a 15-ml. centrifuge tube. The alcohol is evaporated off under a stream of N_2 or air while heating the tube to 40° C. Any solid material adhering to the side of tube is washed

down with a small quantity of alcohol and this solvent is again blown off. The tubes containing the samples are left overnight in a vacuum desiccator over conc. sulphuric acid.

Oxidation of extract The sample is wetted with 0·10 glacial acetic acid and after standing 10 minutes, 1·50 ml. 0·003 M periodic acid reagent is added with shaking. The oxidation is allowed to proceed for one hour at room temperature at the end of which time the reaction is stopped with 0·40 ml. of the stannous chloride solution.

Determination of the formaldehyde by microdiffusion. This is described in the previous chapter.

The calculation may be referred to μg. deoxycorticosterone by oxidising known quantities of deoxycorticosterone under the same conditions as the urinary extract, and plotting a curve of the optical densities obtained. Hollander *et al.* (1951) find this to be a straight line passing through the origin.

The μg. deoxycorticosterone from their curve may be expressed by the relation—

$$\mu g.\ \text{DOC in the microdiffusion unit} = X \times 37{\cdot}4$$

where X is the optical density.

Special reagents (for the Hollander *et al.* (1951) determination)

Glacial acetic acid—distilled over chromic acid.

Chloroform. Redistilled reagent grade chloroform.

Ethanol (aldehyde free)—45% alcohol distilled over 0·1% 2, 4-dinitrophenylhydrazine.

Periodic acid—0·003 M solution stored in ice box.

Stannous chloride reagent. This is prepared fresh daily by dissolving 2 g. stannous chloride in 5 ml. 0·1 N hydrochloric acid. Solution is effected more readily if the stannous chloride has previously been ground in a mortar and pestle.

Notes on method

1. The average normal excretion over 24 hours using this method is found to be 0·66 ± 0·05 μg. expressed as deoxycorticosterone (from 28 results, 14 subjects ; giving standard deviation of the mean ; the standard deviation of the single result is approximately 0·28).

2. It would appear that formaldehydogenic steroids added to urinary extracts are not quantitatively recoverably by this method (due to some inhibition of the periodate oxidation), whereas formaldehyde so added is recoverable. A partial purification of the extract by Girard separation into ketonic and non-ketonic fractions showed that the recovery of such steroids from the ketonic fraction is 86–100%.

Procedure of Wilson using microdiffusion (1953)

Essential differences between this and the previous procedure is the use of ether to remove interfering substances from the urine extract and the use of higher temperatures (85° C.), in the transference of formaldehyde in the microdiffusion units.

Hewitt, Gladding and Purves (1945) indicated the suitability of sodium bismuthate for clearing glycols.

Oxidation. To the extract containing 30 to 100 μg. of cortisone equivalents in 0·10 ml. absolute ethanol, is added 0·60 ml. 0·005 M periodic acid and well mixed. One hour is allowed for the oxidation and then 0·20 ml. 6% stannous chloride added.

Ether extraction. 3 ml. of pure ether saturated with water is run in from a burette and briskly shaken about 30 times with the mixture, then allowed to settle.

Diffusion. Conway units (No. 1) are prepared with a little silicone stopcock grease applied evenly to the rims. 0·3 ml. of the aqueous layer is pipetted into the central chamber using a 1 ml. serologic pipette graduated to the tip. 1·5 ml. of chromotropic reagent containing stannous chloride is added to the outer chamber.

The units are immediately sealed with their covers and placed in the oven at 85° for one hour.

The 0·30 ml. in the central chamber will have distilled over completely.

Colorimetry. The units are removed and cooled for 20 to 30 minutes. The contents of the outer chamber are then withdrawn for spectrophotometric examination using 570 mμ. wave length.

The amounts in the unknown may be calculated, using cortisone standards in absolute ethanol suitably treated throughout (10 to 30 μg. cortisone in inner chamber).

It may be noted that the aliquots analysed as 0·30/0·90 equal one third of the original amounts used. Suitable reagent blanks are also carried out.

Reagents

Periodic acid. 0·005 *M*—114 mg. dissolved in 100 ml. 0·15 *M* H_2SO_4.

Stannous chloride. 6%—600 mg. dissolved in 10 ml. 0·1 *N* HCl before use.

Chromotropic acid reagent. 2 g. of the sodium salt dissolved in 250 ml. of distilled water. While keeping the mixture cool, 850 ml. of conc. H_2SO_4 is added. The reagent is stored in a brown bottle away from light. Just before use 3 ml. of 6% stannous chloride reagent may be added per 100 ml. of chromotropic acid reagent. Although not essential, this addition tends to give lower blank readings.

Notes on method

1. The use of ether in the manner described allows recoveries of added cortisone within 80 to 100% which is comparable to the results obtained by Hollander *et al.* where the ketonic fraction of the urinary extract is separated.

FORMALDEHYDOGENIC STEROIDS

Sodium Bismuthate as Oxidant

BROOKS and Norymberski (1953) and Edwards and Kellie (1954) have used it in the formation of formaldehyde from the formaldehydogenic steroids prior to the determination of the formaldehyde using a microdiffusion procedure.

Brooks and Norymberski (1953) write, ' In its effect, selectivity and rate of reaction, sodium bismuthate closely resembles periodic acid and lead tetracetate.' (In contradistinction to periodic acid, however, sodium bismuthate was found to cleave hydroxy acids, and lead to 17-ketosteroids, where the steroid had a dihyroxyacetone side chain, an analytically useful property.)

For the present purpose, the insolubility of sodium bismuthate offers the technical advantage that a large excess of the reagent can be used, and the surplus removed simply by centrifuging.

Bismuthate oxidation (*Brooks and Norymberski,* 1953)

The steroid residue is dissolved in 2·0 ml. of aqueous acetic acid (5% *v*/*v*). 5 mg. of sodium bismuthate (A.R.) is added and the tube shaken 30 mins. with protection from daylight. The mixture is centrifuged and 0·40 ml. samples taken and the formaldehyde determined as in the previous method.

Bismuthate oxidation as carried out by Edwards and Kellie (1954)

The steroid to be assayed (5·25 m-moles) is transferred in solution to the oxidation tube and dried below 40° in a stream of N_2. The dry residue is then washed down to the tip of the tube with $CHCl_3$ and the solvent removed under N_2. The residue is dissolved in propionic acid (10 μl.) by rotation of the tube and, after standing 10 min., diluted with water (150 μl.). Some urine extracts give a cloudy solution at this stage but if the solution is stirred during dilution, duplicate determinations agree closely.

The solution is cooled to 10° by immersion in a chilled-water bath and finely powdered sodium bismuthate (A.R. 0·5 mg. approx.) is added. (A convenient spatula may be made by flattening the end of a No. II hypodermic needle.) The temperature is maintained and the mixture, stirred at intervals, is allowed to react in the dark for 45 min. The stirrer is then withdrawn and the tube is centrifuged at 2000 rev./min. for 5 min. to remove the $NaBiO_3$ and any cloudiness present. The amounts of reagents and steroid may be increased in proportion if required.

The formaldehyde is now determined as described in Chapter XXXIV. Edwards and Kellie use modified Conway units as already described.

Notes on methods

1. Two kinds of blank determinations may be suitably carried out with the above determination, one in which kieselguhr is substituted for the bismuthate and the other where the full procedure is adopted but omitting the steroid. These may be termed 'unoxidised' and 'oxidised' blanks.

2. High blank values may be obtained if light is not excluded during the oxidation.

3. With freshly prepared propionic acid the difference between the 'oxidised' and 'unoxidised' reagent blanks at 10° is so small that the latter can be dispensed with.

The propionic acid keeps well at 0° under N_2 but an 'oxidised' reagent blank should be carried out with each batch of determinations.

4. If the 'unoxidised' blanks are high due to volatile chromogenic substances, these may be removed or minimised by standing the preparations over concentrated H_2SO_4 and P_2O_5 in a vacuum dessicator ($<$1 mm. Hg.).

Reagents

Water. All water used in the preparation of reagents and in the determinations is distilled from alkaline $KMnO_4$.

Chloroform. $CHCl_3$ (laboratory reagent grade) is washed with a solution of NH_2OH, HCl (5%, *w/v*), made alkaline with ammonia, and then washed three times with distilled water. After drying overnight over anhydrous $CaCl_2$, the $CHCl_3$ is distilled, rejecting the first and last 10% of the distillate. Ethanol (2%, *v/v*) is added and the reagent is stored in amber glass bottles.

Propionic acid. Propionic acid treated with about 1% (*w/v*) CrO_3 is warmed to complete solution and allowed to stand overnight. It is distilled under reduced pressure below 55° and the distillate is subsequently redistilled twice at atmospheric pressure through an efficient fractionating column. The constant-boiling fraction distilling at about 142° is collected, again treated with CrO_3, and distilled under reduced pressure. The purified propionic acid is stored in the dark under N_2 at 0°.

CHAPTER XXXVII (Da 4)

GLYCINE (FORMALDEHYDE PRODUCED BY NINHYDRIN OXIDATION)

The Determination of Glycine in Plasma

Principle. Schwartz, Robertson and Holmes (1955) have adapted the principle of Alexander, Landwehr and Seligman (1945), (as modified by Christensen, Riggs and Rae, 1951) to a microdiffusion procedure.

The principle depends on the conversion of glycine to formaldehyde by the action of ninhydrin, and the subsequent reaction of formaldehyde with chromotropic acid (Eegriwe, 1937). In the present method the formaldehyde diffuses into the chromotropic acid in the central chamber.

Procedure of Schwartz, Robertson and Holmes (1955)

One ml. of heparinised plasma (or serum) contained in a 15 ml. centrifuge tube is diluted with 4 ml. of picric acid solution (0·044 *N*). The mixture is stirred and centrifuged: and 0·5 or 1·0 ml. of the supernatant fluid is pipetted into the outer well of a Conway microdiffusion unit (No. 1 type). The fluid is placed near the outside rim of the outer well to prevent creeping into the central well during evaporation.

Evaporation to dryness may be effected in about one hour by placing the units under an exhaust fan in a hood.

The rims of the Conway vessels are greased with a rather thick layer of petrolatum. This can be conveniently done with a syringe previously filled with melted petrolatum and cooled to room temperature. Five-tenths millimetre of ethyl phosphate is added to the outer chamber near the outside rim of each Conway vessel. A pipette with a rubber two-valve bulb should be used to avoid contact of this reagent with the skin or lungs. The unit is gently rotated so that the ethyl phosphate is uniformly distributed over the area previously covered by the picric acid solution.

The evaporation, greasing, and addition of ethyl phosphate is carried out serially on multiple Conway units. The next step, prior to storage in the dark, must be completed for each individual unit before proceding to the next unit because of the light sensitivity of the chromotropic acid reagent. We have found that mixing of the contents of inner and outer wells is easily avoided during handling if units are placed in an individual wooden holder (Fig. 57). Using a 50 ml. amber-coloured burette, 2 ml. of the chromotropic acid reagent is added to the centre well. Next, 0·5 ml. of the ninhydrin reagent is placed in the outer well. A glass

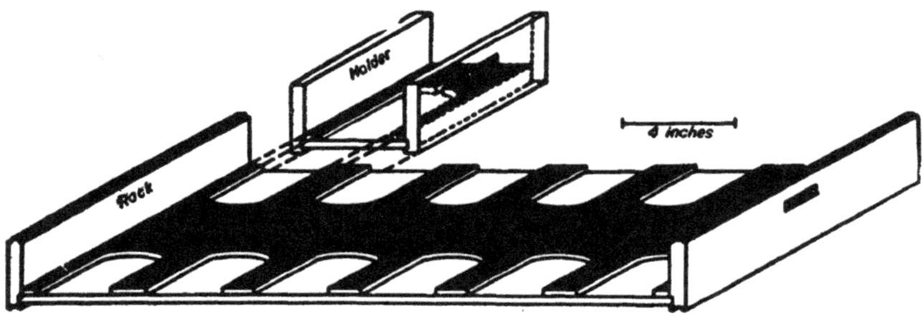

FIG. 57. Rack and holder used for handling Conway units.

cover is quickly placed over the greased rim with gentle tapping to expel air bubbles left in the petrolatum seal. An airtight seal is, of course, essential. To ensure uniform mixing of the ninhydrin reagent, the unit is rotated slowly, with care, so that the chromotropic acid does not spill out of the inner well. The unit is transferred to the large wooden rack, without jarring by sliding the unit forward by means of a pencil placed in the small groove in the wooden holder. The units are permitted to stand overnight in the dark while the formaldehyde diffuses over into the centre well.

On the following morning, the covers of the Conway vessels are removed and prior to transfer to 13 by 100 mm. test tubes, the contents of the inner well are stirred gently with the pipette tip. The test tubes, protected from direct light, are heated in a boiling water bath for thirty minutes. After cooling, the solutions are transferred to calibrated test tube cuvettes and read at 570 mμ in the spectrophotometer. A glycine standard containing 0·10 μM glycine per millilitre and a reagent blank consisting of 1 ml. of 0·044 N picric acid are run in duplicate through the entire procedure concurrently with the unknown. The reagent acid blanks are read against a distilled water blank and the standard and unknowns are read against a reagent blank. 'Creeping' of picric acid from the outer into the inner well will produce a pink colour in the inner well after overnight standing. Such blanks or samples give fallaciously high readings and must be discarded.

Reagents

1. *Picric acid solution,* 0·044 N. Twenty-two grams of picric acid are dissolved in 2,000 ml. of distilled water. This solution is titrated against 0·1 N NaOH using phenolphthalein as an indicator.
2. *Ethyl phosphate.*
3. *Chromotropic acid* reagent.

Stock solution. 5% (*w*/*v*) aqueous sodium chromotropate (sodium 1, 8-dihydroxynaphthalene-3, 6-disulphonate). This solution is stored in the refrigerator in a glass-stoppered brown bottle, thawed as needed, and then refrozen.

Working reagent. Six millilitres of the stock solution are diluted to 100 ml. with concentrated H_2SO_4. The volumetric flask is placed in cool water, and H_2SO_4 is added slowly with mixing to prevent overheating. Stored in a refrigerator in a glass-stoppered brown bottle, this solution remains stable for one week.

4. *Citrate buffer,* *p*H 5·0 : Forty-two grams of citric acid are dissolved in 400 ml. of 1 N NaOH, diluted to 2,000 ml. with distilled water and then stored in the refrigerator.
5. *Ninhydrin reagent.* Two grams of ninhydrin are diluted to 100 ml. with citrate buffer. This solution is stored in a brown bottle. Fresh solutions are prepared weekly.
6. *Glycine standards.* Stock solution containing 1 μM glycine per millilitre is prepared by dissolving 37·5 mg. glycine in 500 ml. of 0·044 N picric acid solution.

Working standard contains 0·10 μM glycine per millilitre. Ten millilitres of the stock solution are diluted to 100 ml. with 0·044 N picric acid solution.

Notes on the method

1. The determinations are made in duplicate. With the procedure described, Schwartz *et al.* consider that 72 glycine determinations may be conveniently carried out simultaneously.

2. They suggest the use of simple plywood racks, illustrated in Fig. 62 to assist the efficient manipulation of large numbers of Conway units.

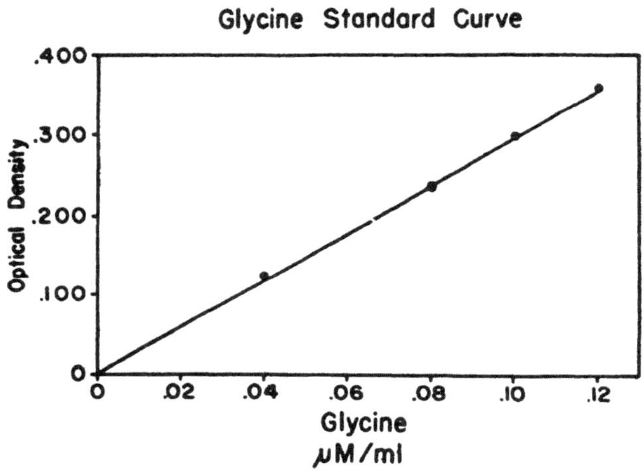

FIG. 58. Demonstration of the linear relationship between glycine concentration and colour yield

3. The standard curve (Fig. 58) shows that Beer's law applies up to 0·12 μM per ml. Calculations of glycine content are based on the glycine standards run concurrently with the samples to be analysed.

4. Recoveries of 98·7 ± 1·4% are obtained when glycine is added to normal plasma. As little as 0·02 μM of glycine can be determined by this procedure and the method is described as 'simple, accurate, specific and highly reproducible'.

5. The method may be also adapted to the determination of glycine in tissues and reference is made to its application to rat diaphragm. (Schwartz, Holmes and Robertson—to be published.)

CHAPTER XXXVIII (*Db* 1)

Db. ACETALDEHYDE GROUP

ACETALDEHYDE

Semicarbazide Absorption

Determination of acetaldehyde in blood

THE principle of the method used by Burbridge, Hine and Schick (1950) is to transfer the acetaldehyde by microdiffusion into a semicarbazide solution contained in the central chamber of a standard unit. After the diffusion is complete, the extinction of the diluted semicarbazide solution is determined using a Beckman DU spectrophotometer, at wave length 224 mμ and slit at 0·86 millimetre.

Procedure of Burbridge, Hine and Schick (1950)

Three ml. of the buffered semicarbazide HCl solution are pipetted into the central chamber. The top of the outer wall of the cell is greased with lubriseal. Two ml. each of 2/3 *N* sulphuric acid and 10% sodium tungstate are added to the outer chamber just prior to addition of the blood sample. Two millilitres of well-shaken fresh blood are added to the outer chamber and the cover is immediately put in place. The blood and protein precipitants are mixed by gently rotating the cell. The cells are then placed in an oven with the temperature controlled between 28° and 30° C. and the diffusion is allowed to proceed for 90 minutes. At the end of this time the contents of the central chamber are quantitatively transferred to 10 ml. volumetric flasks and brought to volume with distilled water. The optical densities of these solutions are read on a Beckman DU spectrophotometer at wave length 224 mμ and with slit width of 0·86 millimetre. The reference blank for these readings is 3 ml. of the semicarbazide solution diluted to volume in a 10 ml. volumetric flask. From the optical densities obtained, the corresponding micrograms of acetaldehyde can be read from the standard curve. The final value is then multiplied by five in order to correct for dilution.

Formation of a standard curve

Into nine 10 ml. volumetric flasks are pipetted 3 ml. of buffered semicarbazide solution. Into three of these are added 1, 2 or 3 ml. of the buffered semicarbazide solution or 3 ml. of a 15·6 μg. per ml. cooled solution of acetaldehyde. All nine flasks are allowed to warm to room temperature, diluted to volume and the optical densities determined on the Beckman DU spectrophotometer, using as a blank 3 ml. of the buffered semicarbazide solution diluted to 10 ml. At the wave length and slit width given, a linear relationship exists between the optical density and concentration.

Special absorbing reagent

Semicarbazide hydrochloride solution. This is made up as 0·0067 molar solution in a phosphate buffer (pH$=$7·0; 0·06 M NaH_2PO_4 and 0·14 M Na_2HPO_4).

Notes on method

1. The authors note the fact that large numbers of the analyses can be carried out in a relatively short time. Thus, one analyst completed as many as forty-eight determinations in a five-hour period.

2. For use with the standard Unit it would seem advisable to reduce the total volume in the outer chamber to 2 to 3 ml. and that in the central chamber to 1 to 2 ml.

3. Acetaldehyde recoveries of 3·1 to 15·6 mg. per ml., from water or blood, were quantitative within the sampling error.

Burbridge *et al.* note that when a 2 ml. sample of blood is analysed, acetaldehyde in concentrations as low as 0·2 μg./ml. can be determined with an accuracy range of $\pm$3·5%.

4. Normal acetaldehyde blood levels found by this method ranged from 40 to 110 μg. per cent in fourteen healthy persons both fasting and non-fasting subjects. (Stotz, 1943, had reported the normal free acetaldehyde content of blood to be from 22 to 37 μg. per cent in seven fasting individuals.)

5. Semicarbazides could also be formed by other volatile aldehydes and ketones. Measurable amounts of acetone in normal blood have not been reported ; also the authors showed that acetaldehyde determinations fell within the normal range even when strongly positive tests for acetone were given by the urine.

7. From their standard curve and using their procedure the micrograms acetaldehyde per ml. is given by the approximate relation, μg. per ml. $= 3 \cdot 0 \times E$ where E is the extinction. (This is inserted here only for guidance.)

CHAPTER XXXIX (*Db* 2)

ACETALDEHYDE FROM LACTIC ACID AND THREONINE WITH BISULPHITE ABSORPTION

Lactic Acid and Threonine

Lactic acid in tissues Winnick (1942a) used a microdiffusion method based on that of Gordon and Quastel (1939), who found that lactic acid is oxidised quantitatively to acetaldehyde by ceric sulphate in acid solution at 50° C. These investigators employed a current of nitrogen gas to carry the acetaldehyde into bisulphite solution, where the bound aldehyde was determined iodometrically, as in the method of Friedemann, Cotonio, and Shaffer (1927).

Winnick lists the following differences of his method from the above :

(1) Single lactic acid determinations are performed on 1 ml. blood samples or on approximately 0·5 g. of tissue.

(2) Zinc hydroxide is used instead of trichloroacetic acid to precipitate blood proteins. The recoveries are low by 5 to 10 per cent. if either trichloroacetic or tungstic acid is used.

(3) Blood filtrates are freed of sugar by treatment with cupric hydroxide prior to analysis (McCready, Mitchell, and Kirk, 1939).

(4) Determinations are usually conducted at room temperature instead of at 50°.

(5) The acetaldehyde which results from the oxidation of lactic acid is determined by a microdiffusion procedure, using the bisulphite reaction as described for acetone.

In using the Winnick method different observers have found the determinations of standard solutions and recoveries too low for accurate determinations.

In the previous edition a modification was described which appeared more satisfactory, though this may have been related to special conditions and the level of lactic acid analysed. Other workers here obtained results which were in general below the theoretical. It is possible from the work of Long (1946) and of

Elsden and Gibson (1954) that this may have been related to some oxidation of the resulting acetaldehyde to acetic acid. Further study of the method (Doolin, 1955 ; Ryan, 1956) has led to the adoption of the following procedure.

Lactic acid determination in blood, applicable also to tissue (Ryan, 1958).

Principle

The method is a modification of that described by Burbridge, Hine and Schick (1950) for the determination of acetaldehyde in blood. The principle depends on the oxidation of lactic acid to acetaldehyde, by means of ceric sulphate, in the outer chamber of the unit (No. 1). The acetaldehyde so produced is absorbed in a buffered solution of semicarbazide in the central chamber. After the diffusion is complete the extinction of the diluted semicarbazide solution is determined using a Beckman DU spectrophotometer at wave length 224 mμ.

Ryan's procedure.

Treatment of blood sample. Into each of two centrifuge tubes 2 ml. of 15% perchloric acid is introduced. Into one of the tubes 7 ml. of distilled water is added and into the other 6 ml. of distilled water + 1 ml. of a standard lactic acid soln. (containing 20 mg. % lactic acid), are added. 2 ml. of the freshly collected blood is then added to each tube and, after thoroughly mixing, the suspensions are centrifuged. The supernatant fluids are poured off and a measured quantity from each tube, (8·0–8·5 ml.), is treated with sufficient 40% KOH (0·2–0·4 ml.) to precipitate the excess perchloric acid. The $KClO_3$ precipitate settles out rapidly leaving a clear solution which is then ready for analysis.

Microdiffusion procedure.

2 ml. of the semicarbazide absorbing reagent is introduced into the centre chamber of each unit. Into the outer chamber of each unit (except blanks) 0·4 ml. of 6N sulphuric acid saturated with ceric sulphate is added, followed by 2 ml. of the protein free blood extracts. The units are sealed by means of lids smeared with a paraffin wax-vaseline fixative (Conway, 1947), and rotated in the

usual manner. A blank estimation using 6N sulphuric acid instead of the ceric sulphate soln., is also carried out. The units are incubated at 37° C for a period of two hours. At the end of this time 1 ml. is removed from the centre chamber of each unit, and diluted to 10 ml. with dist. water. The extinctions of the diluted solutions are then read at a wavelength 224 mμ, using a Beckman spectrophotometer.

Reagents.

Semicarbazide hydrochloride solution. This is made up as 0·0067M semicarbazide HCl in a phosphate buffer (pH 7·0 ; 0·06M NaH_2PO_4 and 0·14M Na_2HPO_4).

6*N Sulphuric acid.* This solution may be prepared with sufficient accuracy by diluting 162 ml. conc. H_2SO_4 to 1 litre.

Saturated ceric sulphate. The ceric sulphate is freshly prepared before using by grinding to a fine powder in a clean dry mortar. Sufficient of the 6N H_2SO_4 is taken in a test tube and excess of the powdered ceric sulphate added and well shaken. The undissolved ceric sulphate is removed by centrifuging before use.

Standard Lactic Acid. A stock solution, equivalent to 100 mg.% lactic acid, is prepared by dissolving 1·065 gm. of pure lithium lactate and 1 ml. of conc. H_2SO_4 in a litre of water.

15% *Perchloric acid.* 40% *Potassium hydroxide.*

Notes on method.

(1) Recoveries of added lactic acid by the above method were found to be quantitative and the determinations of blood lactic acid were very similar to those obtained by the method of Elsden and Gibson (1954) on the same blood samples.

(2) No appreciable error results from glucose in filtrate.

(3) An ordinary trichloroacetic acid filtrate is now found quite suitable and the internal standard is unnecessary.

Threonine

Winnick's microdiffusion method (*1942b*). This is an adaptation of Shinn and Nicolet's procedure (1941) whereby this amino acid is oxidised to acetaldehyde at room temperature by neutral periodate solution. Martin and Synge (1941) obtained only 70% of the maximum theoretical quantity of acetaldehyde on treatment of dl-threonine with neutral periodate solution, using a complicated apparatus. Winnick had no difficulty in confirming Shinn and Nicolet's finding that 1 mole acetaldehyde results from the oxidation of 1 mole of threonine.

In Winnick's procedure a sample of protein hydrolysate is treated with neutral periodate solution in the outer chamber of the standard ' unit ' and the acetaldehyde resulting from the oxidation of threonine passes by gaseous diffusion into the central chamber, where it is absorbed by bisulphite solution. Phosphate buffer is used to maintain a *p*H of 7·0 in the outer chamber.

Procedure.

(1) *Preliminary protein hydrolysis.* An accurately weighed protein sample (usually about 0·5 g.) is hydrolysed for at least 24 hours with about 20 ml. of 3*N* HCl. The hydrolysate is neutralised to a *p*H of approximately 7·0 with 3*N* sodium hydroxide. If HCl more concentrated than 5 to 6*N* has been used, it is preferable to concentrate the hydrolysate first *in vacuo* to a volume of 5 to 10 ml. in order to remove most of this acid. The neutral solution is made up with water to a suitable volume (100 to 250 ml.) and analyses are performed on aliquots. The preceding operations can be conducted on a much smaller scale if the supply of protein is limited.

(2) *Oxidation and microdiffusion procedure.* 1 to 1·5 ml. of 0·25*M* sodium bisulphite is pipetted into the central chamber of the microdiffusion ' unit '. A 2 or 3 ml. aliquot of the solution to be analysed is pipetted into the outer chamber. 1 ml. of approximately 0·1*M* tripotassium phosphate is delivered also into the outer chamber. This solution is adjusted so that 1 ml. will just neutralise 1 ml. of 0·2*M* periodic acid (to be added next) to a *p*H of about 7·0. The greased glass cover-plate is placed over the ' unit ', leaving a narrow opening. Through this slit 1 ml. of 0·2*M* periodic acid is pipetted quickly into the outer chamber. The cover is slid over immediately to seal the vessel completely, and the latter is rotated to mix the solutions in the outer chamber.

After the ' unit ' has stood for at least 4 to 5 hours at room temperature, the cover is removed and the excess bisulphite is oxidised with 1N iodine. Powdered disodium acid phosphate is added to dissociate the acetaldehyde-bisulphite compound, and the liberated bisulphite is titrated with standard 0·005N iodine. The titration procedure is the same as that described for acetone and lactic acid determinations.

Calculation. This is based on the fact that 1 ml. of 0·005N iodine is equivalent to 0·298 mg. of threonine.

Further notes on the method.

(1) As shown by Winnick, proline, histidine, cystine, alanine, aspartic acid, isoleucine, tryptophane and methionine do not interfere with the method, these compounds being representative of nearly all the different structural types of amino acids found in proteins.

A similar observation made by Martin and Synge (1941) showed that no significant quantity of acetaldehyde resulted when a mixture of 14 different amino acids (not including threonine) was oxidised with periodate.

(2) Glucose likewise does not interfere.

Solutions required.

3N HCl *and* 3N NaOH.

Periodic acid solution. Crystalline periodic acid is used, a 0·2M solution in water being made up. It is not feasible to prepare a neutral periodate reagent, since a precipitate results if the periodic acid is mixed directly with the potassium phosphate solution.

CHAPTER XL (E1)

E. VOLATILE KETONES

ACETONE (INCLUDING A RAPID CLINICAL METHOD USING THE NESSLER SOLUTION)

Acetone.

Winnick's method for acetone in blood or urine. In this the acetone in the blood (acidified) or urine diffuses into sodium bisulphite in the central chamber. Klein (1940) has shown that the bisulphite reaction can be used for the quantitative determination of acetone plus acetoacetic acid in blood filtrates free of sugar. The dissociation of the carbonylbisulphite compounds is repressed by using a large excess of $NaHSO_3$ and lowering the temperature to 5° C. After the addition reaction is complete, the excess of bisulphite is removed by concentrated iodine solution, and the solution is made alkaline with Na_2HPO_4, thereby dissociating the carbonyl-bisulphite compounds. The bisulphite thus liberated is then titrated with standard 0·005*N* iodine solution. In Klein's method with blood filtrates most of the pyruvic acid present in the blood is also determined as well as acetoacetic acid. With the microdiffusion procedure there is a high degree of specificity for acetone.

Procedure. Blood samples are collected in tubes which contain oxalate or citrate and are tightly fitted with rubber stoppers. The blood (or urine) can stand for a number of hours in the ice box without appreciable change in the acetone concentration.

Into the central chamber of ' unit ' No. 1 is run roughly 2 ml. of approximately 0·15*M* $NaHSO_3$, and when blood is being determined about 0·5 ml. of 2*N* H_2SO_4 into the outer chamber (this is omitted with urine). The lid is suitably greased (Winnick uses stop-cock lubricant) and placed in position, and 1 or 2 ml. sample of the blood or urine is then quickly introduced into the outer chamber, the lid being displaced a little to allow the insertion of the pipette. After delivery the lid is immediately replaced in position.

If blood is being analysed the ' unit ' is rotated at once to mix the blood and the acid.

After six hours or more at room temperature (which Winnick mentions as being 25° for his analyses) the ' unit ' is cooled for at least 30 minutes in a shallow dish of water containing ice cubes. The lid is removed and a small drop of starch solution is added to the $NaHSO_3$ solution in the central chamber. Using a fine glass stirring rod, the large excess of $NaHSO_3$ is removed with $1N$ iodine solution added very slowly, dropwise from a burette. Nearing the end-point, small fractions of drops of iodine are transferred from the burette tip to the solution with the tip of the stirring rod.

When a permanent light purple colour is reached, the ' unit ' is removed from the ice bath and warmed to room temperature. Then about 0·3 to 0·4 g. of powdered Na_2HPO_4 is added to the central solution, and the latter is stirred well. The $NaHSO_3$ which is liberated from combination with the acetone is titrated with standard $0{\cdot}005N$ iodine solution to the same end-point as before.

Calculation. 1 ml. of $0{\cdot}005N$ iodine is equivalent to 0·145 mg. acetone, so that the mg. of acetone per 100 ml. of blood or urine can be calculated directly.

Notes on the method.

(1) 2 ml. Bang burettes would appear suitable for the above titrations, but the use of the horizontal burette previously described, at least for the final titration, using $N/20$ or $N/40$ iodine, would appear very suitable. Each large division of $N/20$ iodine is the equivalent of 0·0145 mg. acetone.

(2) To hasten the microdiffusion, and as pointed out by Winnick, the ' units ' may be incubated at 40–50°, the time required for complete diffusion being then reduced to about 3 hrs. A fixative of higher melting point such as 2 parts vaseline to 1 part paraffin wax may then be used.

(3) The microdiffusion may also be hastened by keeping the volume in the outer chamber as low as possible consonant with the complete covering of the surface. Thus where the ' units ' are well cleaned, and the outer chambers rinsed before drying with a little saponin solution, 1 ml. will be found quite sufficient to cover

the surface. With blood 0·5 ml. may be used with the 0·5 ml. acid and 1 ml. of urine without acid. (The $NaHSO_3$ in the central chamber could also be reduced to 1 ml.) Such a procedure may be expected to reduce the time to 3–4 hours at 25° C.

(4) According to Winnick the method is accurate to 2–3% for acetone levels of 10 to 30 mg./100 ml.

Solutions, etc.

$2N$ H_2SO_4.
0·15M $NaHSO_3$.
1N iodine.
0·005N (with Bang burette) or $N/20$ or $N/40$ with Conway burette.
Powdered Na_2HPO_4.
1% starch solution.

A rapid clinical method for determining acetone in blood and urine. (Werch's procedure.) This depends on the formation of a white precipitate with Nessler's reagent (Deniges, 1930; Fearon and Webb, 1936). The procedure is as follows:

The ' units ' with their covers are placed on a table spread with black paper (a dark background helps in the reading of the end-point), and the edges of the outer wall smeared with vaseline. The reagents are then introduced. First, 2 c.c. of Nessler's solution are run into the central chamber, and 1 c.c. of dilute acid into the outer chamber, being distributed evenly around. A transparent glass lid is next placed so as to cover the entire ' unit ' except for a small slit, permitting the introduction of the end of the pipette used to run in the unknown solution. A stop watch held in one hand, and a pipette containing 3 c.c. of the unknown in the other, is started as soon as the pipette stops flowing freely. The hand holding the stop watch next slides the glass plate so that it completely covers the ' unit ' as an immediate gesture after starting the watch. When acetone is present precipitation of the Nessler's solution occurs at the periphery of the inner chamber in the form of threads, like spokes of a wheel, or as a cloud. It is that second when the spokes or cloud extend inward 1 mm. that is taken as the end-point.

Calculation. The rate of formation of the precipitate depends on the concentration of acetone present and on the temperature.

The relations are expressed by Werch as

$$D = \frac{179t^2}{(24 - \theta)^2}$$

where t is the time in seconds and θ the temperature in degrees C, while D is the dilution. For convenience in determining D, Werch has constructed the nomograph given in Fig. 59.

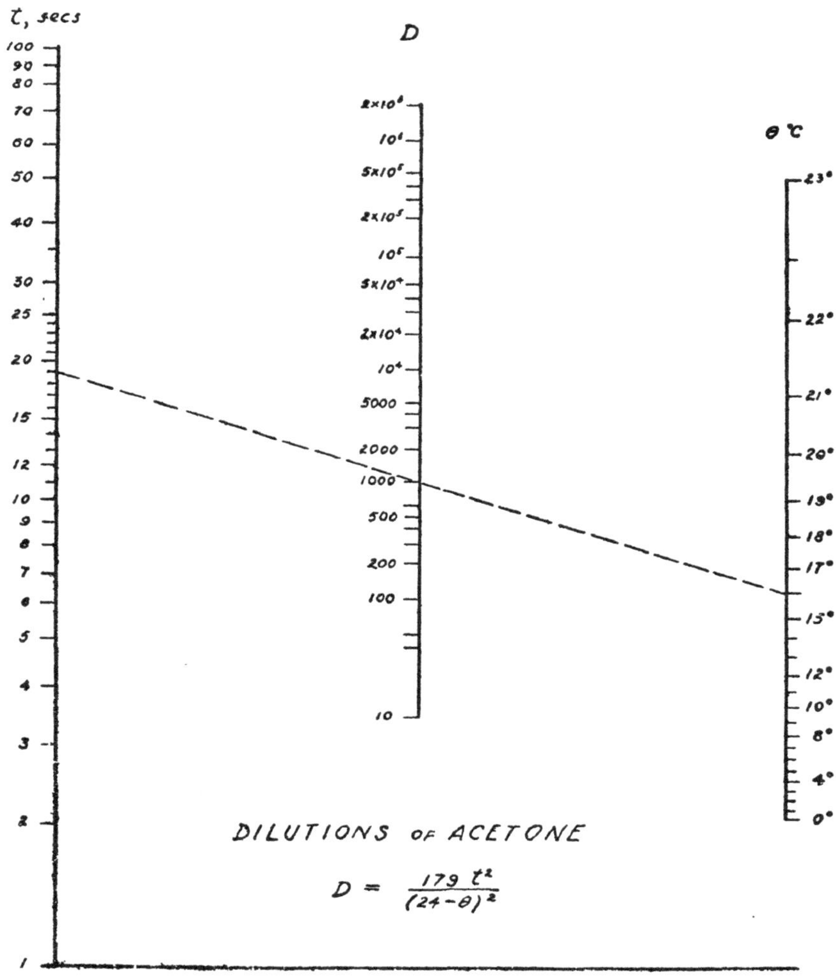

FIG. 59. Nomographic Chart for use in the determination of acetone by Werch's method. (From the *Jour. of Lab. and Clin. Medicine*, **25**, p. 414, 1940.)

Notes on the method. (1) It is clear that compared with Winnick's bisulphite procedure this will give only very approximate results, which, none the less, may be of clinical value. They are very easily and rapidly obtained.

(2) The temperature effect is considerable, and for this reason and as will appear from the equation, determinations made above 20° C. are not reliable. It seems to the present writer that the temperature effect could perhaps in general be better allowed for and direct calculation simplified (apart from the nomograph) by carrying out duplicate determinations of the t value on a standard 1 in 10,000 solution of acetone at the same time as that of the unknown. The acetone in the latter, as milligrammes per cent., is then given by $10 \times R^2$ where R is the t value of the standard divided by that for the unknown. At the same time, Werch's procedure (using the nomograph) has the merit of being the simplest in practice.

(3) With normal blood and urine, precipitation of the Nessler solution occurs only after several hours and in a very thin film. Whenever acetone is present reasonable time readings are obtained. In a number of cases of diabetes mellitus, with clinical evidence of deep acidosis, the t values for blood samples, as examined by Werch (at 16° C.), were around one minute. This corresponds to a dilution of 1 : 10,000.

CHAPTER XLI

THE HALOGENS (INTRODUCTORY)

A principle used in the determination of the halogen salts consists in oxidation to the free element in the outer chamber, by oxidants such as acid permanganate or dichromate, and the absorption of these by diffusion in the central chamber in 1 ml. of 20% potassium iodide, where it liberates an equivalent amount of free iodine.

The high percentage of iodide is selected to prevent subsequent diffusion of iodine from the central chamber.

Subsequent to the complete absorption, the free iodine when it exceeds a concentration of $N/1000$ (35 μg chlorine) is titrated with $N/20$ thiosulphate, a distinct end-point change occurring with less than a small division of the micro burette and sufficiently sharp for obtaining a coefficient of variation of 0·5% with the single determination, using amounts of the order of 0·3 mg. chlorine.

It may be noted that the presence of the 20% iodide introduces a purplish colour into the starch iodide end-point instead of the usual blue, but this introduces no practical difficulty in the titration as described.

Below a level of 35 μg chlorine (1 ml. of $N/1000$) No. 2 ' unit ' may be used, with 0·1 ml. 20% KI in the central chamber, titrating with suitably dilute thiosulphate. For the lowest levels (less than 7 μg Cl) the starch iodide colour is determined after starch addition. Though the method has been worked out on the above principle, it is capable of wide variation with respect to possible absorbents and modification of the colorimetry or the titration method, but such need only arise when it is desired to exceed the high accuracies already found.

With the colorimetry at the lowest levels, it is advisable to plot one's own curve of recoveries with respect to the theoretical quantities, as slight and possibly unsuspected changes in procedure may alter a little the values already described.

Another principle used in the microdiffusion procedure has been introduced by Gordon (1951). In this the free halogen liberated by oxidation in the outer chamber oxidises in turn Fast Green in solution in the central chamber, the amount oxidised being determined colorimetrically.

The titration with thiosulphate. The thiosulphate used is $N/20$ and keeps well when a trace of carbonate is added. If controls of 1 ml. of $N/100$ HCl are occasionally carried out the thiosulphate need not be otherwise standardised. The starch solution used is 0·2% dissolved by boiling, with a minute quantity of mercuric iodide as described in text. Other observers may prefer different solutions, such as cold extract of soluble starch cleared by centrifuging. Any starch solution used, however, should be clear and give a marked end-point change with less than a single small division on the burette.

In the region 35 to 7 μg Cl, the strength of the thiosulphate may be $N/200$–$N/400$ with the ordinary bore graduated tubing of the Conway burette but more concentrated solutions can be used with narrower bore.

Delicacy of the microdiffusion method of halogen analysis. The possible delicacy of chloride determinations depends on how far we can determine photometrically the concentration of free iodine in the iodide using starch solution, or some equally sensitive reagent for testing free chlorine (e.g. *o*-tolidine) or free iodine.

With 0·5 ml. of 0·1% starch addition to 1 ml. 20% KI we get for the free iodine equivalent of 1 μg. Cl absorbed an extinction of 0·94 with 50 mm. stratum layer. With 0·7 ml. iodide and 0·3 ml. starch this would become 1·4. The extinction values were determined with a Leitz instrument using filter No. 3, but higher values can be obtained with filters No. 2 or 1, giving wavebands in the purple, but not so convenient to work with owing to the weakness of this band in the usual light source. The extinction may in short be increased to 2·0.

Since accurate photometric determinations can be carried out with extinctions of 0·25, it is possible therefore to determine chloride accurately down to 0·1 μg. chlorine with the standard 'unit' (No. 1). This would mean, however, special preparation of 'units' to decrease the blank value further, though this difficulty scarcely arises with bromide; also a further investigation of the slight initial lag in free iodine formation, as described in the subsequent account of methods, would be desirable.

The above considerations apply to a limit of 50 mm. stratum using volume of the order of 1 ml. Further possibilities exist using the micro cells of the Spekker or Pulfrich instruments.

Gordon, using the oxidation of Fast Green by the procedure to be described later, gives the limits of chlorine detection as 0·1 μg., but much lower levels could possibly be reached by suitable modification of the colour determination and volume adjustments.

CHAPTER XLII

CHLORIDE (BY OXIDATION TO CHLORINE AND ABSORPTION INTO IODIDE)

Principle and Introductory. The principle used in the micro-determination of chloride consists in oxidising the chloride to chlorine gas by a suitable acid permanganate mixture in the outer chamber, the chlorine so formed being absorbed by potassium iodide in the central chamber, where it liberates an equivalent amount of iodine.

The formation of the chlorine gas takes place quite readily at room temperature, and its absorption is complete in one hour and a half.

For a range of chloride down to 35 μg (0·035 mg.) the liberated iodine is best titrated with sodium thiosulphate solution, most suitably by 0·05N solution from the horizontal burette described in Chapter VI. Chloride quantities of the order 0·3–0·4 mg. chlorine (corresponding to the amount in 0·1 ml. plasma) can be determined with a coefficient of variation of 0·5%.

(Below 35 μg Cl, and down to 7 μg Cl, the No. 2 ' unit ' is convenient, the thiosulphate being $N/200$ or $N/400$ as previously mentioned.)

Below 7 μg chlorine the iodine liberated in the central chamber is determined colorimetrically or spectrophotometrically after developing the starch-iodine colour.

Procedure.

(*a*) **Chloride determination down to 35 μg chlorine.** Into the central chamber of a ' unit ' is run approximately 1 ml. of a 20% potassium iodide solution. Into the outer chamber is introduced about 0·2 g. of pure potassium permanganate, taking care that none of the grains are spilled into the central chamber ; this may be done without any difficulty when using a spoon spatula. The quantity of permanganate need be only very roughly judged and weighed but once to form an idea of the bulk. 1 ml. of the fluid to be analysed is introduced from an Ostwald pipette into the outer chamber. One of the square glass lids for covering the 'unit' is now lightly smeared over its surface with the special fixative (see below). The lid is placed so as to just allow the introduction of the tip of a pipette. The 'unit' is slightly tilted—this most conveniently by resting it on another lid—and 1 ml. of a sulphuric acid solution containing 24–30% by volume of pure sulphuric acid (s.g. 1·84) is run into the outer chamber. The lid is immediately fixed in position and the fluid in the outer chamber rotated gently about 10 times to ensure adequate mixing. The 'unit' is then left aside on the bench after being first examined to see if it is sealed all around—which is evident by the transparency of the contact. After 1½ hours the lid is removed and the contents of the central chamber are titrated with 0·05*N* thiosulphate from the burette previously described, using a drop of starch indicator towards the end. The iodine starch colour is purple or violet in the presence of the iodide. For this titration the Bang 2 ml. standard burette may also be used containing 0·005*N* thiosulphate.

A blank determination with the reagents is carried out at the same time.

Calculation of the result. After subtracting the blank value, the amount of thiosulphate is multiplied by its chlorine equivalent for the determination. This latter may be determined with fresh thiosulphate solution by making an initial duplicate determination of 1 ml. of 0·0141*N* HCl, each ml. of which contains 0·5 mg. of chloride as chlorine. The burette reading here multiplied by 2 gives the figure required for 1 mg. of chlorine. In the Conway burette 0·05*N* thiosulphate is used, and when made up with a trace of sodium carbonate keeps for a long period. (See reagents below.)

If the thiosulphate is directly standardised by titration of standard iodine in 20% KI, the result, calculated as chloride, is multiplied by 1·03. This factor does not arise from the oxidation and absorption of the chlorine not being quantitative, but is due to a slight diffusion of the liberated iodine from the central chamber. The factor 1·03 applies for 1½–2½ hours after the addition of the sulphuric acid to the outer chamber. (This point is again referred to below.)

The above method is independent of the presence of protein to the extent of 1 mg., i.e. 0·1% in 1 ml.

If it is found desirable to carry out a determination with greater or lesser volumes than 1 ml., the total acidity in the outer chamber should lie between 12 to 15 volumes % of sulphuric acid. This and other conditions are discussed subsequently.

If there is not much organic matter present it may be found more convenient to add the permanganate as 1 ml. of the saturated solution to 1 ml. of the solution for analysis in the outer chamber. The acid subsequently added should be 0·5 ml. of 60–75% (by volume) sulphuric acid.

Reagents.

1. 20% *potassium iodide.* This solution does not require to be accurately made up. It should be stored in a brown bottle and is best made in small quantities at a time. A small amount of free iodine will not affect the result since it will be allowed for in the blank. For chloride determinations below 35 μg chlorine the potassium iodide should, however, be freshly made up, or at least give no perceptible colour when viewed in a thick layer.

2. 24–30% *(by volume) sulphuric acid.* The purest sulphuric acid should be used with s.g. 1·84, 24–30 ml. being run into about 60 ml. distilled water in a beaker and made up to 100 ml. after cooling.

3. 0·05*N thiosulphate.* This is made up in the usual way by dissolving 12·42 g. to make 1 litre of solution. To preserve the solution 50 mg. of sodium carbonate are added per litre.

4. *Potassium permanganate.* Only the purest analytical reagents should be used. The crystals are ground to a powder in a clean mortar and preserved for convenience in a small wide-necked stoppered bottle.

5. *The fixative.* This is formed by melting 50 g. of paraffin wax (M.P. 49° C.) in 80 ml. of pure liquid paraffin in a beaker and cooling the mixture ; more or less of solid paraffin may be used in accordance with the prevailing temperature of the room. The paraffin wax supplied by the British Drug Houses is suitable for use, though some samples supplied at higher melting points were not sufficiently free from impurities, when carrying out the finest determinations.

6. 0·2% *starch solution.* A suitable solution is that of Mutnianski as described by Kolthoff (1931). In this method 2 g.

of soluble starch and 10 mg. mercuric iodide are ground to a paste with a little water and poured into 1 litre of boiling water. The solution is clear and does not alter over a long period if kept in a brown bottle.

Accuracy, etc., of method. A coefficient of variation of 0·55% may be expected. In a typical group of 25 determinations 23 gave results within 1·2% of the theoretical.

It may be noted that the method can deal, practically speaking, with the largest amounts of chloride, but at the same time—as above mentioned—it is advisable for convenience in titration that the fluid be diluted if necessary, so that 1 ml. contains not more than about 0·7 mg. chlorine (or not more than twice the strength of a 1 in 10 blood filtrate). Smaller volumes than 1 ml. may also be used, with a slightly reduced accuracy. The total acid strength in the outer chamber, however, should always be within 12–15% by volume.

(*b*) **Micro-determination of chloride below 35 μg chlorine.** This is carried out in a similar way to the above using No. 2 standard ' unit '. Into the central chamber is pipetted 0·1 ml. 20% KI, and into the outer 1 ml. of the solution after adding a knife point of powdered pure permanganate. 0·2 ml. of 75% H_2SO_4 is then introduced into the outer chamber after placing the lid in position as above. After two hours the inner chamber is titrated with standard thiosulphate ($N/200$ or $N/400$, or stronger depending on the bore of the graduated tube used in the Conway burette). A small drop of starch solution is used to indicate the end-point.

(*c*) **Micro-determination of chloride below 7 μg Cl.** This is carried out as above, but it may be more satisfactory to add the permanganate and acid in one solution, 0·5 ml. of equal volumes of 75% sulphuric acid and saturated permanganate.

The final determination is carried out colorimetrically or spectrophotometrically using the starch-iodine colour.

In the determination 2 ml. of 0·1% starch are pipetted into a small tube and as much as possible of the central chamber contents pipetted into the tube and the contents mixed. The central chamber is then washed out with the mixed fluid, the washings being returned each time to the tube.

Finally the extinction of the fluid is obtained using a light filter in or towards the red end of the spectrum, and the amount of Cl read off from a curve drawn for standard solutions similarly determined. The relation of extinction to quantity is a straight line which may not pass exactly through the origin but cut the abscissa slightly to the right.

(*d*) **Application to chloride determination in blood, urine and tissues. Chloride in blood.** The chloride is here determined in protein-free filtrates, the most suitable being the tungstate filtrate of Folin and Wu (1919). Of this filtrate 1 ml. is a suitable volume giving in a subsquent titration a reading of about 20 large divisions on the burette containing 0·05*N* thiosulphate. The method described for quantities down to 35 μg chlorine may be followed in detail. It may be found more convenient, however, to add the acid and permanganate together as 1 ml. of a mixture containing equal parts of sulphuric acid—60% by volume—and saturated permanganate. This mixture should be freshly made up in the required amount for a group of determinations, the acid and permanganate being kept separate as stock solutions. A blank determination should also be carried out with the reagents in the same dilution.

Chloride in urine. The chloride in urine may be determined for convenience in 0·1 ml. of undiluted urine (a slightly higher accuracy will be obtained using 1 ml. of a urine diluted one in ten). The method described for chloride determinations above 35 μg chlorine is followed. Into the outer chamber is pipetted 0·1 ml. urine from an Ostwald pipette after the addition of approximately 0·2 g. of powdered permanganate, the acid subsequently added to the outer chamber being 2 ml. of 15% (by volume) sulphuric acid (where 1 ml. of a diluted urine is used 1 ml. of 30% sulphuric acid should be added). The remaining procedure is the same as before, the titration being carried out two hours after the addition of the acid.

Accuracy of method with blood and urine. The accuracy with blood is the same as that for determining the chloride in 1 ml. of 0·01*N* HCl, and may be expressed as a coefficient of variation of 0·5%.

Whereas one might have expected the chloride determinations

on protein-free filtrates to proceed as with standard solutions, the large amount of organic matter in urine would at first appear likely to introduce a disturbing factor. The recovery of added chloride, however, has been found quantitative even with albuminous urine containing 1% protein. From a large number of recoveries and duplicate determinations the coefficient of variation for the single determination using 0·1 ml. urine has been found to be 0·7%.

Chloride in tissues.

Amphibian tissues. About 0·200 g. of muscle (for example) is accurately weighed out on a watch-glass and transferred to a small mortar. For convenience of calculation the tissue weighed may lie in the range 0·190–0·210 g. A small quantity of pure quartz sand (Merck) is added to the mortar and from a graduated 2 ml. pipette or micro-burette a volume of distilled water which is calculated as follows. The weight of the tissue is multiplied by 20 and 2·36 subtracted. This with 0·2 g. tissue, for example, will mean the addition of 1·64 ml. (this volume is so chosen that the 2 ml. taken from the subsequent fluid cleared by centrifuging contain the chloride in 100 mg. tissue). The mixture is thoroughly ground to a fine paste, which should require no longer than 2 min. 2·0 ml. of 0·0667*N* sulphuric acid are added and 0·2 ml. of 10% tungstate, the mixture being further ground for half a minute. It is then transferred to a small centrifuge tube and centrifuged along with a similar mixture from a blank determination which will contain no tissue but 1·80 ml. distilled water and the remaining reagents as before. If it is found difficult on centrifuging to obtain a perfectly clear fluid, it is better to filter.

2·0 ml. of the clear fluid are transferred to the outer chamber of a ‘ unit ’ containing about 0·2 g. of powdered permanganate and 1 ml. of 20% potassium iodide in the central compartment. The determination is then made in the usual manner, 0·5 ml. of a 60% (by volume) sulphuric acid solution being used. The ‘ units ’ are left aside for 2 hours at room temperature and subsequently titrated or estimated colorimetrically by the method described for chloride in the range 35–7 μg chlorine.

Calculation. Using the above method where 0·2 g. tissue is used, the total fluid volume is 4·0 ml. and the chloride corresponding to 100 mg. of tissue is contained in the 2·0 ml. of the clear fluid after centrifuging. When 0·21 and 0·19 g. of tissue are used in accordance with the above scheme it may be readily calculated that the 2 ml. of the clear fluid contain the chloride corresponding to 99·8 and 100·2 mg. of tissue respectively, which for the purposes of the determination are equivalent to 100 mg.

The chloride equivalent of the thiosulphate used in the titration expressed as mg. chlorine gives the g. chlorine per 100 g. tissue.

Where the direct colorimetric method is used, which will be found very convenient, the plunger in the iodide is set at 9·2 mm. and the millimetres of grey solution, using Leitz filter No. 3, multiplied by 4 gives the mg. chlorine per 100 g. in the tissue.

Accuracy of the method. The method gives a 99% recovery of added chloride, and the variable error is similar to that already described.

Mammalian or amphibian tissues (*Conway and Hingerty*, 1943).

Taking muscle as an example, one part by weight is added to 20 times this figure in volumes of sodium sulphate solution (1·3% anhydrous for amphibian tissue, and 2·0% for mammalian).

The thickness of the muscle section or sections should not exceed about a few mm. The muscle is stirred in the suspending fluid by shaking or bubbling with air for one hour, or is left overnight in small conical flasks in which the level of liquid is not greater than about 0·5 cm.

With amphibian muscle 1 c.c. of this extract is taken directly into the outer chamber of ' unit ' No. 2 and the chloride determined by the method already described for levels 35 to 7 μg. Cl.

With mammalian muscle, the fluid is deproteinized by adding to 10 vols. one vol. of 10% sodium tungstate and 1 vol. of 2/3 *N* sulphuric acid and filtering or centrifuging.

Calculation. The Cl obtained is multiplied by 20·3 for amphibian muscle, and by 24·4 for mammalian muscle, to give the Cl per gram of tissue.

Alternative procedure using the standard ' unit '.

A similar procedure is adopted in preparing the extract. The central chamber of the ' unit ' contains 1 ml. of 20% KI, and before closing the 'unit' 1 ml. of saturated permanganate is added to the 1·0 ml. extract. After placing the lid in position 0·5 ml. of 75% sulphuric acid (by volume) is introduced, etc. 0·5 ml. of the 20% KI from the central chamber at the end is added to 5 ml. of 0·2% starch, and the extinction determined after mixing, using a filter in or towards the red end of the spectrum.

The procedure is standardised by suitable controls.

(*e*) **Factors influencing the formation and absorption of chlorine in the above determinations.** If the conditions are varied from the above descriptions it is essential to know how changes in these conditions will affect the rate of formation and absorption of the chlorine. The factors determining these rates will therefore be very briefly discussed.

(1) *The time-absorption curve at constant temperature.* The process of oxidation by the acid permanganate takes appreciable time, so that the curve of absorption depends on two factors, namely the rate at which the chlorine is formed and the rate at which it is absorbed at unit concentration. The curve is nearly linear up to about 60% of complete absorption as shown in Fig. 60. It may be noted that the absorption of the last few % is very slow compared with the initial rate and is not explicable by the decreased concentration in the outer chamber. It is probably due to the slow release of a little dissolved chlorine from the fixative. Keeping to the conditions of the method described above, the absorption of the formed chlorine with its consequent iodine formation corresponds after 90 minutes at room temperature to 97·0–97·5% of the chloride added to the outer chamber.

Within the next hour the absorption increases slowly until all the chloride is absorbed as chlorine, but at the same time there has been a slight diffusion of iodine from the central chamber. This iodine diffusion is a constant amount of 1·2% per hour of the iodine in the central chamber. The net result is that from $1\frac{1}{2}$ to $2\frac{1}{2}$ hours after the addition of the acid to the outer chamber the iodine present in the central chamber corresponds to 97·0–97·5% of the theoretical equivalent of the chloride in the analytical sample. A factor of 1·03 is therefore required in an absolute calculation of the chloride from the thiosulphate.

(2) *The effect of the permanganate concentration in the outer chamber.* This effect is shown in Fig. 60. The rate of formation and absorption of chlorine from chloride is in practically linear relationship to the permanganate concentration in the outer chamber—at least from about 2% downwards. The outer

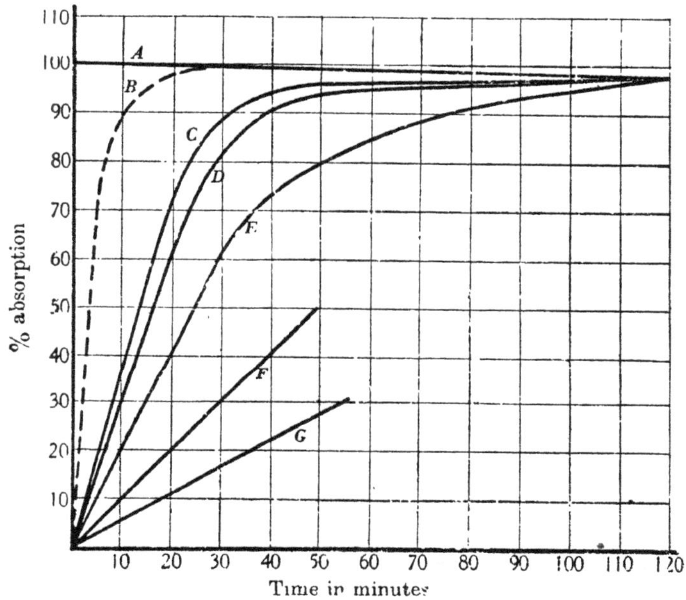

FIG. 60. Curves *D*, *E*, *F* and *G* represent the chlorine absorptions from the outer chamber containing 0·1 ml. of 0·1*N* HCl, 1 ml. of 25% (by volume) sulphuric acid and 1 ml. of 5, 2, 1 and 0·5% $KMnO_4$ respectively. Curve *C* is similar, solid permanganate being used and 2 ml. of 12·5% (by volume) sulphuric acid. Curve *B* represents the absorption of chlorine from 2·1 ml. of chlorine water in outer chamber. Curve *A* represents the fall in percentage value of free iodine in the outer chamber introduced before a blank determination (allowance being made for blank value). Experiments conducted at room temperature of about 18° (from the *Biochemical Journal*).

chamber in these experiments contained 0·1 ml. of 0·1*N* HCl, 1 ml. of permanganate solution (5, 2, 1 or 0·5%) and 1 ml. of sulphuric acid, 25% by volume (see curves *D*, *E*, *F*, *G* in Fig. 60). With one of the curves (*C*) solid permanganate was used, the volume being made up to 2·1 ml. by using 2 ml. of 12·5% (by volume) sulphuric acid. The figure shows also the percentage absorption rate of chlorine (curve *B*) from 2·1 ml. of acidified chlorine water.

Where excess solid permanganate is used (about 0·2 g. added to the outer chamber) the rate of chlorine formation is but little greater than where the concentration of the permanganate in the outer chamber is 2·4%. Water saturated with potassium permanganate at 18° contains about 6%, but even with the small quantity of fluid in the outer chamber appreciable time may be taken for full saturation.

(3) *The effect of varying the acid concentration in the outer chamber.* Fig. 61 illustrates the effect of changing the acid concentration.

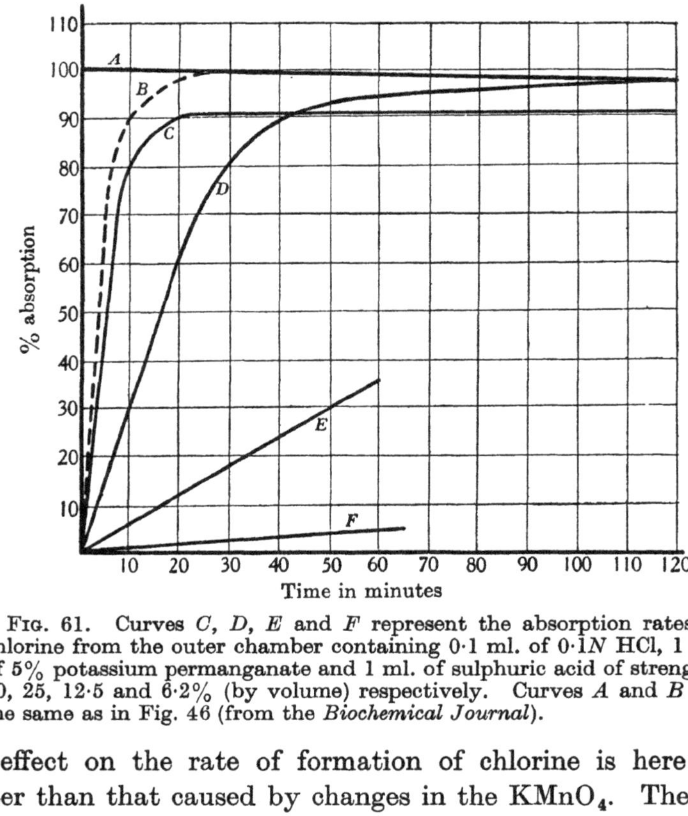

FIG. 61. Curves *C*, *D*, *E* and *F* represent the absorption rates of chlorine from the outer chamber containing 0·1 ml. of 0·1*N* HCl, 1 ml. of 5% potassium permanganate and 1 ml. of sulphuric acid of strengths 60, 25, 12·5 and 6·2% (by volume) respectively. Curves *A* and *B* are the same as in Fig. 46 (from the *Biochemical Journal*).

The effect on the rate of formation of chlorine is here much greater than that caused by changes in the $KMnO_4$. The figure shows that when the acid concentration reaches 29% by volume of sulphuric acid (see curve *C*) the absorption of the chlorine in the central chamber is very similar to that from chlorine water (curve *B*). At the same time a fall occurs in the total quantity absorbed. Below about 15% sulphuric acid this effect becomes negligible—the theoretical equivalent of iodine to added chloride being present in the inner chamber after two hours (see curve *D*), allowing for the small quantity of iodine diffused during this period from the 20% potassium iodide.

For such chloride determinations as here described, many other 'redox' systems with high characteristic potentials were also tried, but the acid permanganate was found the most suitable.

(4) *The effect of the fluid volume in the outer chamber.* It was shown for ammonia analyses as described in Chapter III that the percentage absorption rate of a given quantity of ammonia, other things being equal, is inversely proportional to the volume of the fluid in which it is contained. After the addition of alkali to the outer chamber all the ammonia is immediately liberated as gas, its tension being in inverse ratio to the fluid volume. With the chloride determinations, however, the absorption rate is much faster than the rate of formation with the acid permanganate concentrations used, and the amount of chlorine formed from a given quantity of chloride, under similar conditions, is independent of the volume in which it is contained. Consequently the absorption rate of chlorine is practically independent of the fluid volume in the outer chamber. However, it is found that when the volume exceeds about 3 ml. the last traces of the chlorine are more slowly delivered so that it is advisable to keep the volume below this figure.

(5) *The effect of temperature.* The determinations previously described were conducted at an average room temperature of about 18°. Judging from a few experiments where the temperature was increased to 36°, it would appear that there is an increase in the velocity of the whole process of 3 or 4% per degree rise in temperature.

The time for a determination could therefore be halved by conducting the process at 38° instead of at 18°. The fixative described above, however, cannot be employed for this purpose as it becomes too fluid at such raised temperatures.

On account of the small temperature effect the time for a determination as given above need not be altered if the temperature of the room falls to 14 or 15°.

(6) *Effect of changes in the dimensions of the absorbing apparatus.* The above times and quantities are for the standard 'unit' No. 1. With 'unit' No. 2 practically the same times hold, since the

vapour pressure of chlorine in solution is relatively high, and the diffusion rates will be largely independent of the surfaces, but will be much influenced by the depth of the layer in the outer chamber. Also in the present determination the rate of the reaction is much slower than the diffusion rate of chlorine gas, so that it largely determines the time of the analysis.

CHAPTER XLIII (*Fa* 3)

CHLORIDE (BY OXIDATION TO CHLORINE AND ABSORPTION INTO FAST GREEN)

THE use of the triphenylmethane dye, Fast Green (Gordon, 1952) makes the colorimetric procedure, in the lowest ranges more sensitive than that previously described above (Chapter XLII). The method is applied to chlorinated insecticides.

Procedure of Gordon

Conway microdiffusion units of borosilicate glass 30 mm. diameter with Lucite covers are used. The units are used with a plastic cell holder and an automatic shaker (obtainable from Microchemical Specialities Co., Berkeley, California). The use of this rocking shaker makes for quantitative diffusion of chlorine or bromine in 30 to 45 minutes, and ‘ results are highly reproducible even when reagent volume, concentration, viscosity, etc., are varied within wide limits ’. The fixative used is Chlorinated Petrolatum.

Chlorinated Petrolatum. A mixture of 9 parts of U.S.P. White Petrolatum and 1 part white paraffin wax is dissolved in benzene at 50° to 60° and chlorine gas is bubbled through for 30 minutes. The solution is then heated on a hot plate in a fume hood, to a final temperature of 150° to 160° to remove benzene and volatile products of the chlorination. The residue is cooled and used as fixative for the covers of the Conway units.

Six different analytical procedures are described by Gordon depending on the type of material to be analysed.

METHOD A.

Chloride in solutions—absence of other substances reacting with acid permanganate.

0·1 to 0·4 ml. (depending on the Cl range which is suitably 0·5 to 2·5 μg.) is pipetted into the outer chamber. The fixative is applied to the outer edge of the units, and 0·15 ml. of the 0·0167% Fast Green reagent is pipetted into the central chamber. Approximately 0·2 ml. of the acid permanganate reagent is then pipetted into the outer chamber and the units sealed. Gordon then uses the special shaker, for 30 to 45 minutes. A dozen units are usually employed together including reagent blanks, and estimations in duplicate. At the end, the contents of the inner chamber are transferred with washing to a graduated test tube and made up to the 5 ml. mark. The remaining unoxidised dye is determined by suitable colorimetry (Gordon uses a Klett-Summerson colorimeter with a set of 15 exactly matched, calibrated 5 and 10 ml. tubes, and a special filter with high transmittance above 600 mμ).

Calculation. A standard curve is given by Gordon in which the total micrograms NaCl is plotted against the colorimeter readings The total μg. NaCl is first determined for an unknown solution and that for the reagent blank is subtracted.

Reagents

Fast Green FCF, 0·0167% in 0·05 *M* sulphuric acid. 0·15 ml. contains 25 μg. of dye which is convenient for use.

Acid permanganate. Equal mixture of saturated potassium permanganate and 75% H_2SO_4 (mixed before a series of analysis) may be used as already described. If traces of Cl are present, evacuation by connecting with the water pump for 5 minutes or longer will greatly reduce such contribution to the reagent blank.

Notes on method

1. The colour of the dye solution fades somewhat in the first few days after making, but then remains stable for months.
2. The final solutions for colorimetry could, if necessary, be left in the tubes for 1 or 2 days, also after stopping the shaker the units could be left untouched overnight, if so required.
3. The range could be suitably extended if necessary by increasing the strength of the Fast Green solution.
4. If the dye solution in any unit turns yellow in 5 to 10 minutes on the shaker it should be replaced using less chloride in outer chamber.
5. Clearly the procedure is applicable to the No. 2 Standard

units, and shaking need not be used if longer times are allowed as already described.

METHOD B.

For chloride in solutions with the use of other substances reacting with acid permanganate.

After introducing the chloride solution as in Method A. 0·2 to 0·4 ml. of the 20% chromic–30% sulphuric acid is then pipetted into the outer chamber of each unit. Let stand 15 minutes to 1 hour. With organic matter present, the colour changes from orange to red-brown ; if it changes to green more chromic acid is added.

The subsequent procedure is as in Method A.

Reagent

Chromic acid solution. C.P., 20% (*w*/*v*) in 30% (*v*/*v*) sulphuric acid (low chloride). The chloride is removed by slowly heating to the boiling point in a fume hood.

Methods applicable to Organically Bound Chlorine

Alkali-labile chloride

It may be noted that about 10 to 30% of the weight of chlorinated insecticides is alkali-labile chlorine.

METHOD C.

(*a*) Into 10 ml. beakers pipette 1 ml. of the chlorinated compound in Cl-free benzene or other organic solvent. Add to each beaker 0·5 ml. of 2 *M* sodium propoxide in 1-propanol. Mix the solutions.

For the complete or nearly complete theoretical yield, the reaction is conducted at 95° for about 30 to 40 minutes. (The relative yield at room temperature at 60 to 90 minutes may be useful in characterising the insecticides.)

To obtain suitably the temperature of 95° C. a metal plate over a water bath is suggested, and the covering of each small beaker with a large one slows sufficiently the escape of the solvent (through the pouring lip).

Before the beakers go completely dry remove and allow to cool. Add 0·55 ml. of 1 *M* sulphuric acid which brings the *p*H to about 3 to 4. Wash down any dry sodium propoxide on the walls of the beaker.

Benzene extraction of organic solvent or other organic material

This may be carried out with small volumes of added benzene with shaking then filling to the brim with benzene, siphoning off and repeating.

The residual benzene and some of the water are removed by setting on the water bath for a few minutes. Residual volume is preferably 0·40 to 0·45 ml. This is transferred with a little washing to make a volume of 0·5 ml. (this may be done into a pipette with controlled suction). After suitable mixing an aliquot is taken and the procedure then is as for Method A or B.

Reagent

Sodium n-propoxide, approximately 2 *M* in 1-propanol is prepared by dissolving clean sodium metal in anhydrous 1-propanol (redistilled over sodium from Eastman, Kodak No. 848). The solution contains chloride ; most of this will settle out on standing. The clear supernatant is siphoned off ; this has a very low chloride blank. The solution gradually turns yellow, and becomes cloudy after repeated exposure to air, but does not interfere with its analytical use.'

Oxidant-labile chloride

It is pointed out (Gordon, 1952) that all the chloride can be liberated from an organic compound by oxidation with sodium peroxide in a Parr bomb or by other such procedures. At the same time some chlorinated compounds are readily oxidised by acid permanganate.

Gordon describes a Method D in which a measured volume of the substance in benzene or other solvent is evaporated (by exposure and shaking, assisted also by an air current) in the outer chamber and then oxidised with acid permanganate : as in Method A. Different strengths of oxidant have been investigated, one of which contained also perchloric acid.

Organic chlorine in extracts of biological material

For this Method C may be applied. The possibility of various losses has been discussed by Gordon, and the necessity of checking the process by recoveries of added chlorinated insecticides.

If more than 50 mg. of plant extractives is present in the 1 ml. aliquot of benzene solution for analysis, shaking four times with sulphuric acid in separatory funnels, followed by two washings with water is necessary. Method C can then be applied.

Analytical procedure for bromide

The principles described in the subsequent chapter have been utilised with some variation with respect to the differential use of oxidants.

CHAPTER XLIV (*Fa* 4)

BROMIDE

Principle and Introduction. The principle and general technique of the chloride determinations may also be applied to that of other halogens. Both free bromine and iodine are sufficiently volatile to require about the same absorption times as chlorine. Following the micro-bromide technique described in the present chapter bromide can be determined to an ultimate standard deviation of 0·08 μg. Br, which in milli-equiv. is about one-third that found with chloride. The standard deviation of the blank value may be expressed as 0·037 μg. Br, which is only about one-seventh of the equivalent value found with chloride. The difference probably arises from minute traces of chloride likely to be present in water and on glassware. The early papers on micro-determination of bromide have been reviewed by Bernhardt and Ucko (1925). Accounts of later developments may be found in the papers of Quastel and Yates (1934), Dixon (1935), Bertram (1933), Stoll and Brenken (1934), Indovina (1935), Leipert (1935), and Ucko (1936). Here reference will be made only to such previous work having immediate bearing on the present method and the results obtained for normal blood and urine.

Such micro-bromide determinations consist essentially of three stages. Firstly, the ashing of the material. Secondly, the oxidation of the bromide to bromine by a suitable oxidant. (These two stages may be combined in wet ashing methods, such as that of Leipert and Watzlawek, 1934.) Thirdly, the determination of the liberated bromine after extraction, aeration, etc.

In the present method these three stages are also followed for blood and urine, ashing being obviously unnecessary for purely inorganic mixtures or such as contain only traces of organic matter (e.g. sea water). The stages, however, are here carried out in a special way and permit the least absolute error of bromide with the least manipulation. Also, it is possible with this method to carry out large numbers of determinations at the same time.

Into the outer chambers of the ' units ' is introduced a dichromate oxidation mixture and in the centre 1 ml. of 20% KI in which the formed bromine is absorbed, liberating its equivalent of free iodine.

The ashing procedure. The ashing of the blood is carried out after preliminary removal of the proteins with methanol and subsequent evaporation of the cleared fluid. This reduces the incineration time to a few minutes and was introduced here for the reason that it is now practically certain that all the bromide in normal blood is present in inorganic form. If a complete ashing procedure is desired, the technique of Indovina (1935) seems very suitable. In this method magnesium oxide is employed to prevent bromide losses on heating.

The oxidation procedure. The oxidant found most suitable for bromide determinations is dichromate and sulphuric acid. The dichromate is used in powdered form and the sulphuric acid as 1 ml. of a solution containing 40 vol. % of sulphuric acid to 1 ml. of the ash extract. (Upwards of 48 vol. % may be used after bromide administration.) Acid dichromate of this strength has a scarcely detectable influence on 1 ml. of 1% NaCl even after 20 hours' action, whereas all the bromide is oxidised and absorbed in 2–2½ hours.

The use of a suitable acid permanganate mixture would have the advantage of rendering the method independent of the presence of both chloride and iodide. It was found, however, to have the disadvantage of causing marked losses as bromate, when the concentration of bromide for determination is very small.

The determination of the bromine formed on oxidation. In the present method the bromine diffuses into the central chamber of the ' unit ', liberating its iodine equivalent from 20% KI. The iodine is then determined as for chloride over the corresponding ranges.

In the method used by Leipert and Watzlawek (1934) for the determination of the bromide content of normal blood the formed bromine is oxidised to bromate after absorption into alkali, the bromate being then determined iodimetrically. An advantage of this procedure is that six times the equivalent of the original bromine is finally present as iodine. This variation could also be used with the ‘ unit ’, with possible advantage for the minutest bromide determinations.

The delicacy of the method selected, however, is so great that it is doubtful if a substantial increase in accuracy would result, and further operations with numerous reagents would be required. Also errors in the bromine formation and absorption are likewise multiplied by six.

Concerning specific colorimetric reactions for bromine, mention may be made of the use of the reagent of Guareschi (1913). This was used, for example, by Oppenheimer (1921), by Bernhardt and Ucko (1925) and by Indovina (1935). The reagent is basic fuchsin decolorised by sulphurous acid and is in fact the Schiff aldehyde reagent used by Guareschi for bromine detection. (Indovina decolorises with sulphuric acid.) It gives a violet colour with free bromine.

From Bernhardt and Ucko's account the reagent would appear to be far from satisfactory, and it may be noted that Ucko (1936, 1934) has abandoned it in favour of the fluorescein reagents (Swarts, 1889, as cited by Bernhardt and Ucko, 1925 ; Baubigny, 1897), which with free bromine form eosin, or tetrabromofluorescein. Using this latter reagent for the determination of bromide in normal blood and tissues, Ucko (1936) has obtained markedly different values from his previous results (1925).

It is likely that the fluorescein reagent could be also used with the ‘ unit ’, but does not seem to offer any advantage over the present method.

Procedure.

Inorganic.

(*a*) *Above 80 μg* Br. Into the outer chamber of a ‘ unit ’ is placed a small quantity of pure powdered potassium dichromate —roughly about 0·2 g.—by means of a spoon spatula, and into the central chamber 1 ml. of 20% KI. Into the outer chamber is now introduced 1 ml. of the bromide solution and the lid smeared with the fixative (see below) and placed in position. The ‘ unit ’ is slightly tilted by resting it on the edge of a spare lid, the lid is slightly displaced and 1 ml. of a solution containing 40% (by vol.) sulphuric acid is run in quickly without blowing—best from a simple tube with a mark at 1 ml. approximately. A solution containing 45% acid (by vol.) may be used when the Br/Cl ratio exceeds 1/500. After the introduction of the acid the lid is quickly replaced. The fluids in the outer chamber are well mixed by rotation. After 2 hours the free iodine in the central chamber is titrated with 0·05*N* thiosulphate from the special burette, the number of divisions on the burette being multiplied by the bromine equivalent or factor which may be determined as for chloride by making some initial determinations with standards or titrating standard iodine in 20% iodide in the central chamber. In this latter case the thiosulphate required is multiplied by 0·97 to allow for the slight constant loss of iodine from the central chamber during a determination.

(*b*) *Below 80 μg* Br. The procedure is carried out in a similar way to the previous method. 0·1 ml. 20% KI is used in the central chamber of ' unit ' No. 2. A little powdered dichromate is introduced into the outer chamber, and 1 ml. of the bromide solution. On placing the lid in position 0·5 ml. of 60% sulphuric acid is run into the outer chamber. After about 2 hours the content of the central chamber is titrated carefully with *N*/200 or *N*/400 thiosulphate from the Conway burette. Stronger solutions can be used with finer bored graduated tubes.

(*c*) *Below 16 μg* Br. The same procedure is adopted as in previous method, but at the end the procedure follows that described for Cl below 7 *μg* Cl. The determination is controlled by a curve made out for standard solutions under similar conditions. Like the Cl curve the relation between extinction and quantity is a straight line, but one which does not pass quite through the origin.

Ashing and extraction procedure for blood and urine.

Normal blood (*1 ml.*) Into a 15 ml. centrifuge-tube is pipetted 1 ml. of blood, 12 ml. of 95% methanol are added and the whole is well mixed. The tube is then capped and centrifuged for 5 minutes. 10 ml. of the clear fluid are pipetted into a small crucible (20–30 ml. capacity) and 0·2 ml. of a standard KBr solution containing 2 *μg* Br is added. The fluid is dried in the oven at

90–100° and then the crucible is heated in the Bunsen flame until the bottom and sides have been brought for about 1 minute to a dull red heat. It is then cooled and 0·5 ml. of distilled water is pipetted into the crucible, mixed and rotated around the walls. The solution is then pipetted into the outer chamber of a 'unit'. Another 0·5 ml. water is added and the process repeated. The presence of some carbon particles does not interfere. The determination then follows the inorganic procedure (above outlined) for quantities under 16 μg Br. The 2 μg Br are added to bring the amount analysed on to the linear curve. A blank determination is carried out in an exactly similar manner, including the small bromide addition.

Blood after bromide administration (*0·2 ml.*). A similar procedure is adopted. 5 ml. of 95% methanol are used and 3 ml. of the clear fluid pipetted off for drying. No bromide need be added. The mode of determining the free iodine in the central chamber will depend on the order of magnitude of Br expected. It may be stated that the ingestion of 2 g. KBr by the human subject will produce about 16–32 μg Br in 0·2 ml. of whole blood.

Urine. The same procedure for ashing and extraction may be adopted for normal urine as for urine after bromide administration. Since the bromide in urine very closely follows the chloride it is always advisable to carry out a preliminary chloride determination. The bromide concentration has in fact little significance apart from this chloride figure. Where the urine contains the average quantity or more of chloride, 2 ml. are used for ashing with correspondingly larger volumes with reduced chloride concentrations. 2 ml. urine are pipetted into a crucible and dried in the oven. The crucible with dried contents is now held over the Bunsen flame and cautiously incinerated. The contents char at first, then fumes are emitted, the residue finally melting as a rule when the bottom of the crucible turns red. At this stage the crucible is withdrawn from the flame and cooled. The whole procedure does not require more than a few minutes, and the bottom of the crucible is brought only for about 1 minute to a dull red heat. After cooling, 2 ml. of distilled water are added and mixed with the contents. After a short time 1 ml. is pipetted into the outer chamber of a ' unit ', the standard method being then followed. If it is found difficult to obtain 1 ml. free from many charred particles, 2 ml. of extract are centrifuged in a small tube, which procedure may be followed as a routine. After bromide administration the method above 80 μg Br can be used and the free iodine present in the central chamber titrated with the $N/20$ thiosulphate from the horizontal burette.

Notes on method and solutions required. No special care need be taken to free the analytical reagents used from bromide, since it is found that they contain none. But it is advisable to clean the ' units ', after the usual cleaning, with a little alcohol and ether, subsequently rinsing them with distilled water. For the smallest bromide quantities the ' units ' should not be exposed to direct sunlight during the absorption, and may be covered with a light cloth ; under these conditions the blank values will show no alteration even after 24 hours or more.

40% sulphuric acid. This is made up as a volume concentration, 40 ml. of the purest sulphuric acid (sp. gr. 1·84) being made up to 100 ml. with water. A certain latitude (of a few % at most) is permissible in the actual concentration, but the purest sulphuric acid must be used. Forty-five vol. % is advisable when the Br/Cl ratio exceeds 1/500.

20% potassium iodide. The purest reagent should be employed, and for the minutest bromide determinations it should be freshly prepared. When over 80 μg Br are being determined, a slight coloration is of no consequence as it will be allowed for in the blank.

0·2% starch. This is best prepared as described for chloride determinations in the previous chapter.

The fixative. This is also the same as for chloride.

0·02N thiosulphate. This is made up as a stock solution from N or 0·1N thiosulphate and stored in a brown bottle in the dark.

Standard bromide. 1·489 g. of pure powdered and dry KBr are dissolved in 1 litre of water. Then each ml. contains 1000 μg Br.

Dichromate. The purest potassium dichromate to be used ; for the lowest levels, a little chromate may be mixed (to about 1%).

Accuracy of the method. The accuracy for the higher quantities resembles that already found for chloride, 0·6% as a coefficient of variation being found in a large series of determinations of

800 μg Br. For the lowest quantities the bromide method is somewhat more accurate than the chloride owing to the smaller blank values—which arise only from the light absorption by the added starch with occasional traces of free iodine in the 20% iodide.

The effect of chloride and iodide on the bromide determinations. For biological determinations it is essential that chloride should have no appreciable influence since it will almost always be present in largely preponderating concentrations. With the above determinations in the 2 hours of bromide absorption the chloride oxidation is scarcely measurable when the chloride concentration in the outer chamber is of the order of 0·5–1·0% NaCl (i.e. 1–2% in the fluid examined). Even after 24 hours' action the amount represents less than 2 μg Br.

Iodide, on the other hand, interferes and will increase the free iodine in the central chamber. It is, however, readily removable. If to 1 ml. of the fluid in the outer chamber 0·2 g. dichromate be added and 0·1 ml. of N sulphuric acid, then by mere exposure of the 'unit' on the bench—the central chamber being empty—all the iodide will have disappeared in one and a half hours, the free iodine diffusing into the atmosphere. If the iodide present much exceeds 100 μg iodine, this exposure time should be extended.

Note on the bromide content of normal blood and urine. Using the above method, values for normal blood bromide varied between 227 and 572 μg Br/100 ml., the mean value being 372 μg/100 ml. The corresponding urine samples gave a mean value of 656 μg Br/100 ml. with a range of 297–855. The chloride of the urine samples was determined as well as the bromide, and the ratio of chloride to bromide (expressed as milli-equivs.) found to be 2150. The chloride in the blood samples was not determined, but if we assume that the mean ratio of bromide in the whole blood to bromide in the plasma is 1 : 1·12—a figure derived from Leipert's data (1935)—and that the concentration of chloride in plasma is rather constant and 10·35 milli-equiv./100 ml., then we may derive the mean value of Cl/Br for the plasma as 1990 (this corrects the original figure of 2180). Comparing this value with the corresponding ratio of 2150 for urine, it appears that, compared with chloride, the bromide in urine shows a very similar ratio with that in blood.

Pincussen and Roman (1929), supported by Zondek and Bier (1931), advanced the theory that normal blood bromide consisted largely of bromide combined in organic linkage and functioning like thyroxine as an important catalyst in the organism. The pituitary was thought to be the site of the formation of this substance. Its passage from the pituitary into the mid-brain was believed to be associated with the advent of normal sleep and the content of bromide in the blood to be much reduced in maniacal depressive states. The views of Pincussen and Zondek were based, however, on faulty methods as shown by Fleischacker and Scheiderer (1932).

Ucko (1934), who with Bernhardt (Bernhardt and Ucko, 1925, 1926) had obtained very high bromine values for the pituitary, has later stated that using a revised method he could not find more bromide in the pituitary than in other glands, and that no valid evidence exists for assuming the presence of an organic bromide catalyst or for the assumption that a gland is associated with its production. From the work of Quastel and Yates (1934), Dixon (1935) and in particular the recent detailed studies of Leipert (1935), and of Leipert and Watzlawek (1934), there are no valid grounds for supposing that the normal blood bromide is other than the inorganic substance. Leipert has shown that protein precipitation, dialysis or ultrafiltration gives no indication of a bromoprotein compound. Moreover, the bromine of blood is completely precipitated by silver nitrate, whereas organic bromide compounds of the type of dibromotyrosine are not precipitated by this reagent.

Dibromotyrosine is one of the two organic bromine compounds known to occur in nature, being found in corals, etc. (Mörner, 1913). The other is 6·6 dibromo-indigo, found to be linked with protein in *purpuridae* (Friedlander, 1907, 1922), and which appears to be identical with the ancient Tyrian purple.

Concerning the origin of the normal blood bromide, the greater part of this must come from the bromide associated with the natural chloride of the food, although this does not often exceed 20% of the total ingested. The Br/Cl ratio in commercial chloride has been found here not to exceed 1 : 7000 for four samples, whereas the Br/Cl ratio of sea water and of ordinary drinking water (e.g. Dublin Vartry water) is of the order 1 : 700 to 1000. Presumably the natural foodstuffs possess a similar ratio to the latter with regional variations. The commercial chloride taken would therefore dilute the ' natural ' ratio—the general limiting value of which may be presumed to be that of the sea.

The failure of the evidence hitherto presented for an organic bromide compound in the blood by no means removes biological interest from micro-determination of bromide. Among other considerations the study of the Br/Cl ratio may throw considerable light on the chloride metabolism as well as on the functioning of such organs as the kidney. From the clinical side also an easy and accurate method for carrying out micro-bromide determinations on very small quantities (finger blood after a single therapeutic dose) is of considerable interest.

The effect of varying the conditions in the above bromide method.

Varying the acid concentration. The effect is illustrated in Fig. 62, the rate of absorption from bromine water being also given for comparison (dotted line). It will be seen that changing the acid concentration has a marked effect of a kind similar to that shown for chloride. 20 minutes after adding the 40% sulphuric acid (as described above) 50% of the bromide has been oxidised and absorbed. When 20% acid is added only 4–5% of the bromide is recovered in the same time. In a similar way to the action of the permanganate system on chloride, a marked deficiency appears in the total amount recovered when the concentration of acid exceeds a certain limit (see curve *C*, Fig. 62). This deficiency, however, is not evident until the concentration of acid in the mixed fluids exceeds 24%. The concentration of acid should in fact be maintained within 24–19%, since a higher concentration than 24% causes a loss in the total amount recovered and less than 19% gives an unduly slow absorption.

In Fig. 62, curve *A* shows that a slight constant loss of iodine occurs from the central chamber (approximately 1% per hour of the total amount present).

The effect of changing the dichromate concentration in the outer chamber. The effect is much less than changing the concentration of acid. Similar results were obtained with the permanganate system and chloride ; but whereas the rate of chloride oxidation

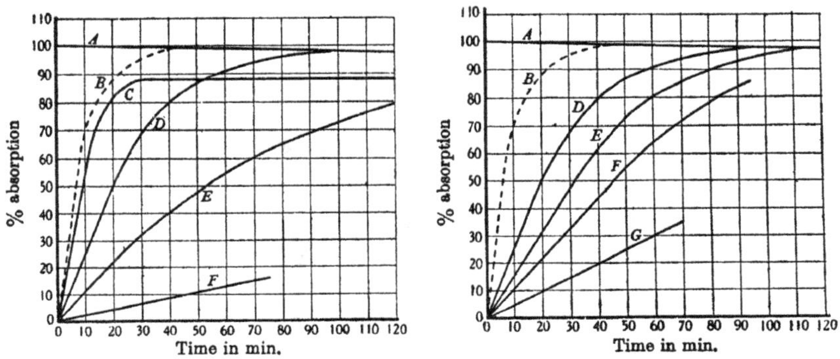

FIGS. 62 AND 62*a*.

FIG. 62. Curves *C*, *D*, *E* and *F* represent the absorption rates of the bromine from the outer chamber ('unit' No. 1) containing 0·1 ml. of 0·1*N* KBr, 1 ml. of saturated dichromate and 1 ml. of sulphuric acid of strengths 75, 40, 30 and 20% (by vol.) respectively. Curve *A* represents the fall in percentage value of free iodine in the outer chamber introduced before a blank determination (allowance being made for blank values). Curve *B* represents the absorption of bromine from 2·1 ml. bromine water in the outer chamber.

FIG. 62*a*. Curves *D*, *E*, *F* and *G* represent the bromine absorption from the outer chamber containing 0·1 ml. *N* KBr, 1 ml. of 40% (by vol.) of sulphuric acid and 1 ml. of saturated, half-, quarter- and one-sixteenth-saturated dichromate respectively. Experiments conducted at room temperature of about 18°. Curves *A* and *B* are the same as in Fig. 62.

seemed to be proportional to the permanganate concentration, the influence of varying the concentration of dichromate appears to follow a square root relation.

The effect is shown in Fig. 62*a*. It may be stated that saturated potassium dichromate at room temperature (18°) contains 9·8 g. dichromate per 100 ml. solution.

IODIDES AND HALOGEN MIXTURES

Iodides. Free iodine diffuses from the outer chamber of the unit into 20% potassium iodide in the central chamber in a similar manner to bromine or chlorine liberated by oxidation. In the presence of the other halogens the microdiffusion method may be found of value in determining iodide.

Procedure. Depending on the range of iodide (similarly to the other halogens), No. 1 'unit' or No. 2 or No. 2A 'units' may be used. The outer rim of the 'units' are smeared with a little liquid paraffin. Into the inner chamber is run 1 ml. 20% KI, or 0·1–0·2 ml. for the small 'units'. Into the outer chamber is pipetted 1 ml. of the fluid (0·5 ml. with the smaller 'units'), about 0·2 g. powdered dichromate and 0·1 ml. of *N* sulphuric acid. Immediately after adding the acid the lid is placed in position on the 'unit'.

The surface of the lid smeared with oil or fixative should be reduced as much as possible.

With minute quantities the use of the 'unit' No. 2A permits of an easy limitation of surface, the area around the holes in lid and on which the little glass cylinder reposes being smeared but not the cylinder itself.

After two hours on the bench the contents of the inner chamber are treated as for the other halogens.

Halogen mixtures. In a mixture of iodide and chloride the following appeared at first an advisable procedure for estimating the three halogens. (It must now be revised in the light of the findings of Öbrink and Ulfendahl.) The initially suggested technique was as follows.

Iodide is determined as in the above method. After the titration or removal of the central contents, the inner chamber is cleaned and dried with some filter paper and a fresh quantity of 20% KI added. The lids are placed again in position and 1 ml. of

22% sulphuric acid added to the outer chamber. The bromide is determined then as already described.

In determining the chloride with the standard 'units', 1 ml. vols. of the fluid are pipetted into the outer chamber of separate 'units', 0·5 ml. of saturated potassium permanganate added and then 0·1 ml. of 2*N* sulphuric acid and the contents mixed and exposed for two hours on the bench at room temperature. 1 ml. of 20% KI is then added to the central chamber, the lid placed in position and the acid in the outer chamber raised to 14–15% (by vol.) of sulphuric acid by adding 0·5 ml. of 60% sulphuric acid, and determining the chloride as already described.

Öbrink and Ulfendahl's Method for determining halogen in halogen mixtures using the microdiffusion and isotopic dilution techniques. These workers made a detailed study of the microdiffusion method on solutions containing only one halogen species and then on various mixtures using the Öbrink modification of the No. 2 Standard Unit (*vide* Chapter II). With regard to the method for single halogens they confirmed fully the microdiffusion method as described here. If two halogens were present their results, using the suggested procedure outlined above for halogen mixtures may be summaried in the following table.

Halogens present	*Results*
Iodide and chloride.	Both accurately determinable.
Iodide and bromide.	Iodide accurately determinable. Bromide gave low results.
Bromide and chloride.	Bromide accurately determinable. Chloride gave low results.

In mixtures of all three halogens bromide alone was accurately determinable using the procedure for halogen mixtures given above. Such difficulties could be overcome by combining the microdiffusion procedure with an isotopic dilution technique.

Principle of Method:

The principle of the isotopic dilution technique may be stated in the words of Öbrink and Ulfendahl, 'If the component that is going to be followed is thoroughly mixed with a component identical in all respects with it except for the substitution of a normal for a radioactive atom, then the fraction of radioactivity present in the final sample is identical with the fraction of the stable compound that has been recovered '.

Procedure. A known amount of a mixture of radioactive iodide, bromide and chloride was added to every sample of mixtures containing all three halogens. This extra volume (0·3 ml.) was thoroughly mixed with the test sample in the unit before the sulphuric acid was added. Each sample was analysed for iodide, bromide and chloride by the microdiffusion method. Thus with $K_2Cr_2O_7$ in a final H_2SO_4 concentration of 0·47 % the estimation of iodide was attempted. With $K_2Cr_2O_7$ in 20 % H_2SO_4 bromide should be oxidized together with some iodide (see Fig. 2) and with $KMnO_4$ in 12·5 % H_2SO_4 both chloride and bromide were expected to leave the middle ring. Because of the radioactivity it was not considered advisable to use the procedure described above, where the bromide was first allowed to escape into the air with the unit uncovered.

After two hours the inner ring was titrated as usual and then its content quantitatively transferred to plastic tubes for radioactivity measurements.

Calculation. The radioactivity measurements give the fractions (f) of each halogen that had diffused into the inner ring. The titration with $Na_2S_2O_3$ give the total number of μ-equivalents (n), which were absorbed in the KI-solution. Thus the following equations could be formulated where x represents the number of μ-equivalents of the halogen in the original sample. The superscripts refer to the I^-, Br^- or Cl^--units, the subscripts to the radioisotopes referred to.

$$n^{I} = f_{I}^{I} \cdot x_{I} + f_{Br}^{I} \cdot x_{Br} + f_{Cl}^{I} \cdot x_{Cl}$$
$$n^{Br} = f_{I}^{Br} \cdot x_{I} + f_{Br}^{Br} \cdot x_{Br} + f_{Cl}^{Br} \cdot x_{Cl}$$
$$n^{Cl} = f_{I}^{Cl} \cdot x_{I} + f_{Br}^{Cl} \cdot x_{Br} + f_{Cl}^{Cl} \cdot x_{Cl}$$

From these equations x_I, x_{Br} and x_{Cl} can be solved.

Special Solution:

A mixture was made of the radioisotopes by taking 1 part of I^{131} containing 1 μC/ml. and 1 μeq I^-/ml., 1 part of Br^{82} containing 1 μC/ml. and 0·8 μeq Br^-/ml. and 1 part of Cl^{36} containing 0·05 μC/ml. and 10 μeq Cl^-/ml. or 1 part of Cl^{38} containing 16 μC/ml. and

3·6 μeq Cl^-/ml. From this mixture 0·3 ml. were added to each sample in the units.

Note on fraction of I^-, Br^- and Cl^- from mixtures recovered in microdiffusion analysis:

(In the microdiffusion procedure used, bromine was not allowed to escape into the atmosphere. The two halogens, chlorine and bromine, being trapped together in the central chamber as given in the Procedure above.) These fractions are shown in the following table from Öbrink and Ulfendahl's paper.

The fractions of I^-, Br^- and Cl^- from mixtures recovered in microdiffusion analysis.

Number of μ-equivalents added to each unit.			*Fraction recovered in the inner ring.*			*The unit analyzed for*
I^-	Br^-	Cl^-	I^{131}	Br^{82}	Cl^{36}	
0·17	0·33	0·50	0·96	0·00	0·00	I
0·17	0·50	0·33	0·76	0·00	0·00	I
0·33	0·17	0·50	0·94	0·00	0·09	I
0·33	0·50	0·17	0·97	0·00	0·20	I
0·50	0·17	0·33	1·00	0·00	0·18	I
0·50	0·33	0·17	0·99	0·00	0·38	I
0·17	0·33	0·50	0·59	0·71	0·00	Br
0·17	0·50	0·33	0·62	0·88	0·00	Br
0·33	0·17	0·50	0·37	0·78	0·08	Br
0·33	0·50	0·17	0·61	0·75	0·00	Br
0·50	0·17	0·33	0·31	0·52	0·00	Br
0·50	0·33	0·17	0·44	1·00	0·07	Br
0·17	0·33	0·50	0·00	0·92	0·90	Cl
0·17	0·50	0·33	0·00	0·81	0·72	Cl
0·33	0·17	0·50	0·00	0·82	0·92	Cl
0·33	0·50	0·17	0·00	0·85	0·62	Cl
0·50	0·17	0·33	0·00	0·73	0·77	Cl
0·50	0·33	0·17	0·00	0·85	0·78	Cl

CHAPTER XLVI

SERIAL DETERMINATION OF ORGANICALLY BOUND HALOGEN

THE description here follows closely that given by Pirt and Chain (1952).

Principle. The compound containing halogen is heated in sealed tubes in a micro Carius furnace with fuming nitric acid in the presence of $AgNO_3$.

The silver halide so formed is dissolved in aqueous ammonia (or potassium cyanide in the case of the iodide) and reduced by hydrazine hydrate (or zinc powder with the iodide) to metallic silver and ionized halogen.

The latter is determined by microdiffusion in accordance with the principles already stated.

In the method as described the metallic silver was also determined. The two procedures give very similar results, the microdiffusion determination being slightly lower.

At the same time it was pointed out that the latter had the advantage of providing a means of estimating separately chloride, bromide and iodide in mixtures of all three.

Procedure of Pirt and Chain

(The determination of the metallic silver is omitted.)

Combustion of the organic substances

The combustion tubes were 90 mm. long by 7 to 8 mm. internal diameter, tapered for about 20 mm. at the bottom. Into these 1 ml. of the solution was introduced and rapidly evaporated by a current of air distributed through a manifold over the surface— the tubes being immersed in a water bath.

The halogen analysed is suitably of the order of 1·5 to 15 μmol.

Solid $AgNO_3$ (about twice the required amount) is added to each tube and 0·1 ml. fuming nitric acid added (with iodoacetate 0·1 ml. ordinary concentrated nitric acid is used).

By suitably tilting, the walls of the tubes are washed down with the acid. The tubes are then sealed with a blow-pipe and heated for 6 hours at 270–300° C. in a furnace placed in the fume cupboard. (It would appear that 0·5 hours at 300° C. is enough for

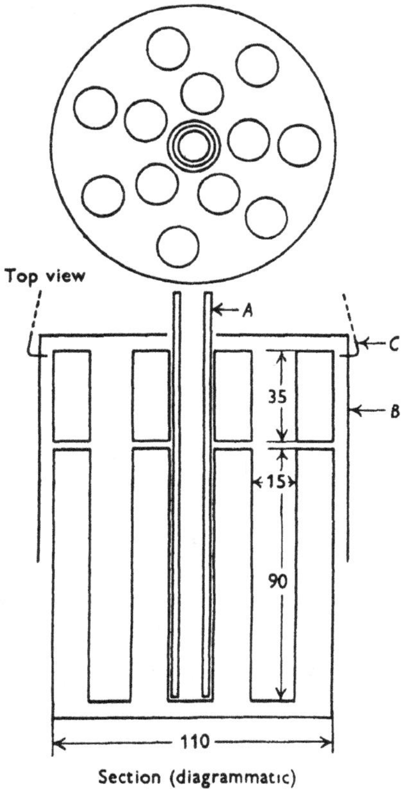

Fig. 63. Micro-Carius furnace. *A*, support for thermometer tube: *B*, tinplate shield ; *C*, wire support. All dimensions in mm.

most halogen compounds ; for precautions the reader is referred to Gatterman and Wieland, 1946.)

The type of furnace used was a bored aluminium block (Fig. 63) heated by a Bunsen burner with a disused tin plate container held above the block by a bucket handle type of suspension. (This acted as a safety device.)

Collection of Silver Halides

Silver chloride and bromide. After cooling and releasing the pressure the tubes are opened just below the shoulder by making a scratch and touching it with a hot glass rod. The HNO_3 is carefully diluted with water (1 ml.), stirred (using a thin drawn-out glass rod) and the tubes are heated on a water bath for 5–10 minutes to decompose the silver halide–$AgNO_3$ complex. After the heat treatment, the silver halide precipitate, which should be in a loose powdery form, is centrifuged into the tapered part of the combustion tube, first adding to the contents of the tube a drop of sodium laurylsulphonate solution which serves to prevent particles of the precipitate being held on the surface of the liquid. The supernatant is removed with a Pasteur pipette and the precipitate washed on the centrifuge in the same way with water acidified with HNO_3, stirring each time and adding a drop of sodium laurylsulphonate.

Silver iodide. The AgI precipitate and HNO_3 are diluted with water (1 ml.). To decompose the AgI–$AgNO_3$ complex which is much more stable than that of Cl^- and Br^-, a double-heating treatment was required. The tube is heated on a boiling-water bath until the precipitate is soft enough to be crushed by a thin glass rod with a ball end; the powdered precipitate was then heated for a further 15 minutes. The tubes are removed from the water bath and to the hot solution is added a little Na_2SO_3 (about 15 mg.) to reduce to I^-, any IO_3 which might have been formed. The precipitate is collected by centrifuging with sodium laurylsulphonate addition and the supernatant removed. Water acidified with HNO_3 is added and, after stirring, the precipitate is heated for a further period of 15 minutes on the boiling-water bath. The AgI is collected, washed once more with the dilute HNO_3 and finally with water to remove traces of acid.

Reduction of silver halides

Silver chloride and bromide. The halide is dissolved by stirring with conc. NH_3 (0·3 ml.). To the solution at room temperature a solution of $N_2H_4 \cdot H_2O$ (0·05 ml.) is added. (It is important to carry out the reduction at about 20°; if the solution is warm the Ag is not all precipitated but tends, in part, to form a colloidal solution). The solution was mixed by rotating the tube between the palms of the hands and after 2 minutes standing, water (about

0·5 ml.) and a drop of sodium laurylsulphonate solution are added. The precipitated metallic Ag is centrifuged down. The supernatant is carefully removed and collected in a dwarf tube with a Pasteur pipette. The Ag precipitate is washed once with water (1 ml.), without stirring, collected by centrifuging and the washing added to the first supernatant.

Silver iodide. The precipitate is dissolved in the aqueous solution of KCN (0·2 ml.). The reduction of the AgCN complex is carried out by addition of Zn dust (about 30 mg., the large excess being necessary to provide a large surface for the reaction) and stirring for about 1 minute. After the reduction, water (1 ml.) and a drop of sodium laurylsulphonate solution are added and the solution centrifuged. The I-containing supernatant was separated by means of a Pasteur pipette, the precipitate containing metallic Ag and excess Zn was washed once by stirring with water (1 ml.) and after centrifuging the wash-water is added to the first supernatant.

Determination of the ionized halogen by microdiffusion procedure.

Chloride ions. The supernatant solutions with washings are evaporated to dryness (using currents of air as mentioned above). The Cl^--containing residues were transferred to the outer rim of the Conway units in two portions of 0·5 ml. of water and the halogen was determined in the manner described by Conway (1947). A silicone fixative was used as small losses of halogen occurred with the normally used paraffin fixative.

Bromide ions. To the solution containing the Br^- was added 0·05 ml. 10% (*w*/*v*) KOH. The solution was evaporated to dryness and the residues were then heated in the oven at 100–110°, for 1 hour to remove all traces of excess $N_2H_4 . H_2O$ and then transferred to the Conway units in two portions of 0·5 ml. of water. If the heating is omitted the presence of excess $N_2H_4 . H_2O$ in the solution leads to N_2 production and the formation of a foam on the surface of the $K_2Cr_2O_7$ solution. This interferes with the diffusion of Br_2 from the outer rim into the inner chamber of the Conway unit. The KOH was added to convert NH_4Br to KBr, since it was found that heating the former salt alone caused a significant loss of Br^-, presumably due to volatilisation of the salt.

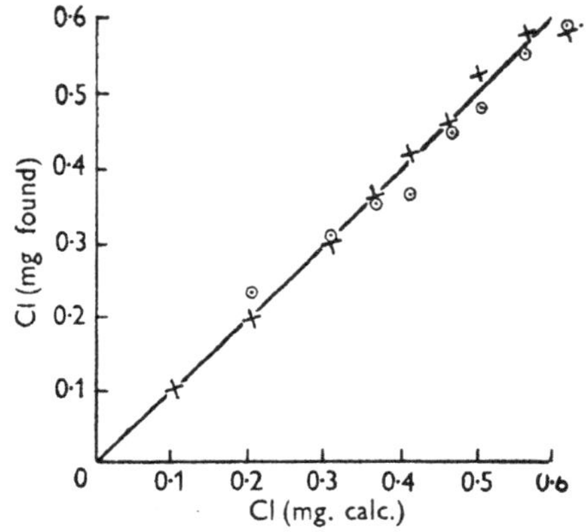

FIG. 64. Chlorine determinations, ×, *o*-Chlorobenzoic acid (argento-metric method); ⊙ *o*-Chlorobenzoic acid (Conway method). The line drawn corresponds to that for theoretical values.

The microdiffusion determinations were carried out as described in Chapter XLII–XLV. A silicone fixative was used.

Iodide ions. The tubes containing the I^- solutions are clipped in a water bath placed in the fume cupboard, 0·8 ml. 0·066 *M*-

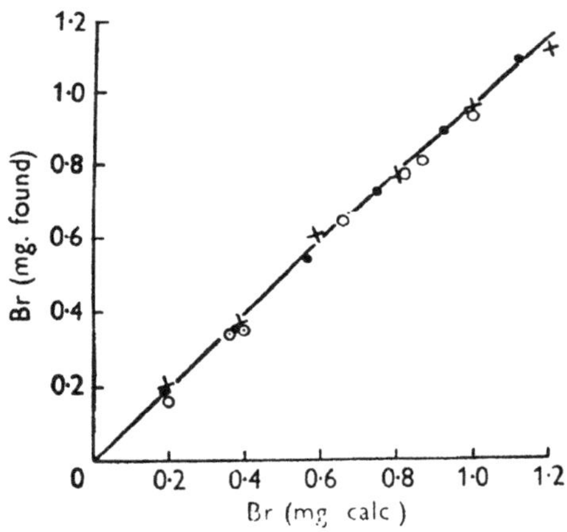

FIG. 65. Bromine determinations. ●, Bromoterephthalic acid (argento-metric method); ×, potassium *p*-bromophenylacetate (argentometric method); ⊙ potassium *p*-bromophenyl acetate (Conway method). The line drawn corresponds to that for the theoretical values.

phosphate buffer (*p*H 7) is added and the solution evaporated to dryness by the air-stream method. Thus the CN^- is removed by volatilizing it as HCN. The residues are extracted with two successive 0·5 ml. portions of water and transferred to the outer chamber of the Conway unit. In order to liberate the I_2 0·1 ml. 4·2 *N* H_2SO_4 is added, and the determination then carried out as described in Chapter XLV except that a silicone fixative is used. A control experiment showed that under these conditions Br_2 is not liberated from a solution containing Br^-.

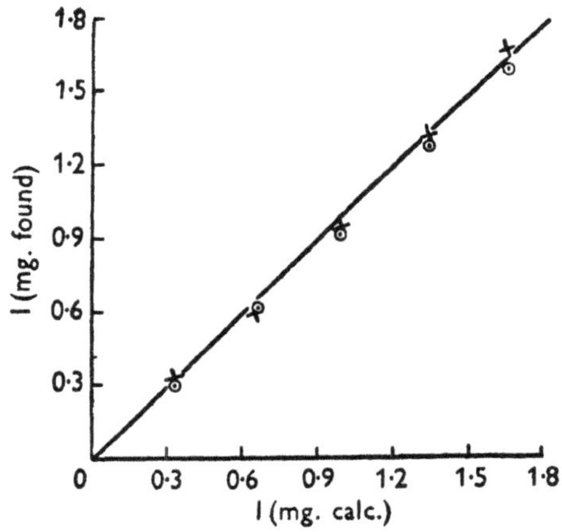

FIG. 66. Iodine determinations. ○, Potassium iodoacetate (argento-metric method); × potassium iodoacetate (Conway method). The line drawn represents the theoretical values.

Results of method

These are summarised in Figs. 64 and 66. In these the results of the metallic silver determinations, after reduction of the halides, are also included.

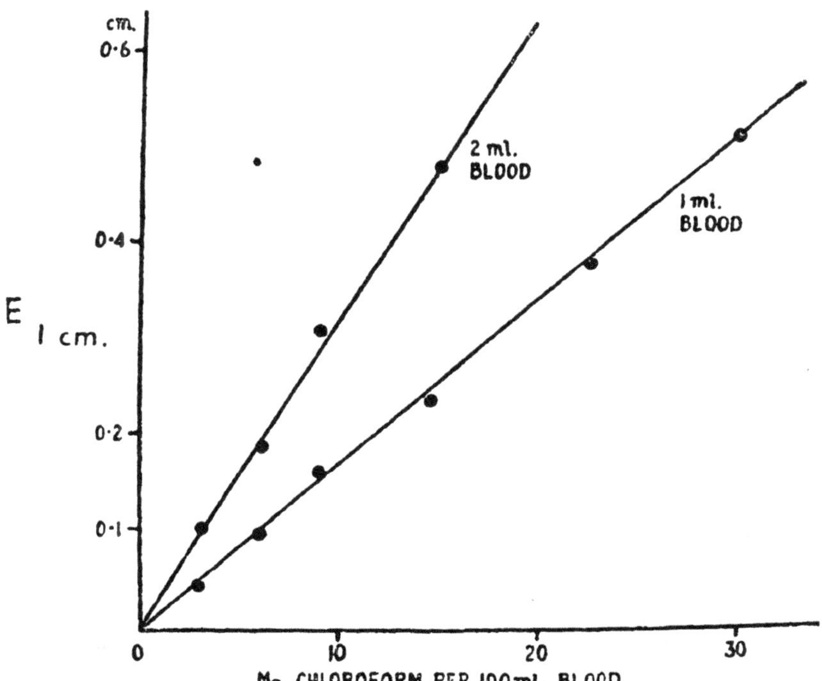

FIG. 67. Extinction values (in Burgen's mothod, 1948) against mg. chloroform per 100 ml. blood, (from Burgen, 1948). Courtesy of the British Medical Journal. (See page 324.)

CHAPTER XLVII

VOLATILE HALOGENATED HYDROCARBONS (CHLOROFORM, TRICHLORETHYLENE AND CARBON TETRACHLORIDE)

Determination of chloroform in blood. The account of this method follows the description of Burgen (1948) who showed that the microdiffusion procedure could replace the recovery of chloroform from blood by steam distillation with advantages as to time saving and economy of apparatus. Only 1 to 2 ml. blood are required for a determination. The method is applicable without modification to estimations in urine, cerebrospinal fluid and tissue extracts. (See also Feldstein and Klendshoj's system for the determination of volatile poisons as given in Chapter L).

Principle of Method : The chloroform diffuses from 1 to 2 ml. undiluted blood in the outer chamber into toluene in the central compartment. The vapour of chloroform is 100 times more soluble in toluene than in blood and when equilibrium is reached 96–97% of the chloroform has been transferred. For 1 mg. blood at 20° C., the transfer is 95% complete in 3 to 4 hrs. and in 15 mins. at 37° C. At the latter temperature it is 90% complete in 45 minutes and 98% complete in the same time with 2 ml. blood.

After the diffusion period the colorimetric reaction with pyridine and caustic alkali (Fukiwara, 1914) Daroga and Pollard, 1941 ; Habgood and Powell, 1945) is utilised.

Burgen's Procedure :

Toluene 0·8 ml. is pipetted into the centre compartment of a standard microdiffusion unit (No. 1) ; 1 or 2 ml. of whole blood collected with suitable precautions (no oil in the syringe ; narrow tubes filled to the top, and closed with a cork covered with silver foil) is rapidly pipetted into the outer compartment and quickly spread with the tip of the pipette. The cover, which has been smeared with fixative, is applied and the whole unit placed in an incubator at 37° C. for 50–66 minutes. The unit is then removed from the incubator, uncovered and 0·5 ml. of toluene from the centre compartment transferred to 5 ml. of pyridine in a test tube. Now 2·5 ml. of 20% sodium hydroxide are added, the tude agitated and then placed in a boiling water bath for precisely 5 minutes. It is then cooled in iced water, and when cool, 5 ml. of the supernatent pyridine layer are transferred to a clean tube and 1 ml. of water added to remove the turbidity. The resulting colour is stable for about one hour.

The colour is measured in a colorimeter with an Ilford micro-3 green filter, setting to zero with a blank. The intensity of absorption has a linear relation to concentration (see Fig. 55). Normally 1 cm. cells are used, but for concentrations below 5 mg. per 100 ml., cells of 2 to 4 cm. should be used, if available. The absolute values for the absorption vary somewhat with different batches of pyridine, but owing to the linearity of the absorption curve the slope of the line can readily be checked by running a few standards.

The colour developed with chloroform is purplish-red, that with trichlorethylene is orange-red, with an extinction about one-third that for chloroform. The blood levels during anaesthesia (6 to 12·5 mg. of trichlorethylene) are also lower than for chloroform and the use of 4 cm. cells are essential.

From the above description it is evident that the extinctions are readily determinable by such instruments as the Spekker Absorptiometer and the Pulfrich Photometer.

Accuracy of Method : Chloroform in concentrations from 5 to 35 mg. per 100 ml. (the concentrations likely to be found during and following anaesthesia) can be determined with an accuracy of plus or minus 5%.

Reagents :

Toluene : Pure toluene is shaken with concentrated sulphuric acid to remove thiophen, and then twice distilled, the middle fraction boiling at 115° C. being retained.

Pyridine : Analytical grade or redistilled. It must be quite clear and colourless.

20% *Sodium Hydroxide.*

Fixative : 5 g. of gum tragacanth are ground with 70 ml. of distilled water to a smooth jelly and 25 ml. of glycerol added.

Fb. CARBON MONOXIDE

CARBON MONOXIDE in blood was determined by Christman and Randall (1933) by absorption of the gas in palladium chloride solution, after the addition of ferricyanide to the blood to liberate the gas in free form. The reaction of the CO with the $PdCl_2$ may be represented as :

$$PdCl_2 + CO + H_2O = CO_2 + 2HCl + Pd$$

1 ml. of $N/50$ $PdCl_2$ solution being equivalent to $22{\cdot}4/100 = 0{\cdot}224$ ml. of CO at N.T.P.

After the absorption Christman and Randall flocculated the colloidal palladium with aluminium sulphate and filtered off, the excess palladium chloride being determined colorimetrically by addition of gum ghatti and excess potassium iodide and comparison of the resulting red colour with a suitable standard.

Wennesland (1940) described a method for CO determination in blood in which the absorbent was also palladium chloride after liberation of the free gas by dilute sulphuric acid. For the transport of the CO a diffusion apparatus was used consisting of two conical flasks, the mouths of which were joined by a short piece of rubber tubing. The excess palladium chloride was determined iodimetrically.

Gray and Sandiford (1946) adapted these two earlier methods for use with the standard unit.

The quantities used here are in accordance with their description, but instead of the iodimetric technique involving prior filtration and washings, a simpler titration method was used with more accurate results.

It will be noticed that in the equation above, two moles of hydrochloric acid are set free for each mole of carbon monoxide. Titrating this free HCl should give a ready method for determining the carbon monoxide absorbed.

In titrating palladium chloride solutions the most suitable end-point *p*H lies in the region 4·0–4·5 *p*H. Near to a *p*H of 5·0 a steep rise begins in the titration curve with considerable buffering, the magnitude depending on the concentration of the palladium chloride.

Among many indicators investigated the most suitable was found to be bromophenol blue.

Since the reduced palladium tends to form a dark colloidal solution, this seriously interferes with titration unless flocculated.

Gray and Sandiford flocculated with aluminium sulphate, but for the purpose of titration, magnesium chloride is much more suitable.

Procedure for determining carbon monoxide in blood (Conway and Ryan, 1948).

Into the outer chamber is pipetted 0·2 c.c. of 5*N* sulphuric acid, and 1 c.c. of palladium chloride of suitable strength (*N*/100 in *N*/1000 HCl, or *N*/200 in *N*/1000 HCl, both solutions incorporating 0·5% magnesium chloride; for choice of palladium chloride strength, see below).

The lids with fixative are placed in position and 2 c.c. of the blood diluted 1 in 4 (1 in 8 if over 15 vols. CO per cent are anticipated) are introduced in the usual fashion, into the outer chamber, the units being tilted slightly so that the blood does not come into immediate contact with the acid. On sealing the lids the blood is mixed with the acid. The units are left aside for $1\frac{1}{2}$ hours.

On removing the lids the units are tilted to one side and the tip of an 0·8 c.c. pipette is inserted under the film of palladium on the surface of the inner chamber, 0·8 c.c. are sucked up and transferred to a small flat-bottomed tube; 0·1 c.c. of 0·25% bromophenol blue solution is added and the yellow solution titrated with *N*/50 KOH from the micro-burette. As the titration proceeds the yellow colour fades, and at a *p*H of 4·0 the end-point is taken as the first appearance of a definite reddish-purple tint. A corresponding blank titration is made of the 0·8 c.c. of the palladium chloride solution used as absorbent.

Calculation.

If '*A*' c.c. represents the difference between the analytical and blank titrations in c.c. of *N*/50 alkali, then $A \times 0{\cdot}224$ gives the c.c.

of carbon monoxide (at N.T.P.) in the $2 \times 1/4 \times 0{\cdot}8$ (or 0·4 c.c.) blood. Hence $A \times 56{\cdot}0$ gives the vols. carbon monoxide per cent.

Solutions.

Palladium chloride mixtures : To make a litre of the $N/100$ palladium chloride solution 0·88 g. of palladium chloride is dissolved with heating in 100 ml. of $N/100$ HCl, 5 g. of $MgCl_2$ are added and the mixture made up on cooling to 1 litre, a similar procedure being used for the $N/200$ palladium chloride solution—0·44 g. of the salt being then used.

Brom-phenol blue solution : This is made up in 0·25 per cent strength.

$N/50$ KOH solution.

Notes on the method.

(1) *Choice of palladium solution* : If it is anticipated that the blood does not contain more than nine vols. per cent of CO the mixture containing $N/200$ palladium chloride should be used. If under 15 vols. per cent $N/100$ strength is used. With the usual blood dilution, and if over 15 vols. per cent the $N/100$ mixture but with a blood dilution of 1 in 8.

(2) *Diluting the blood* : For the 1 in 4 dilution, 6 c.c. of distilled water are introduced into a 10 c.c. test tube and 2 c.c. of the blood (shaken gently before removing the cork to ensure an even distribution of the red cells) are introduced under the surface of water in the test tube. This is then corked, but it is better not to mix the contents until ready for use. The mixing may then be done by gently inverting a few times.

(3) *Accuracy of the method* : Using the above method carbon monoxide in blood can be determined with a standard deviation for the single estimate of only 0·06 vols. per cent. It is doubtful if in practice the manometric method, which is much more troublesome can give a smaller variability than this. A number of determinations using the manometric procedure of Horvath and Roughton (1942) and the microdiffusion procedure on the same blood specimens (with approximately 10 vols. CO per cent) showed a slight difference in the absolute mean values. The manometric results were 0·17 vols. per cent higher.

(4) *Sealing the units* : It is important for the most accurate results that the units be well sealed with the fixative, and any not showing the transparent contact line all around between unit and lid should be discarded.

(5) When the lid of the unit is removed and the surface of the palladium chloride solution examined in a good light against a white background, if the CO of the undiluted blood exceeds approximately 0·5 vol. % a thin grey film of metallic palladium may be seen on the surface of the palladium chloride. With increasing quantities of CO in the original blood the film of palladium becomes more and more pronounced until with about 3 to 5 vols. % the palladium forms a definite metallic mirror on the surface.

(6) *Berka's modification* (1955). In this 0·2 ml. blood sample is used and the central chamber holds a small container holding the palladium chloride solution (see Chapter II). The excess palladium chloride is estimated colorimetrically as a coloured complex with p. nitrosodiethylaniline or else polarographically by reduction of Pd^{++} to Pd. $2M$ NH_4OH and $2M$ NH_4Cl were used as electrolytes in the polorographic method. Results obtained in trials of the method with both laboratory and field samples confirm its usefulness and practicability and compare favourably with those obtained by the Wennesland method on control samples.

Use of the reaction between CO and palladium chloride as a qualitative test for CO in blood.

From the preceding paragraph it will appear that the film of metallic palladium formed on the surface of the central chamber can be used as a qualitative test. If this only is required, the blood need not be diluted and using the small (No. 2) units with 0·2 c.c. of the $N/100$ palladium chloride mixture in the central chamber, and 0·5 c.c. of $5N$ sulphuric acid with 0·5 c.c. blood outside the appearance of a grey film of palladium on the central surface is evident (when the units are examined after $1\frac{1}{2}$ hours) with only 0·1 vols. CO per cent.

Oxygen Capacity Determination of Blood.

The oxygen capacity of blood can be determined by the well known method of measuring the carbon monoxide combining power. In the Van Slyke and Hiller manometric method pure carbon monoxide is used as the equilibrating agent, at a pressure of about 25 mm. Hg, the oxygen and nitrogen being extracted during the process. The carbon monoxide going into physical solution being corrected later in a blank analysis (the correction is very small for such pressures).

With the microdiffusion procedure pure carbon monoxide is most conveniently used at atmospheric pressure for the equilibration. This involves an allowance for the physically dissolved CO as high as 1·7 vols. per cent. Ordinary coal gas containing about 7% CO or 53 mm. Hg pressure, and no appreciable oxygen, was found more suitable. The amount physically dissolved may be neglected.

Procedure. About 4–5 c.c. of blood are introduced into a separating funnel of 200 c.c. capacity. The end of the funnel is connected with the coal gas mains, the gas being passed over the surface of the blood, the funnel being rotated during the passage. At intervals the gas is turned off, and the funnel corked but the rotating continued. After 4 or 5 minutes the funnel is finally corked, placed vertically and the blood allowed to drain down the walls for a few seconds, after which the blood is allowed to run quickly into a small tube previously filled with coal gas, and the tube corked.

The vols. CO per cent in this blood are now determined by the method described above (using $N/100$ palladium chloride mixture and 1 in 8 dilution of blood).

Calculation. If ' A ' is the number of c.c. of $N/50$ alkali difference between the analytical and blank units, then the oxygen capacity is given by

$$\text{Oxygen capacity} = 112 \times A \text{ vols. per cent.}$$

Comparison of results with the manometric method (**Van Slyke and Hiller, 1933**).

Sixteen determinations were carried out on seven different blood specimens both with the manometric and microdiffusion procedure. The agreement was excellent, the total mean values being the same and the maximum divergences for a given blood specimen being 0·5 vols. per cent.

A RAPID CLINICAL METHOD FOR CARBON MONOXIDE DETERMINATION

Principle. The time required for a visible blackening of the rim of a filter paper moistened with palladium chloride solution in the central chamber, with acidified CO containing blood in the outer chamber is noted. The approximate CO percentage of the blood is then read off from a nomogram (Lehmann, 1944).

Procedure of Lehmann. Into the outer chamber is run 2 ml. of 2 *N* H_2SO_4. Then on a filter paper which fits well the central chamber is dropped exactly three drops of *N*/100 palladium chloride solution. Care is taken that the moistened filter paper lies free of any folds or tuckering and without air bubbles underneath.

The lid, smeared with vaseline around a circle corresponding to the zone of contact with the unit, is placed in position and 1 ml. of citrated blood is run into the outer chamber. The lid is at once firmly closed, the unit rotated three times, and then left on the bench. After the rotation the stop watch is set going and the time required for a definite blackening of the rim of the filter paper is noted.

Calculation. The percentage saturation with CO is observed from Fig. 68.

Temperature correction for 20° C. The following table indicates this with sufficient accuracy.

Carboxyhaemoglobin determination according to Lehmann

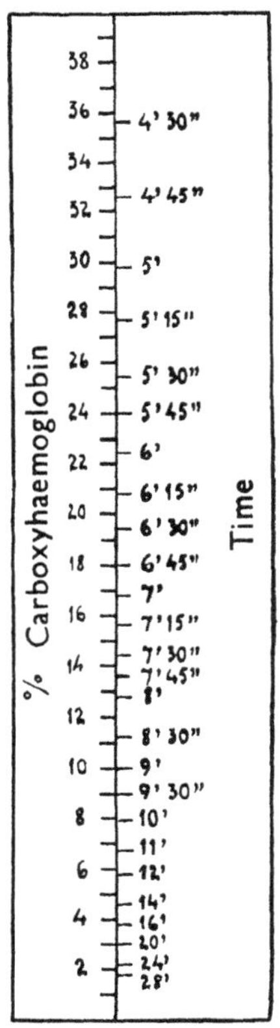

FIG. 68. Diagram showing the relation between time taken for blackening of the filter paper and the carboxyhaemoglobin content.

% CO *Saturation*	*Allowance for* 1° *C.* (% CO *saturation*)
30	5·0
25	3·1
20	2·5
15	2·0
10	1·4
5	0·7
2·5	0·4

Example. Blackening was visible after 7 minutes. This gives 17% CO saturation. Room temperature = 22° C. Hence

$$\text{allowance} = (22 - 20) \times 2 = 4$$

so that the approximate CO saturation is 13%.

Notes on method

1. The observation should be carried out under good lighting conditions.

2. The degree of error involved in the determination may be judged from the following table.

% CO *Saturation*	*Error range* (% *saturation*)
30	±5·0
25	±4·3
20	±3·8
15	±3·0
10	±2·0
5	±1·0
2·5	±0·5

This degree of accuracy is usually all that is necessary for clinical purposes, but if more is required the full procedure described in Chapter XLVIII may be carried out.

3. Symptoms of chronic poisoning appear when the percentage saturation is over 8 to 10%.

4. Tobacco smoking can produce 1 to 2% saturation.

CHAPTER L (G1)

G. VOLATILE SUBSTANCES OF TOXICOLOGICAL INTEREST

FELDSTEIN AND KLENDSHOJ'S SYSTEM FOR THE DETERMINATION OF VOLATILE POISONS BY MICRODIFFUSION (1954)

THE volatile poisons have been divided into four groups, based on their diffusion characteristics.

Group I. Volatile substances which reduce an acid dichromate solution.

This group includes :

Ethanol	Acetaldehyde
Methanol	Formaldehyde
Isopropanol	

A negative test based on the absence of colour change in the dichromate solution is evidence of the entire absence of the group. Ethanol can be accurately determined in the group when the presence of the others in any appreciable amount is excluded. For this purpose, the group has been subdivided :

Group Ia Volatile aldehydes and ketones absorbed in sodium bisulphite, and tested for separately.

This group includes :

Acetaldehyde	Formaldehyde
Acetone.	

Group Ib. Volatile alcohols (methanol and isopropanol) absorbed by dilute sulphuric acid.

Negative tests for members of groups I*a* and I*b* render the reduction of dichromate by ethyl alcohol more specific.

Group II. Volatile acidic substances absorbed in sodium hydroxide solution.

These include :

Cyanide	Sulphide
Phenols.	

Group III. Volatile substances which reduce an acid palladium chloride solution. This group is specific for carbon monoxide.

Group IV. Volatile substances which are more soluble in toluene than in aqueous medium. This group includes the halogenated hydrocarbons :

Chloroform	Carbon tetrachloride
Trichlorethylene.	

'Recovery of members of each group added to blood, urine and tissue slurry (1 in 5 dilution homogenised in the Waring blender) is excellent ; and the microdiffusion procedure has been shown to be rapid, convenient and specific. A complete analysis for the volatile unknowns may be completed in 2–3 hours with as little as 5 ml. of blood. Most of this time is taken up by the diffusion process, leaving the analyst free for other work.'

Group I

Dichromatic absorption.

Ethanol. This is determined as described in Chapter XXXII. 0·8 ml. blood or urine in the outer chamber, 2·0 ml. dichromate solution (3·70 g. in 150 ml. water, 280 ml. concentrated H_2SO_4 and dilute to 650 ml.) in the inner chamber and 1·0 ml. saturated K_2CO_3 as liberating agent in the outer chamber.

Diffusion time. 2 hours at room temperature for blood and urine ; 3 hours at 37° C. for tissue (4 ml. slurry).

(Though Feldstein and Klendshoj have got very good results using 4 ml. tissue slurry the use of 2 ml. would seem more advisable because of the faster diffusion rate and the less danger of spilling.)

Group Ia

Bisulphite absorption

Outer chamber - - - -	3 ml. blood or urine.
Inner chamber - - -	3 ml. 0·15 *M* $NaHSO_3$.
Liberating agent - - -	3–4 drops of 10% H_2SO_4.
Diffusion time - - - -	2 hours at room temperature.

(Tissue slurry—similar treatment if used in same volume.)

In each of 3 test tubes is placed 1 ml. of the $NaHSO_3$ solution from the central chamber.

Acetaldehyde. To the first of these tubes is added 1 drop of 4% $CuSO_4$ solution and 6 ml. concentrated H_2SO_4. This is followed by 0·2 ml. of *p*-hydroxybiphenyl with constant shaking. The tube is placed in a boiling water bath for 1·5 mins. and then cooled. In the absence of formaldehyde a violet colour is produced by acetaldehyde. If no colour is produced the absence of acetaldehyde and formaldehyde may be inferred. For the quantititative determination of acetaldehyde the optical density of the violet colour is read in a spectrophotometer at 540 mμ with a suitable reagent blank. The density is proportional to concentration over the range 2 to 10 μg. of acetaldehyde.

Formaldehyde. To the second tube is added 0·2 ml. of chromotropic acid (0·5% freshly prepared) and 4 ml. of concentrated H_2SO_4. After shaking, the tube is placed in a boiling water bath for 15 mins. and then cooled. The formation of a pink colour is due to the presence of formaldehyde. For the quantitative determination, the contents of the tube are diluted to 10 ml. with water and the optical density of the solution is determined in a spectrophotometer at 580 mμ against a reagent blank set at zero intensity. As with acetaldehyde the density is proportional to the concentration over the range 2 to 10 μg./ml. of $NaHSO_3$ solution.

(Recoveries of acetaldehyde or of formaldehyde by these methods are excellent.)

Acetone. To the third tube is added 4 ml. of 40% NaOH and 1 ml. of salicylaldehyde solution (0·1% in absolute alcohol). The tube is heated at 50 to 60° for 3 mins. and then cooled at room temperature. In the presence of acetone a red colour is produced. For quantitative determination the contents of the tube are diluted to 10 ml. with water and the optical density is measured in a spectrophotometer at 520 mμ. The optical density is proportional to the concentration of acetone over the range of 5 to 50 μg. Recovery of added acetone is excellent.

Notes on Groups I and Ia

When ethanol is being quantitatively determined the tests for formaldehyde, acetaldehyde and acetone may be done qualitatively but quantitative results for these may be carried out if desired.

Note that chromatropic acid reacts only with formaldehyde; whereas *p*-hydroxybiphenyl gives a colour with both acetaldehyde and formaldehyde.

Group Ib

Sulphuric acid absorption

Outer chamber - - - -	1·0 ml. blood or urine (or suitable volume of tissue slurry).
Inner chamber - - - -	2·0 ml. of 10% (*v*/*v*) H_2SO_4.
Liberating agent - - -	1·0 ml. of saturated K_2CO_3.
Diffusion time - - - -	2 hours at room temperature.

(If 2.5 ml. tissue slurry are used as given by Feldstein and Klendshoj, allow $3\frac{1}{2}$ hours.)

Methanol. The method for determining this is based on the oxidation to formaldehyde with permanganate solution and colour development with chromotropic acid (see Chapter XXXI).

Isopropanol. One ml. of the absorbing solution is placed in a test tube graduated at 10 ml. to which is added one drop of 5% $KMnO_4$. The tube is allowed to stand for 10 mins. after which saturated $KHSO_3$ is added dropwise until the permanganate colour disappears. One ml. of salicylaldehyde solution (0·1% in absolute alcohol) is then added, followed by 2 ml. of saturated KOH. The tube is heated in a boiling water bath for 2 mins. and then cooled. A red colour is an indication of isopropanol in the absence of acetone and acetaldehyde. For the quantitative determination see Chapter XXXI.

Group II.

Absorption in 0·1 *N* NaOH

Outer chamber - - - -	3·0 ml. sample.
Inner chamber - - - -	3·0 ml. 0·1 *N* NaOH.
Liberating agent - - -	3 to 4 drops of 10% H_2SO_4.
Diffusion time - - - -	2 hours at room temperature.

Cyanide. The method is based on the conversion of the absorbed cyanide to cyanogen chloride and reaction with methyl-phenyl-pyrazolone to form a blue colour. (See Chapter XXX.)

Sulphide. To one ml. of the absorbing solution is added 1 ml. of bismuth reagent (4·28 g. bismuth nitrate pentahydrate in 300 ml. glacial acetic acid and add 1·5 litres of water). Nitrogen gas is bubbled through the solution to remove oxygen. The absence of any brown colour is evidence for the absence of sulphide. If present, it may be quantitatively determined (see Chapter XLVII).

Phenols. To one ml. of the absorbing solution in a test tube is added 2 to 3 drops of 10% NaOH and 0·5 ml. phenol reagent (Folin and Ciocalteu, 1929). A blue colour shows a positive result. See Chapter XXX for quantitative determination.

Group III

Absorption in palladium chloride solution

Outer chamber - - - -	1·0 ml. blood.
Inner chamber - - - -	2·0 ml. PdCl solution (0·22 g. PdCl in 250 ml. 0·01 *N* HCl).
Liberating agent - - -	1·0 ml. 10% (*v*/*v*) H_2SO_4.
Diffusion time - - - -	1 hour at room temperature.

Carbon monoxide. Black film or precipitate in the central chamber indicates a positive result. For quantitative determination see Chapter XLVIII. The method forms an excellent screening test.

Group IV

Absorption in toluene

Outer chamber - - - -	1 to 4 ml. of biological sample.
Inner chamber - - - -	1·0 ml. toluene.
Diffusion time - - - -	2 hours at room temperature (or 1 hour at 37° C.).

Halogenated hydrocarbons. (Chloroform, trichlorethylene and carbon tetrachloride.) The method is based on the reaction of the halogenated hydrocarbon in the absorbing solution with pyridine and alkali (20% NaOH) to form a pink colour. (For quantitative determination see Chapter XLVI.)

CHAPTER LI (H 1)

H. OTHER MISCELLANEOUS APPLICATIONS OF THE MICRODIFFUSION PRINCIPLE
TOTAL MOLECULAR CONCENTRATION IN FLUID SAMPLES OF ABOUT 3–4 MILLIGRAMS

THE microdiffusion principle may be applied to the measurement of vapour pressure differences, and hence of molecular concentrations (Conway, Brady and Hingerty, 1945). In the method described here changes in the water content of a small amount of fluid in a little receptacle are determined by weighing. The container is placed in the central chamber of a microdiffusion ' unit ' (No. 2) and 1 ml. of a standard saline solution in the outer. To approach equilibrium in about 24–30 hours it is found necessary to incubate to 54° C. using only 3–4 milligrams of fluid. The diffusion of the water is very slow owing to the very small tension differences.

Method.

Containers. The containers for the small quantities of fluid are small cups of Pyrex glass weighing about 0·8 g. and having the following dimensions. The internal diameter is about 9 mms., external diameter 11 mms. and height 8 mms. They can be readily made from glass tubing. Each container is placed in the central chamber of a No. 2 ' unit '. To ensure spreading of the small quantity of fluid over the bottom surface of the cups, after cleaning, these are steeped in 0·05% saponin for about 15 minutes before drying.

Lids. Square pieces of plate glass about 1 cm. thick and 5 cm. square are used. These are preheated in the incubator to the required temperature and smeared with fixative before beginning a series of weighings.

Fixative. A mixture of vaseline and beeswax in proportions of 3 to 4 by weight is used.

Procedure. A number of 'units' containing 1 ml. of standard saline (1% for the calculations described below) are prepared and an equal number of lids are preheated and smeared with fixative in the incubator. Three (or more) 'units' are used for each sample. Each cup is weighed accurately to 0·01 mg. on a microbalance, being introduced for this purpose into a small tared weighing bottle. After weighing, a quantity of 3–4 mg. of the fluid for examination is rapidly introduced from a simple micro glass pipette drawn out to a capillary point, the small sample being blown out gently. The accuracy of the delivery is here of little consequence since the volume delivered will be weighed subsequently. Immediately after the delivery of the fluid, the cup is again introduced into the weighing bottle and re-weighed. After weighing, the little cup is placed in the central chamber of the 'unit' and quickly covered by the preheated lid, and the whole placed on a wooden board which had been previously in position in the incubator.

After 30 hrs. the little cup is removed as quickly as possible and its weight again determined. In removing the 'units' with cups from the incubator, the door of the latter is opened and closed as quickly as possible to avoid any marked fall in temperature. Special care must be taken in the transfer of the cup to the weighing bottle that no undue exposure to the air occurs.

Calculation. The calculation is made in terms of standard saline conditions, which then in turn can be referred to freezing point depressions, if necessary.

TABLE XVI

R.	NaCl *equivalent.*	*R.*	NaCl *equivalent.*
0·95	1·09	1·25	1·50
1·00	1·17	1·30	1·56
1·05	1·24	1·35	1·62
1·10	1·31	1·40	1·67
1·15	1·38	1·45	1·72
1·20	1·44	1·50	1·77
		1·55	1·82

The ratio for the sample—(wt. after)/(wt. before) referred to as *R*, is determined. Then if this ratio falls below 0·9 the saline concentration of NaCl as g.% to which the sample is equivalent is calculated as $R/0{\cdot}9$. Thus, if R is 0·73, the NaCl concentration giving the same vapour pressure as the sample is 0·73/0·9 or 0·81 g.%.

If the ratio is above 0·9, Table XVI gives the relation.

Accuracy of method. In accordance with the above calculation, a standard deviation of approximately 2·7% was found with 36 single determinations on solutions ranging from 0·5 to 2·0% NaCl.

For 11 determinations carried out in triplicate, ten gave values differing by no more than 2·8% from the theoretical and one gave 6·7%.

These figures were obtained using an incubator heated from the bottom, better results are anticipated when an incubator heated also from the sides is used.

Notes on the method.

(1) It is a very curious fact that when the fluid concentration falls below a value corresponding to 1% NaCl (present in the outer chamber) there is *more* than the theoretical loss of water. The reason for this is not clear, but, as will be seen, it does not prevent exact determinations carried out as above.

(2) When beyond 1% and up to 2% with the above method the gain of water is likewise *less* than the theoretical, the process being probably incomplete up to 2% NaCl (or similar concentration of other substance) in the sample.

(3) The accuracy of the determination depends on the uniformity of heating and, as above noted, an incubator heated from the sides as well as the bottom is advisable.

(4) A capillary pipette may be used for delivering the fluid for examination into the cups and, provided the amount lies between 3 and 4 milligrams, it is sufficiently accurate, since the actual amount is then weighed.

CHAPTER LII (H 2)

SEPARATION OF CRYSTALS AND "GUMS" BY MICRODIFFUSION

THIS account of the separation of crystals and 'gums' follows closely that given by A. L. Bacharach in the *Analyst* (1947).

The principle as first used by F. Tutin involved the separation of 'gums' from crystals on a porous plate, using a solvent but not wetting the mixture with the solvent.

In the original technique a fragment of unglazed plate or tile—from about 1 to 6 sq. cm. in area and of any outline, but preferably as flat as possible was used. On this fragment a quantity of the mixed crystals and gum is uniformly spread, in a layer a few mm. thick and the fragment is laid (as in Fig. 69) on a smooth surface, such as a glazed tile or a sheet of glass.

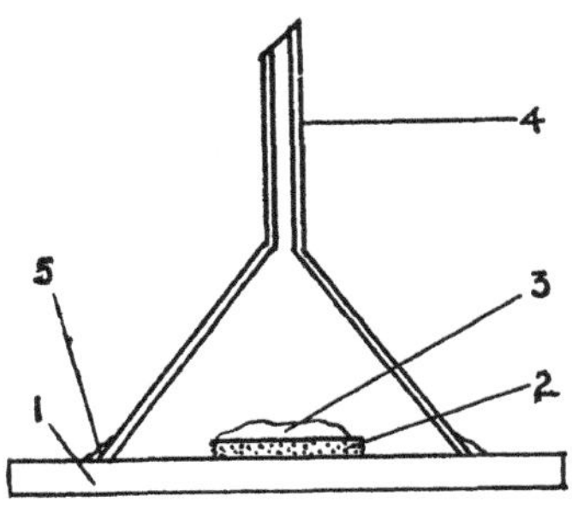

FIG. 69. Separation of 'gums' and crystals, older method (from Bacharach, 1947. Courtesy of the Analyst).

'An ordinary glass funnel is inverted over the fragment, with the stem upwards and open to the air. Solvent from a wash-bottle is then blown round the outside of the funnel where this makes contact with the smooth tile or glass, so as to give a sealing ring of solvent. When this has evaporated beyond a certain point, the ring will be broken and must be replenished with more solvent. It will be found that after a comparatively short time, depending on the solvent chosen the relative amounts of gum and crystals and so on, the gum has penetrated into the porous plate, leaving the crystals more or less " high and dry " on the surface and ready for further manipulative treatment.'

For further purification ordinary recrystallisation may then be applied. Also the gummy material (which may be of special interest) can be extracted in a Soxhlet or other suitable apparatus.

The loss of solvent and the need for continuous attention in its replacement—with consequent interruption of the procedure overnight—are disadvantages. These are avoided by the use of the standard microdiffusion unit or of a macro unit based on the same principle.

Bacharach's procedure with the standard unit.

This is exemplified by the separation of carotene crystals from some commercial ' carotene paste '. Spread on a tile the paste is placed in the central chamber of the unit, and a little acetone introduced into the outer chamber, the unit being sealed in the usual way. Almost complete separation is obvious in 24 hours or less, without any attention for replacement of solvent.

For work on a larger scale Bacharach suggests an arrangement of two glass flat-bottomed evaporation dishes, one inside the other, with an annular space between, to give what is in effect a macro ' unit ' which can be taken apart. Ether was found more rapid and effective than acetone in ' forcing ' the gum into the plate, although evaporation losses with this set-up may relatively, as well as absolutely, be rather greater than with the standard microdiffusion unit (No. 1).

Three or four such arrangements of two dishes and pieces of tile of about 10 sq. cm. would probably suffice for the overnight separation of a gram or more of carotene.

It is pointed out that such microdiffusion procedure provides a simple and rapid method for separating crystalline material from a more or less viscous fluid. Recrystallisation of such mixtures directly, with a view to isolating the crystalline material, may involve very appreciable losses, and possibly a tedious process of evaporation if the crystals happen to dissolve in the chosen solvent.

CHAPTER LIII (H 3)

SOME CONSIDERATIONS ON QUALITATIVE MICRO-DIFFUSION ANALYSIS

THE microdiffusion 'units' may be used for various qualitative tests, arising out of the quantitative methods described. Thus the absorption of ammonia in the central chamber may be observed by the change in colour of a very weak acid solution containing the methyl red-methylene blue indicator, or of the boric acid reagent with indicator as described in Chapter X. Carbon dioxide may be tested for by absorption into barium hydroxide solution producing a white precipitate, which may be made very delicate by using the No. 2 'units'. The halogens may be qualitatively examined by observing—as occurs in the methods described—the yellowing of the solution in the central chamber. The discrimination between chlorides, bromides and iodides can be carried out in a manner readily deducible from the quantitative description.

Other tests that may be performed in this way are as those used by Rossler as described below; namely, the detection of sulphite by iodine-starch solution, nitrate by potassium iodide, sulphide by lead acetate, etc.

In an article on "The Application of Diffusion to Micro-Analysis," Wyatt (1944) gives an account of certain early work on the identification of gases by diffusion. Rossler found that by the use of such apparatus as *A* (Fig. 70) it is possible to detect 0·02 mg. CO_2. He placed the sample at *a* and suspended a drop of baryta water at the base of the upper tube *b*, the outer tube being then dipped in acid. Rossler also used this apparatus for the tests mentioned above.

From Wyatt's account it will be seen that various modifications of Rossler's apparatus have been proposed, of which the simplest is probably that due to Molisch (1921) as described by Emich (1932). In this a small cylinder cut from a glass tube (ends ground plane and parallel) rests on a glass slide and supports another as in Fig. 70 *B*, which may also be replaced by a cover glass. The absorbing drop is at position *b* and the reactant at *a*. The arrangement has been improved by Mayrhofer's inclusion of a capsule *a* (*C*, Fig. 70), to contain the reactants, spray being retained by a glass cover on *a* with four holes for the passage of the gas or vapour. A piece of filter paper has been recommended for this purpose by other authors.

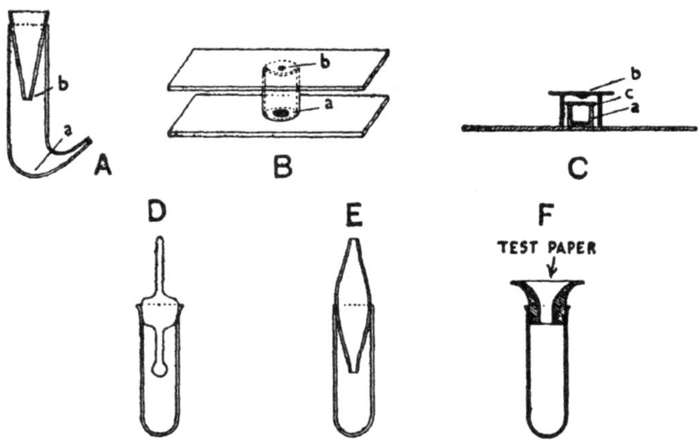

FIG. 70. Cells for qualitative microdiffusion analysis (from Wyatt's article in *Metallurgia*, 1944, p. 329).

Feigl and Krumholz (1929) used an arrangement as in *D*. A similar arrangement had been proposed by Emich in which a small test tube is closed by a cork carrying a platinum wire loop to hold the absorbing solution.

In *D* (Fig. 70) the cork is replaced by a ground glass joint, and the wire loop by a glass tube ending in a small bulb.

E is another variant due to Feigl (1937), while *F* is useful for the recognition of gases and vapours by their action on test papers. These are laid on the small ground-in funnel.

Such apparatus has been found useful in the *qualitative* testing of gases and vapours. On the *quantitative* side the early work of Schlosing and Shaffer has been referred to in Chapter II. The fundamental principles of the method at that time were not worked out and it was abandoned. In some designs of apparatus it was again approached but the transfer was thought of essentially as a distillation (or as a passage of water in which the vapour was entrapped) and not a diffusion.

In the following, an account is given of fine qualitative testing for acetone and alcohol, elaborated by Fearon and Webb (1936) and using the standard ' unit '. The description follows closely their account in the *Analyst* (1936).

The micro-detection of acetone and alcohol.

Fearon and Webb's procedure. Acetone. Two ml. of Nessler's reagent are placed in the inner compartment of the ' unit ' and 2 to 3 ml. of the solution under examination are placed in the outer compartment. The solution must be slightly acid in order to prevent the escape of any ammonia that may be present.

The apparatus is closed, kept at room temperature or incubated at 50° C. The presence of acetone is shown by the appearance of a pale yellow precipitate with Nessler's reagent. The method is extremely delicate. Acetone concentrations down to 0·01 per cent. give a reaction in less than a minute. At 0·002 per cent. the reaction is perceptible in about 5 minutes, and 0·0005 per cent. of acetone is detectable within an hour. When tested in this way samples of normal urines yield no precipitate even after remaining for 12 to 24 hours. If, however, a specimen of normal urine be strongly acidified with concentrated hydrochloric acid before being placed in the ' unit ', a precipitate appears within a couple of hours when incubated at 50° C. This was found to be due to the liberation of volatile mercaptans—a contingency overlooked by some of the previous workers who used Nessler's reagent as a test for acetone in urine (Deniges, 1930). For this reason, the urine should be acidified with a minimal quantity of dilute acid. When the solution under examination contains large quantities of acetone the precipitate in the Nessler reagent gradually redissolves in the excess of absorbed acetone vapour.

Alcohol.

Two ml. of a solution of 2 per cent. potassium chromate in nitric acid previously diluted (1 : 2) are placed in the inner compartment and 2 ml. of the liquid under examination are placed in the outer compartment of the apparatus. The ' unit ' is incubated at 50° C. and the presence of alcohol is shown by the development of a blue colour in the reagent as shown by W. R. Fearon and D. M. Mitchell (1932).

The nitro-chromic reaction is very delicate, and with a suitable dilution of the reagent will—as shown by Webb—detect the alcohol in 0·1 ml. of a 0·025 per cent. solution after 24 hours' diffusion (1933). Under these conditions, all specimens of normal urine examined showed the presence of minute quantities of alcohol.

The nitro-chromic reaction is, of course, not specific for ethyl alcohol, but responds to any compound containing the $—CH_2 . OH$ group. The reagent was originally shown by C. Ainsworth Mitchell to react to the presence of formaldehyde (1896), and for this reason the solution to be tested for alcohol should be treated previously with an excess of 5 per cent. silver nitrate and 20 per cent. sodium hydroxide to destroy any formaldehyde that may be present.

PART III

THE ERROR OF VOLUMETRIC TITRATION

CHAPTER LIV

INTRODUCTORY

THE error of volumetric titration is a composite of the glass and the chemical errors, each of which is in turn divisible into a constant and a variable error.

It is necessary to define exactly what is meant by these two kinds of error—constant and variable.

We may take for example a 10 ml. pipette. Such a pipette is made to deliver 10 ml. at some standard temperature and with certain conditions of delivery and drainage. The *constant error* may be defined as the deviation from 10 ml. of the mean of a large number of deliveries at the standard temperature and under the stated conditions. The *variable error*, on the other hand, may be defined as the standard deviation of any single delivery from the mean of a large number.

The total constant error of volumetric titration will be the arithmetical sum of the constant errors of the glass ware plus the chemical error expressed in the same units. The total variable error as considered below is not the simple summation of the different variable errors.

The manner of expressing the variable error. With uniform temperature and mode of delivery we shall suppose that the individual deliveries from a pipette are weighed up to a large number, the mean taken and the results grouped in small regular intervals around this mean value.

If the number falling within these intervals—as percentages of the whole number observed—are plotted against the midpoint of the intervals, we shall obtain a succession of points falling more or less closely along a special kind of curve as in Fig. 71.

This curve is described by the equation

$$y_x/y_0 = e^{-0\cdot 5\frac{x^2}{\sigma^2}} \tag{33}$$

where y_x is any ordinate and y_0 is the ordinate at $x=0$, at which point we may consider the position of the arithmetical mean to lie. The standard deviation is given by σ, and x is any deviation from the mean. The standard deviation for any distribution may be obtained by summing the squares of all the deviations from the mean, dividing by N the number of observations and extracting the square root. This may be written

$$\sigma = \sqrt{\frac{\Sigma x^2}{N}}. \tag{34}$$

When the number of observations is small (less than 20) N is replaced by $(N-1)$ in this equation.

Fig. 71 is known as the *normal curve*, or the *curve of error*, and its theoretical derivation will be found described in text-books on statistics. The area under the curve in Fig. 71 may be considered to represent the total number of observations and the area under any section as the number falling within the corresponding range of deviation. Normal curves of error of the kind we are considering, distributed symmetrically around the mean value, are defined only by the standard deviation and the magnitude of the mean, but by the standard deviation only when the mean is taken as the unit of measurement.

The variable error, here defined as the standard deviation of an observation from the mean, is not confined to normal distributions, but with the latter—to which the total variable error of titration will correspond reasonably well—it is always possible to say what chances there are of an observation falling outside a given multiple of the variable error or standard deviation.

The variable error so defined has certain properties of which the following are practically important.

(1) *Variable errors are uncorrelated except in certain special instances and will add not as their sum but as the square root of the sum of their squares.* As an illustration, if we were to take the sum of two successive deliveries from a 25 ml. pipette with a variable

error for the single delivery of 6·5 c.mm., the variable error of the sum is not 13·0 c.mm. but 9·2 ($=\sqrt{2}\times 6{\cdot}5$). In general

$$\sigma_s{}^2 = \sigma_1{}^2 + \sigma_2{}^2 + \sigma_3{}^2 + \dots \qquad (35)$$

(For the proof of this equation for uncorrelated deviations the reader is referred to statistical text-books.)

The equation is valid for any type of distribution of the deviations, but requires that one distribution should be uncorrelated

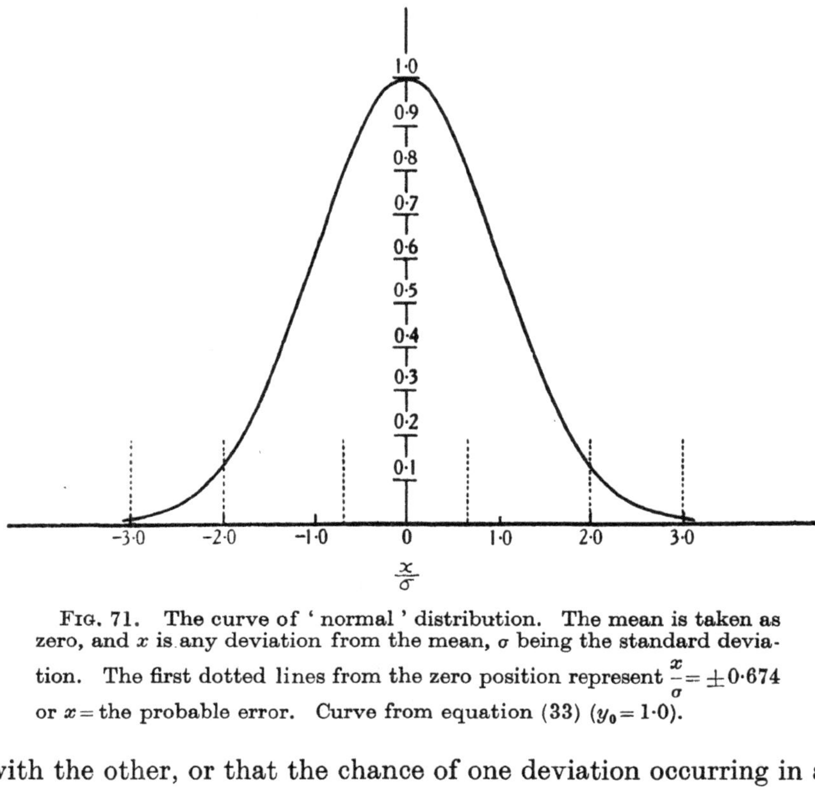

FIG. 71. The curve of 'normal' distribution. The mean is taken as zero, and x is any deviation from the mean, σ being the standard deviation. The first dotted lines from the zero position represent $\frac{x}{\sigma} = \pm 0{\cdot}674$ or $x =$ the probable error. Curve from equation (33) ($y_0 = 1{\cdot}0$).

with the other, or that the chance of one deviation occurring in a particular procedure should not influence the chances of a deviation in a second procedure.

That the pipetting of one volume will not affect the distribution in the pipetting of another volume is obvious enough, but that a pipette deviation should not influence the chances of a *special* burette deviation is not so clear, but may be presented thus. If we were to suppose that the burette delivered with complete accuracy, then the distribution of its delivery on titrating would follow exactly that of the pipette—when expressed in equivalents. When, however, the burette delivers with an error of its own this latter will add quite independently of the first error and will be uncorrelated. The total error as expressed by the burette readings will then be the sum of two uncorrelated errors and their addition will be expressed by equation (35).

Similarly factors causing the variable chemical error will be uncorrelated with those causing the variable glass error.

(2) *The variable error as defined above can be treated with a precision similar to that of the constant error.*

To translate the variable error so expressed into actual limits, we may consider that the total error of titration will be sufficiently close to a normal curve of distribution that the properties of this curve can be applied. We have then the following data :

TABLE XVII

Range of variable error ($x\sigma, \pm$).	*% of total obs. falling outside range.*	*Chances of an obs. falling outside range.*
1·5	13·4	1 in 7·5
2·0	4·5	1 in 22
2·5	1·4	1 in 70
3·0	0·27	1 in 370
3·5	0·046	1 in 2,180
4·0	0·006	1 in 16,700

Plus or minus twice the standard deviation or variable error which includes 95% of the distribution, may be taken as the practical limit of the range in titration (apart from the constant error). Three times the standard deviation may be regarded as covering all the possibilities, for it is to be noted that the curve of error here will not exactly correspond to the normal ; it will be caused by a limited number of factors which are not themselves expressible in their variability by normal curves, but rather, perhaps, by rectangular distributions.

For this reason, very small theoretical chances from the normal curve are to be taken as in fact equivalent to zero chance ; and it is likely that by altering our scale of limits this has a general application in nature. One obvious rectangular distribution we shall deal with here is that of the drop or end-point emergence.

Summation of a rectangular and a normal distribution. This is considered as occurring around the same mean and from uncorrelated factors. Occasionally a distribution can be of a rectangular instead of a normal kind. Such a distribution of error could arise, for example, if every other error were removed in macro-titration except that arising from the size of the drop at the end-point. Over the whole range of the distribution there exists the same probability of an individual observation falling.

The standard deviation of such rectangular distributions is given by

$$\sigma_r = \frac{l}{\sqrt{12}}, \tag{36}$$

where l is the length of the rectangle. The standard deviations of these rectangular distributions add exactly as those of the normal distributions. We have again

$$\sigma_s^2 = \sigma_r^2 + \sigma_1^2 + \sigma_2^2 + \dots, \tag{37}$$

where σ_r is the standard deviation of the rectangular distribution. The only disability that can arise in our treatment of the rectangular distribution is that we could no longer say that 95·5% of the observations would be included in $\pm 2 \times$ the standard deviation, though even here it may be noted that *all* observations in the rectangular distribution are included in $\pm 1{\cdot}73 \times$ the standard deviation. However, the summation of rectangular distributions of uncorrelated deviations around a mean very rapidly approaches the *normal* curve. As will be seen from Fig. 72, it is close to it after three similar summations around the mean, which in each case is represented in the figure as 1·0. Since in actual titrations the rectangular distribution of the end-point emergence will add to the reading error and to the delivery error when the latter is relatively appreciable, as also to the pipette error, and possible variable chemical error (each of which is either a normal curve or similar thereto), it will be seen that the properties of the normal curve may be applied in general to titration results.

Summation of the variable glass errors of pipette delivery and burette titration. All the variable errors in the titration procedure are finally registered as burette volumes. If we had no special burette error the burette readings would follow the variations of the pipette delivery, but as equivalent volumes. If the burette solution were twice as concentrated as the pipetted solution the volume variations observed would be only half those of the

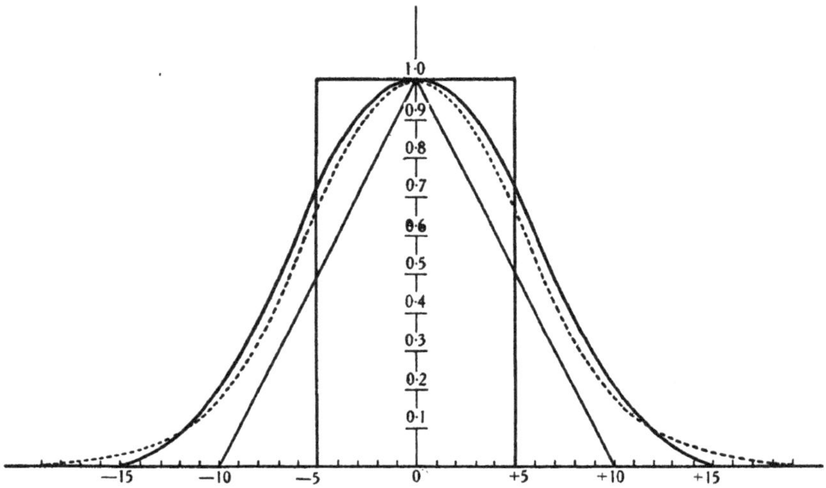

FIG. 72. Curves showing the summation of equal and uncorrelated *rectangular* distributions around the mean. The triangle gives the addition of two such. The dotted line is a curve of normal distribution with standard deviation a little greater than the companion curve for the summation of three rectangular distributions.

pipette. In general if σ_p is the variable error of the pipette delivery, this will appear as a burette volume of $\sigma_p \dfrac{V_b}{V_p}$. If then σ_b is the special variable error of the burette itself we have the total variable error σ_s given by the equation :

$$\left(\sigma_p \frac{V_b}{V_p}\right)^2 + (\sigma_b)^2 = \sigma_s^2 \tag{38}$$

or

$$\left(\frac{\sigma_p}{V_p} \times 100\right)^2 + \left(\frac{\sigma_b}{V_b} \times 100\right)^2 = \left(\frac{\sigma_s}{V_b} \times 100\right)^2, \tag{39}$$

from which it appears that *the pipette and the burette errors can be added as coefficients of variation.*

Also by multiplying the equation (38) across by c_b^2 (c_b = burette concentration), we obtain

$$(\sigma_p c_p)^2 + (\sigma_b c_b)^2 = (\sigma_s c_b)^2, \tag{40}$$

so that the *variable errors may also be added as equivalents.*

In a similar way the variable chemical error—which becomes appreciable when the end-point lacks definition with dilute solutions—is read as a burette volume with uncorrelated pipette and special burette errors, the three summing as equivalents or coefficients of variation.

A numerical effect of variable errors adding as the square root of the sum of their squares. If two variable errors are added one of which is half the other, the addition gives

$$\sqrt{\sigma^2+(0\cdot5\sigma)^2}=\sqrt{1\cdot25\sigma^2}=1\cdot12\sigma, \tag{41}$$

the first error being increased only 12% by the addition. Similarly if the second error is one-third or one-fourth the first the addition gives an increase of only 5 and 3% respectively. *In the treatment of small variable errors in general, one may neglect those constituent errors the sum of which is less than one-third (or even less than one-half) the whole error.*

Precision in the estimate of the variable error from small numbers. So far we have been considering variable errors determined from distributions so numerous that no appreciable error occurred in the determination of the standard deviations. For the small numbers with which it is here only practicable to work—groups of 25 in general—the error in the determination of the standard deviation is appreciable. It becomes of no importance, however, when a multiple of the standard deviation or total variable error is taken in the assignment of limits, since for such limits we may take twice or three times the variable error according to our convention. The error in the determination of the standard deviation of a normal distribution, and expressed itself as a standard deviation, is given by the equation

$$e_\sigma=\frac{\sigma}{\sqrt{2N}}\cdot \tag{42}$$

For 25 results, therefore, this amounts to $0\cdot14\sigma$. Hence two variable errors, determined each for 25 observations and one of which is 1·28 times the other, while very probably different (the odds being about 20 to 1 in favour), cannot be regarded as very significantly so.

In the subsequent account, all variable errors mentioned as actual figures have been determined on distributions of about 25. To avoid undue repetition of the plus and minus sign and of the associated error, this has been omitted, but is to be understood as existing throughout the discussion.

CHAPTER LV

THE VARIABLE GLASS ERROR

THE PIPETTE

In delivering fluid from a pipette of ordinary type, with or without bulb, the chief sources of the variable error are the variation in the amount of fluid left behind on the walls, the variable adhesion of droplets to the delivery tip and the error in the adjustment of the meniscus to the mark. This last error is usually negligible compared with the others. For example, using a 2 ml. standard pipette, a range of ± 1 mm. variation of the meniscus could be allowed before producing a 10% effect on the total standard deviation of the pipette delivery. The meniscus, however, could be adjusted to within one-tenth of this range.

Through the whole range of pipettes, of usual standard type, or simple straight tube, the meniscus adjustment to the mark—provided this is carefully done and against a uniform background—will be therefore regarded as a negligible factor in the variable pipette error.

(A note on the syringe pipettes is given towards the end of the chapter.)

The second error source mentioned—arising from adhesion of droplets to the delivery tip—is important in the case of micro-pipettes, and may produce a relatively large increase in the error if not provided against by a fine delivery tip or nozzle.

Determination of the total variable error of a pipette delivery. The determinations recorded here have all been obtained with pipettes cleaned in the usual way by soaking overnight in strong sulphuric acid and bichromate mixture, well rinsed, and treated as a rule with absolute alcohol and ether before drying in an air current or in the oven.

The error was obtained using the pipette as if in an actual delivery for analytical purposes, the filling being done by sucking and the volume controlled to the mark by the index finger. For efficient control of the meniscus it may be mentioned that the finger should be slightly moist, neither very dry nor definitely wet.

The fluid was sucked to a little above the mark, the pipette tip removed from the fluid and the meniscus allowed to descend slowly to the line—against a uniform background with the eye at the meniscus level. The tip of the pipette was then touched against the glass wall of the vessel containing the stock fluid, and the pipette quickly turned into the horizontal position, the outer glass surface at the end of the pipette being cleaned with a little clean filter paper. The fluid in the pipette was then allowed to run free into a large tared weighing bottle, the pipette being vertically held with the tip above the fluid in bottle. The instant the flow ceased was noted and subsequently either of the following procedures was at first tried :

(*a*) The tip was pressed against the glass wall of the receiving vessel immediately on the cessation of delivery and the drop blown out sharply, rotating the tip in contact with the glass and immediately withdrawing.

(*b*) The tip was held for 15 seconds against the wall of the container, the latter being held so that it was slightly slant to the vertically held pipette.

(*c*) The tip immediately after delivery was held for 15 seconds just touching the fluid received into an originally dry container.

It was found that no appreciable difference exists between (*a*) and (*b*), though it is generally supposed that blowing out the drop gives a higher error—e.g. Kolthoff (1931), Schloesser (1904) (but not Ostwald, 1902)—and this no doubt is true if the drop be blown out without due regard to time and without touching the glass wall in the manner indicated. The third method gave the highest error, though Wagner has not found it so (1898), and the difference is not so great but that it may be held to come within the sampling error.

As will be shown subsequently, there are definite reasons why (*a*) and (*b*) methods should give practically identical results and why the third method may be expected to give a higher value for the variable error.

The investigation of the three methods of delivery was carried out chiefly on the 25 ml. standard pipette, but since method (*a*) is not inferior to (*b*) and has the merit of saving time, it has been adopted in the systematic studies here recorded. It has been found by the author superior to the second method when using micro-pipettes.

(1) *Method of direct weighing.* In this determination of the variable error 25 careful weighings and deliveries are carried out, and after each delivery the fluid is removed by another similar pipette and the receiving flask or weighing bottle weighed again. The determination is always made on a series carried out in succession and under conditions as nearly similar as possible. These latter include attention to uniform water temperature, particularly necessary with large pipettes. From the weights of the deliveries the standard deviation or variable error is given by

$$\sigma_p = \sqrt{\frac{\Sigma x^2}{N}}, \tag{43}$$

where Σx^2 is the summation of the squares of the deviations (x values) from the arithmetical mean, and N is the total number of observations. The use of $(N-1)$ in the denominator instead of N as is advised for small numbers is an unnecessary refinement with 25 values. The variable error where not given as a coefficient of variation (% of the mean) may be conveniently expressed in cubic millimetres, which for this purpose may be taken as equivalent to milligrams of water.

(2) *Method of duplicate differences.* This is a method specially useful with the smaller pipettes and having certain advantages. Two similar weighing bottles of rather large size (30 ml. or more) provided with stoppers are balanced exactly one on each pan of the analytical balance. A pipette delivery is then made into each in turn and the two bottles again balanced, the difference in weight being accurately noted. This difference for the 25 ml. pipette, for example, will rarely exceed 20 milligrams. A number of such observations are made and the standard deviation of the single delivery—the variable error—is determined from the formula,

$$\sigma = \sqrt{\frac{\Sigma x^2}{2N}}, \tag{44}$$

where x is here the difference between the members of any duplicate and N the number of duplicates. The method has the following advantages.

(*a*) The determination is practically independent of small temperature changes and a series may be interrupted for any length of time.

(*b*) It is uninfluenced by small errors in the larger weights—if such errors exist—and the differences are generally covered by the rider on the milligram scale. For demonstration of the variable error to students it is particularly illustrative.

(3) *Method of titration.* This method may be used for very small pipettes, such as the 0·1 ml. For example, successive deliveries of 0·1 ml. of $N/1$ hydrochloric acid are made from such a pipette and titrated with $N/10$ carbon dioxide-free alkali from a standard 2 ml. Bang burette. A similar series is carried out delivering 1 ml. of $N/10$ hydrochloric acid and titrated in the same way. The following equation then applies

$$v_{p_1}^2 - v_{p_2}^2 = v_{r_1}^2 - v_{r_2}^2, \tag{45}$$

or the difference of the squares of the coefficients of variation for the 0·1 and the 1·0 ml. pipettes is equal to the difference between the squares of the coefficients of variation for the two sets of titrations. This must be true since every other factor in the two groups of titrations is the same apart from the pipette errors. At such concentrations also the variable chemical error does not enter.

From the coefficient of variation so determined (v_{p_1}) the standard deviation in c.mm. is obviously given by

$$\sigma_{p_1} = 10 V v_{p_1}, \tag{46}$$

where V is the mean volume of the delivery in ml.

Such a method gives the variable error of the total pipette delivery into a vessel. As mentioned above, small pipettes may show a comparatively large increase in error due to adhesions of minute droplets to the external wall of the delivery tip. This is prevented by the use of fine delivery tips.

The variable pipette error obtained by the above methods.

The error with conventional delivery times. For this the 25, the 5 and the 2 ml. standard pipettes were examined. The pipettes were of a type corresponding to the requirements of Class A test in the National Physical Laboratory, the tip being constructed so that its terminal surface is at right angles to the axis of the tube, being ground down for this purpose and forming a superior finish to the rounding off in the blowpipe flame.

The following table gives a summary of the results obtained :

TABLE XVIII

Pipette.	*Delivery time sec.*	*Internal surface sq. cm.*	*Variable error c.mm. 25 obs.*	*Coefficient of variation.*
25 ml.	29	75	6·3	0·025
5 ml.	22	30	2·6	0·052
2 ml.	17	18	1·8	0·090

It would appear from the table that as a working rule the variable error of the standard pipettes may be written in c.mm. as

$$\sigma_p = 1{\cdot}3\sqrt{V} \tag{47}$$

and the coefficient of variation

$$v_p = \frac{0{\cdot}13}{\sqrt{V}}, \tag{48}$$

V in both equations being expressed as millilitres. Also the variable error may be seen to go approximately as the internal surface.

The effect of drainage on the variable error of standard pipettes. The following table (Table XIX) illustrates the results from three different kinds of procedure after delivery—mentioned at the beginning of the chapter.

TABLE XIX

Pipette.	*Delivery time secs.*	*Variable error c.mm. (25 obs.).*	*Procedure after delivery.*
25 ml.	29	6·3	Drop blown out as described
25 ml.	29	6·4	15 seconds' drainage, tip touching wall
25 ml.	29	9·0	15 seconds' drainage, tip touching fluid

From the results in Table XIX, it would appear that no appreciable difference exists between the variable error after blowing out the drop immediately after delivery, with rotation of tip against the glass wall, and the variable error when 15 seconds' drainage is allowed, with tip against the wall. As will appear subsequently,

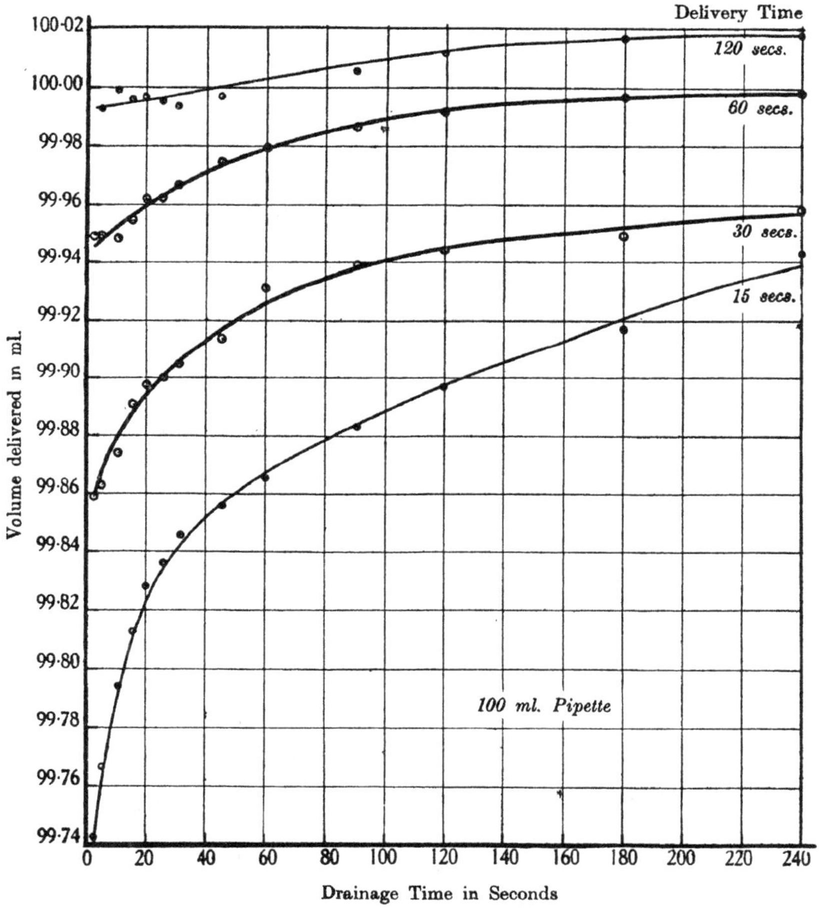

FIG. 73. Drainage curves of a standard 100 ml. pipette from Stott (*Volumetric Glassware*, Witherby, London).

this similarity is only to be expected. On the contrary, the 15 seconds' drainage with the tip touching the delivered fluid would appear to cause an increase in the variable error, this changing from 6·4 to 9·0 c.mm.

In Fig. 73, taken from Stott's book on *Volumetric Glassware*

(1927), the effect of drainage on the total delivery from a 100 ml. standard pipette is given for different delivery times. The figure shows that drainage time cannot be wholly ignored, that is to say, we must proceed in relation to precise times whether immediately after delivery or after some stated period. Using 15 seconds with the 25 ml. pipette the amount drained is 8 c.mm. or about 0·5 c.mm. per second. Since we are not likely to be out by more than one second—and scarcely that from one delivery to another —when we give zero drainage, proceeding sharply from the end of the delivery, it will be seen that this has no appreciable effect on the variable error of 6·5 c.mm. We are more likely to be in error by a few seconds when draining for 15 unless this is taken each time from a stop-watch.

The effect on the variable error of altering the delivery time. *The delivery time, the internal surface and the bore of the tube are the three chief determining influences on the amount of fluid left on the walls and in consequence on the variable error.* The effect of the tube boring will be best considered when dealing with deliveries from straight tubes and consequently in the burette chapter. Before proceeding to the relation between the variable error and the fluid on the walls—the underlying relationship for glassware delivery—the general effect of altering the delivery time from pipettes may be seen from the following table.

TABLE XX

Pipette.	*Delivery time sec.*	*Variable error c.mm.*	*Coefficient of variation.*
25 ml. std.	29·0	6·3	0·025
25 ml. std.	5·0	12·4	0·050
1 ml.	2·5	1·3	0·130
1 ml.	<1·0	2·6	0·258
0·73 ml.	40·0	0·21	0·029
0·73 ml.	5·0	0·70	0·096
0·73 ml.	<1·0	3·53	0·483

The 1 and 0·7 ml. pipettes were simple straight tube pipettes made in the laboratory. In the table is illustrated the surprising fact already commented on in Chapter V and given in Table III, that a simple straight tube 0·7 ml. pipette (length from tip to mark—16·5 cm.) can be made to deliver with the same percentage accuracy as the standard 25 ml. pipette by merely decreasing the delivery rate or the rate of fall of the meniscus (as also by providing a fine delivery tip and using the blow-out procedure at end of delivery in the manner described above). The delivery time of the 25 ml. standard pipette (Table XIX) was diminished from 29 to 5 seconds by filing the tip.

The relation between the variable error and the amount of fluid left on the walls. *This is fundamental to the whole subject of variable error, and goes as a simple proportionality of error to wall fluid.* Provided the meniscus does not fall faster than about 15 centimetres a second (when the variable error cannot be determined with precision from the rapidity of after drainage), it is found that the variable error (σ_p) and wall fluid (F_p) are related as

$$\sigma_p = 0{\cdot}05F_p, \qquad (49)$$

both expressed as c.mm.

The relation is shown in the following table (Table XXI) for the 0·7 ml. straight tube pipette, the 1·0 ml. straight tube pipette and the 25 ml. standard pipette.

TABLE XXI

Pipette.	*Internal surface.*	*Length cm.*	*Time of delivery sec.*	*Fluid on walls c.mm.*	*Fluid per sq. cm. int. surf.*	*Variable error c.mm.*	*Ratio : Error/ wall fluid.*
0·73 ml.	12·65	17·2	40·0	5·0	0·40	0·21	0·042
0·73 ml.	12·65	17·2	4·8	17·4	1·37	0·70	0·040
1·0 ml.	14·1	16·0	2·5	29·0	2·06	1·30	0·045
25·0 ml.	75·0	—	20·0	124·0	1·65	6·30	0·051
25·0 ml.	75·0	—	5·0	268·0	3·57	12·4	0·046

The 0·73 and 1·0 ml. are straight tube pipettes.

It appears from the table that the relation holds as closely as may be expected from 5·0 c.mm. on the wall of the 0·73 ml. pipette to 268 c.mm. on the wall of the 25 ml. pipette ; or from 0·40 to 3·57 c.mm. per square centimetre internal surface.

When the delivery time is very rapid—more than 10 cm. meniscus fall per second in the simple tube pipette—the proportion of variable error to wall fluid tends to rise owing to the very rapid after-drainage.

Relation of wall fluid to delivery time. The wall fluid was determined after delivering *N*/1 hydrochloric acid in the usual way, taking care not to fill the pipette appreciably above the mark. Immediately after the delivery the drop of fluid in the tip was blown out as already described and the pipette then held horizontally and the external surface carefully wiped. The pipette was then rinsed five times with about one-third its volume of distilled water each time, the washings collected and titrated.

There was no difference found between normal sulphuric and normal hydrochloric acid with respect to the amount left on the walls. With one 25 ml. standard pipette five determinations with each acid gave 112 c.mm. left on the walls at room temperature (18·8° C.), 26 seconds being taken with this pipette for the delivery time. It may be noted that Schloesser (1906) got only 0·003 ml. less delivery from a 25 ml. pipette, with *N* sulphuric acid than with distilled water, which amount could arise as sampling error. With the 100 ml. pipette he got 0·003 ml. sulphuric acid in excess, and with the same pipette 0·017 ml. excess of hydrochloric acid, but all such may arise from the normal variations of the pipette delivery. One may conclude that either normal sulphuric or normal hydrochloric acids will suffice for measuring the wall fluid after deliveries.

With a 25 ml. standard pipette, in which large variations in delivery time were made either by removing small portions of the tip or by reducing the opening in the flame, the following results were obtained :

TABLE XXII

Pipette.	*Delivery time sec.*	*Wall fluid c.mm.*	$\sqrt{t}$ × *wall fluid.*
25 ml.	1·4	410	485
,,	5·0	268	598
,,	29·0	118	635
,,	360·0	31	588

The wall fluid goes approximately in an inverse relation to the square root of the time of delivery. It will in fact be shown subsequently that with the burette or straight tubes the relation also applies, and as close as is practically required. It is not surprising here that the amount of wall fluid should have this relation with time of delivery since the fall of a viscous fluid along a surface will obey diffusion relations.

Owing to the fact that the standard pipette consists of wide barrel and narrow straight tubing, no simple relation exists connecting the wall fluid with pipette volume, but for straight tube pipettes a relation does exist between the wall fluid, the delivery time and the bore diameter. This will be considered under the delivery error of the burette.

For each single pipette type, however, the equation

$$F_w = \frac{a}{\sqrt{t}} \tag{50}$$

applies, F_w being the wall fluid after delivery, and a a constant. For the 25 ml. standard pipette a is approximately 580 c.mm., t being expressed in seconds.

As already mentioned, there is no appreciable difference in the variable error when the drop is blown out with tip against side rotated while blowing, and in allowing drainage for 15 seconds after delivery with tip against side. With a 25 ml. pipette it was found that the wall fluid immediately after delivery and after 15 seconds was 118 and 112 c.mm. respectively. Since the variable error goes as 5% of these values it follows that the drainage effect is here quite negligible. It would, of course, be a different matter if no attention was paid to the time after delivery and one, ten or twenty seconds taken indifferently. Whatever procedure or time be taken it must be consistently adhered to.

The effect of temperature on the wall fluid. Temperature affects the amount of fluid left on the walls after delivery, the relation being of an inverse kind, a fall of temperature causing increased wall fluid.

This was investigated for the 25 ml. pipette delivering normal hydrochloric acid, the temperature of the acid being varied. The following table shows the effect :

TABLE XXIII

Pipette.	*Time to deliver sec.*	*Temperature of acid.*	*Wall fluid (titration corrected for density change) c.mm.*	*Wall fluid as % of whole delivery.*
25 ml.	24·0	31·0	94 (4)	0·376
25 ml.	24·6	18·8	111 (5)	0·444
25 ml.	24·8	11·0	124 (3)	0·496
25 ml.	—	6·5	133 (2)	0·532

(The brackets in the fourth column refer to numbers of determinations.)

No drainage time was allowed after delivery, the drop being blown out against the side in the manner already described. The effect of the temperature change from 6·5° to 31·0° may be expressed as a mean change per degree increase of $-0{\cdot}006\%$ of the whole 25 ml. delivered.

Since the variable error will very probably go in the usual relation to the wall fluid we may say that temperature will have no practical effect on this since an effect of $-0{\cdot}0004\%$ per degree is negligible. With regard to the constant error, however, *the effect exceeds that of the glass expansion with temperature.*

Comparison of the variable error of the standard pipettes with the tolerances allowed in construction. We may in this way compare the Class A tolerances of the National Physical Laboratory, Middlesex (and those of the Bureau of Standards, Washington), with the variable errors of the standard pipettes as in Table XXIV.

TABLE XXIV

Nominal capacity ml.	*Minimum delivery time sec.*	*Maximum delivery time sec.*	*Tolerances or maximum errors allowed c.mm.*	*Variable error c.mm.*
100	30	50	60 (80)	13·0
50	25	40	40 (50)	9·2
25	20	35	30 (25)	6·5
10	15	25	20 (20)	4·1
5	10	20	20 (10)	2·9
2	7	15	10 (6)	1·8

The figures in brackets refer to the tolerances of the Bureau of Standards, Washington.

It will be seen from the table that the variable error or standard deviation of the single delivery is very much less than the tolerances. Stated as in the table, however, they are not strictly comparable. Usually a plus or minus three times the standard deviation is taken as covering the range. Even multiplied by three the variable error falls appreciably below the Class A tolerances. *We are really dealing here with two kinds of error, the tolerance being a possible constant error to the limit of the value given. This possible constant error can be satisfactorily eliminated by the simple procedure of weighing five deliveries and taking the mean ; for in this way we also allow for the effect of the variable error. This latter, however, cannot be removed for the single delivery under the fixed conditions in using the pipette.*

With regard to the question of standard glass error there are two points of view, that of the maker and of the laboratory worker. The maker must work to certain tolerances which are governed by the possibilities, the cost and the accuracy demand from the laboratories. *Other things apart, the laboratory worker is constrained to work only in relation to the variable error and its implications.* Usually a certificate is provided with the best standard glassware purporting to give the exact delivery volumes under the conditions, and this bridges the two points of view for the pipette and the volumetric flask, but not for the burette, as will be shown. By determining the five delivery mean the laboratory worker can of course dispense with the certificate, and will proceed in this way for simple pipettes made in the laboratory.

The origin and reduction of the variable pipette error. As already shown, the variable error is proportional to the wall fluid after delivery, and it amounts to only 5% of this. It is somewhat surprising that it should be so small ; and it may scarcely be profitable to inquire further into what may be its ultimate origin. There are some factors, however, which may be eliminated. It is not due, for example, to small changes in the delivery time of the pipette leaving behind different amounts of wall fluid. To test this fourteen deliveries from a standard 25 ml. pipette were timed. The mean time was 24·8 seconds with a standard deviation of the individual time of 0·19 second or 0·8% (maximum deviation found −1·5%). Since the wall fluid goes inversely as the square root of the time, this will account only for 0·4% instead of the total 5% observed. It is also not due to small irregular vibrations imparted by the hand to the pipette since similar errors occur with tubes fixed in position, as with the burettes discussed in the next chapter. Whether superfine cleaning, such as we may expect to result from prolonged contact with hot oxidising reagents, or successive use of such and rinsing with specially prepared water, would reduce it further is a point of scarcely any practical importance, since even if it succeeded the first use of the pipette may be expected to leave it again in the condition of the ordinary well cleaned type. It may be said that it is unlikely that it would further reduce the error, since with increasing thickness of wall film due to more rapid delivery times we get a variable error going proportionally thereto.

The variable error may possibly be related to microscopic irregularities of the glass wall, and to the varying play of air currents on the fluid film as the meniscus descends.

Reduction of error by subsequent rinsing. When the volume conditions of the delivery are not fixed it would appear that we could remove the variable error almost entirely by subsequent rinsing. *It will necessitate about five different rinsings* and must be very carefully done, since washing above the meniscus may result in an error almost if not as great as the usual variable error.

In micro work, where we are often in the region where the variable chemical error is of the same order as the variable glass error, this rinsing may far increase the total error instead of diminishing it, owing to the increased fluid volume at the end-point.

The objections to this washing out procedure are then that it will considerably increase the time of titration when it is carefully done, and that it may lead to a more serious error in micro work than that which it is sought to avoid.

Reduction of pipette error by increase of delivery time. This is a valuable method for micro pipettes constructed in the laboratory. It will be found, where the delivery time is not borne in mind with such pipettes, that the outlet will be made too large as a rule, the pipette delivering very rapidly. With very rapid delivery a variable error of the order of 0·4% may be expected (*vide* Table III of Part I). Such an error can be reduced to 0·029 or less if the delivery time is increased to about 40 seconds for a meniscus descent of approximately 16 centimetres.

This effect is obviously of considerable importance for micro work and shows that a 1 ml. technique can be carried through with the same accuracy as the 25 ml. technique with standard glassware and without any burdensome increase in the time required. The possibility is even present of the same accuracy applying to the 0·1 ml. technique with more extended delivery times, the time required being about 2 minutes for a similar meniscus descent of 16 centimetres.

Reduction of pipette error by the use of syringe pipettes.

In Chapter V a brief account of the micrometer syringe pipette or burette of Trevan (1922 ; 1925) and the syringe pipette of Krogh and Keys (1931) has been given.

In such a pipette as that of Krogh and Keys, delivering between two fixed points, there occurs no pure delivery error arising from a variable degree of wall fluid, provided the syringe piston is working without appreciable entrance of fluid between piston and barrel. Also, the reading error is abolished, and the pipette, for a series of deliveries at the same temperature, might be expected to show no measurable error, other than the very small one that might arise from minute variations in the amount of fluid held at the tip, or minute passages of fluid between piston and barrel.

In effect the Krogh-Keys pipette gives a remarkable accuracy at or above a volume of 1 ml. The level of absolute probable error readily obtainable is given as 0·0001 ml. though 0·00002 ml. can be reached by the most careful working. The first means a coefficient of variation of 0·015% with a 1 ml. delivery.

To reach this, with a straight tube, fine tipped pipette, length 20 cms. and bore diameter of approximately 0·25 cm., would require about three minutes for the delivery. At the same time it would be necessary to adjust the meniscus to within about 0·04 of a mm. to have a negligible extra error arising from this source.

It will appear that accuracies of this order are more easily reached with the syringe pipettes than the straight tube pipettes.

When the volume of fluid required for delivery reaches the level of about 10 c.mm. the ease of producing absolute errors of the order of 0·00002 ml. (p.e.) seems more easily attainable by the pipette principle as applied in the use of the Conway burette (Chapter VI).

Example of the addition of pipette errors.

Three ml. of urine are introduced by a standard pipette into a Kjeldahl incinerating flask, and the ammonia after incineration is distilled into 25 ml. of $N/10$ hydrochloric acid, introduced also by a standard pipette. 12·50 ml. $N/10$ KOH are taken to titrate the acid at the end: it is required to find the variable error due to the pipetting alone.

The variable error of the 3 ml. standard pipette is 0·075% (as a coefficient of variation). The total millimols ammonia in the 3 ml. urine is $(25{\cdot}0 - 12{\cdot}5) \times 0{\cdot}1$ or 1·25, so that the variable error in introducing the urine is $0{\cdot}94 \times 10^{-3}$ millimols. The variable error of the 25 ml. standard pipette is 0·026%, so that in introducing 25 ml. $N/10$ acid this amounts to $0{\cdot}65 \times 10^{-3}$ millimols. These errors add as the square root of the sum of their squares, which gives $1{\cdot}14 \times 10^{-3}$ as the addition in millimols, which when divided by 1·25 and multiplied by 100, gives 0·091%.

In a series of urine nitrogen determinations this is the variable error introduced by the pipetting of the urine and the pipetting of the acid into the receiver, using the same glassware.

When different pipettes are used each time, these must be standardised for their constant errors, and such allowed for in each series.

If we were to proceed in the above calculation simply on tolerances, our error would be computed as follows. The tolerance of the 3 ml. standard may be taken as 15 c.mm., which with the above data will mean $6{\cdot}25 \times 10^{-3}$ millimols. Similarly the 25 ml. pipette will have a tolerance of 30 c.mm. which will likewise mean $3{\cdot}0 \times 10^{-3}$ millimols. These would add directly, so that the sum would be $9{\cdot}25 \times 10^{-3}$ millimols, or 0·74%.

Working on the tolerance system means a great and unnecessary increase in our computed error. It is unnecessary since the constant error—which the tolerance covers—can be eliminated to an extent negligibly small compared with the variable error, merely by determining the volume of the five delivery mean.

Summary of the relationships of the variable pipette error. (delivery by gravity).

(*a*) The variable pipette error is inversely proportional to the square root of the time of delivery.

(*b*) It is directly proportional to the amount of wall fluid per square centimetre surface left after delivery, the relation being

$$\sigma = 0{\cdot}05F.$$

(*c*) It is practically independent of the amount of drainage time up to the first 30 seconds or so provided the same routine times of drainage (if any) are given. Drainage in itself corresponding to the ordinary periods allowed, is of no advantage in reducing the variable error.

(*d*) As shown in the next chapter, the relation of the variable error when expressed as a percentage of the whole delivery and from a straight tube is given by

$$v_p = \frac{0{\cdot}022}{\sqrt{dt}},$$

where d is the diameter (cm.) and t the mean time in seconds for a meniscus fall of 1 centimetre.

With regard to the wall fluid, we have the following relations :

(*e*) The amount of wall fluid after delivery is inversely proportional to the square root of the delivery time.

(*f*) It is dependent on temperature, being decreased by 0·006% of the whole delivery per degree.

CHAPTER LVI

THE VARIABLE GLASS ERROR (*continued*)

THE BURETTE

WITH the burette the significant sources of the variable error in titrating may be given as

(1) The pure delivery error.
(2) The manipulation error.
  (*a*) The error of the end-point emergence.
  (*b*) The reading error.

In addition to the delivery error, which is alone significant with the pipette, there are additional sources of error with the burette. There is the error involved in reading the meniscus level at the end-point and also the error in adjusting the meniscus to the mark which becomes appreciable with the large bore burettes. Besides which there is the error of the end-point emergence, generally taken as a drop when using the 50 ml. burette, a procedure leading to gross and unnecessary inaccuracy with the smaller burettes.

The delivery error. From the previous section on the pipette it may be accepted that the variable error of delivery goes proportionally to the fluid left on the walls, since this applies in general to the barrelled pipettes and to the simple straight tube kind. Here we need only consider how the amount of fluid on the walls varies with the time of delivery and with the bore of the tube, these being the two determining factors.

Effect of delivery time on the wall fluid. The question was investigated with the 50 ml. standard burette of internal diameter 1·11 cm.; also with a straight tube of internal diameter 0·290 cm. and with one of 0·105 cm. diameter, representing each the 25 ml., the 1 ml. and the 0·1 ml. techniques. With the simple straight tubes a piece of extra tubing was attached by a short rubber band and the end of this tube drawn out. This was used to simulate the dead space fluid of the ordinary calibrated burette. The tube was filled from below with $N/1$ hydrochloric acid, and when the outflowing fluid just passed below the rubber junction, the tube was quickly detached, held horizontally and the end wiped clean of external acid or any adherent drop. The tube was then rinsed to the mark with distilled water five times, and the washings titrated with $N/200$ alkali. With the 50 ml. burette a different procedure was necessary. In this case the burette was filled from below up to the 50 ml. mark exactly and the tap closed.

TABLE XXV

Internal *diam.* '*d*' *cm.*	*Fluid* *left on* *walls* *c.mm.*	*Fluid* *left* *per sq.* *cm.* *c.mm.* (q).	*Time* *in sec. for* *meniscus* *descent of* *1 cm.* (t).	$q\sqrt{t}$.	$q\sqrt{\frac{t}{d}}$.
I. 1·11 (50 ml. burette)	349	1·93	0·35	1·14	1·09
	234	1·29	0·72	1·09	1·04
	199	1·10	1·27	1·24	1·18
	144	0·80	2·15	1·17	1·11
	104	0·51	4·12	1·04	0·99
II. 0·290	27·3	2·05	0·110	0·68	1·26
	11·4	0·86	0·453	0·58	1·07
	7·4	0·56	0·876	0·52	0·96
	4·9	0·37	2·670	0·61	1·13
III. 0·105	14·2	1·58	0·048	0·35	1·07
	6·0	0·67	0·223	0·32	0·99
	3·1	0·35	0·630	0·28	0·86
0·105 (held horizontally)	3·0	0·33	0·845	0·30	0·93
	9·9	1·10	0·092	0·33	1·02

Length of I (50 ml. burette, graduated section) - 52·3 cm.
Length of II - 16·0 cm.
Length of III - 27·3 cm.

This acid with several rinsings was collected by inverting the burette and not through the open tap. The total fluid was titrated carefully with $N/10$ alkali. Methyl red was used as indicator. The burette was then filled from below to the zero mark with $N/1$ HCl and allowed to empty completely and filled again with the acid, after which the tap was fully opened and 50 ml. allowed to run out, the tap being closed again with the meniscus exactly at the

50 ml. mark. The total residual acid above the closed tap was rinsed out and titrated as above.

Table XXV gives a summary of the results obtained. It will be seen that here as with the pipette the amount of fluid left on the walls goes inversely as the square root of the time. The horizontal position for the third tube (filling done by the horizontal burette previously described) showed no appreciable difference from the vertical position.

It may be noted that a vertical burette meniscus does not descend quite uniformly throughout the whole length. Mean times with a 50 ml. burette for each 10 ml. section were as follows :

0–10 ml. in	10 secs.	30–40 ml. in	14 secs.
10–20 ,,	10 secs.	40–50 ,,	21 secs.
20–30 ,,	12 secs.		

Thus the fall is practically uniform for the upper two-thirds but falls off markedly for the last 10 ml. Using the average time for descent of 1 centimetre, no practical difference is found between the vertical and horizontal positions.

The effect of internal diameter. The weighted mean values of q for the three tubes where q is expressed as amount of fluid left per sq. cm. and t is the mean time for the meniscus descent of 1 cm., is 1·16, 0·60 and 0·324 c.mm. It will be seen that the bore of the tubing has a marked effect on the amount of the residual fluid. If, for example, we were to consider a thin strip of the internal surface in each tube, of the same width and extending over one centimetre along the tube's length, then the amount left on this strip with the same rate of meniscus descent is quite different for the three tubes. Obviously the difference can only relate to the tube bore. The ratio of the three means of q is as 1·0 : 0·52 : 0·28. At the same time the ratio of the square roots of the internal diameters go as 1·00 : 0·51 : 0·31. The relation of quantity of fluid, time of meniscus descent and tube bore can therefore be expressed with sufficient accuracy for our purpose by the equation

$$q = 1{\cdot}1\sqrt{\frac{d}{t}}, \tag{51}$$

where q is c.mm. of fluid per sq. cm. surface, t the time in seconds for 1 cm. descent of meniscus and d the tube bore in cm.

The amount left on the internal surface of the whole tube is given by

$$q_s = \pi dl \times 1{\cdot}1\sqrt{\frac{d}{t}}$$

$$= \frac{3{\cdot}45d^{1{\cdot}5}l}{\sqrt{t}}. \qquad (52)$$

Considering the variable error, this per sq. cm. surface is given (from equations 51 and 49) by

$$\sigma = 0{\cdot}05 \times 1{\cdot}1\sqrt{\frac{d}{t}}$$

$$= 0{\cdot}055 \times \sqrt{\frac{d}{t}}. \qquad (53)$$

The most useful form of the error is here the coefficient of variation (v_b), which we obtain by multiplying σ by the total internal surface, dividing by the volume in millilitres and multiplying by 100, giving

$$v_b = \frac{0{\cdot}055\sqrt{\frac{d}{t}} \times \pi dl}{0{\cdot}25\pi d^2 l \times 10^3} \times 100$$

$$= \frac{0{\cdot}022}{\sqrt{dt}}. \qquad (54)$$

The coefficient of variation for the single delivery from a straight tube as given by the expression $\frac{0{\cdot}022}{\sqrt{dt}}$ *is in this whole question of glass error the most significant for micro work, for by altering the meniscus descent in an inverse relation to the tube diameter the delivery error remains unchanged.*

From the foregoing section certain points of practical importance in titrating arise for comment. The total fluid left on the walls of a burette as already shown is given by

$$q_s = \frac{3{\cdot}45d^{1{\cdot}5}l}{\sqrt{t}}$$

$$= \frac{210}{\sqrt{t}} \text{ (for the 50 ml. burette).}$$

Here t is the mean time (sec.) for a meniscus descent of 1 cm. A change of t from a mean time of 1·0 to 0·5 seconds means a change in wall fluid of 87 c.mm. or 0·18% of the full delivery, a change much greater than the delivery and manipulation errors combined. Again, if a burette is delivering 50 ml. in a mean time of 50 seconds, a change to 45 or 55 seconds means a change of about 5% of the wall fluid, approximately the same as the variable error of the delivery. Such a constant error, however, will in fact be comparatively small when we apply the usual multiple of 3 to the variable error to obtain the range. It will be found, however, that a 50 ml. burette will deliver with great constancy—to within a second or two—if the tap be fully opened, and it is obvious that this must be done in fine titration, and for such also a rough idea of the end-point position must be already known.

The effect of drainage time on the variable delivery error. The effect of drainage time on subsequent rise of the meniscus has been very fully investigated by Stott (National Physical Laboratory), as given in his monograph on *Volumetric Glassware* (1927). The results obtained were confirmed by Lindner and Haslwanter (1929).

Drainage is very slow at first, then rises to a maximum and falls again in a wave-like occurrence. This is well shown in Fig. 74, giving the drainage rate after varying delivery times from a 50 ml. burette (scale length of 57 cm.). The crest of the wave may appear long after the delivery. This is at first surprising, but is no more than one may expect if we consider the solution on the walls as behaving like a viscous fluid.

It will be seen that drainage has no practical effect on the variable error when the burette is read immediately after the delivery. Even at the very fast rate of 18 seconds for the total delivery from a 50 ml, burette, no drainage occurs in the first 30 seconds. After this a marked drainage sets in amounting to a total of 78 c.mm. in the subsequent 90 seconds. When 40 seconds are taken for the full delivery no appreciable drainage occurs within two minutes.

As we descend into the micro scale drainage becomes more important, for not only is the absolute drainage rate increased per unit surface, but still more so when expressed relatively. When the level of the 0·1 ml. technique is reached and using ordinary delivery times, it is necessary to use the horizontal position to avoid errors from this source, unless another device such as that of Rehberg is used. *This is perhaps the main advantage of the horizontal burette described in Chap. VI.* As an example of this increasing influence of the drainage as the bore of the burette is decreased, we may consider the drainage figures in the tables of Lindner and Haslwanter (1929) for the 50 and 10 ml. burettes.

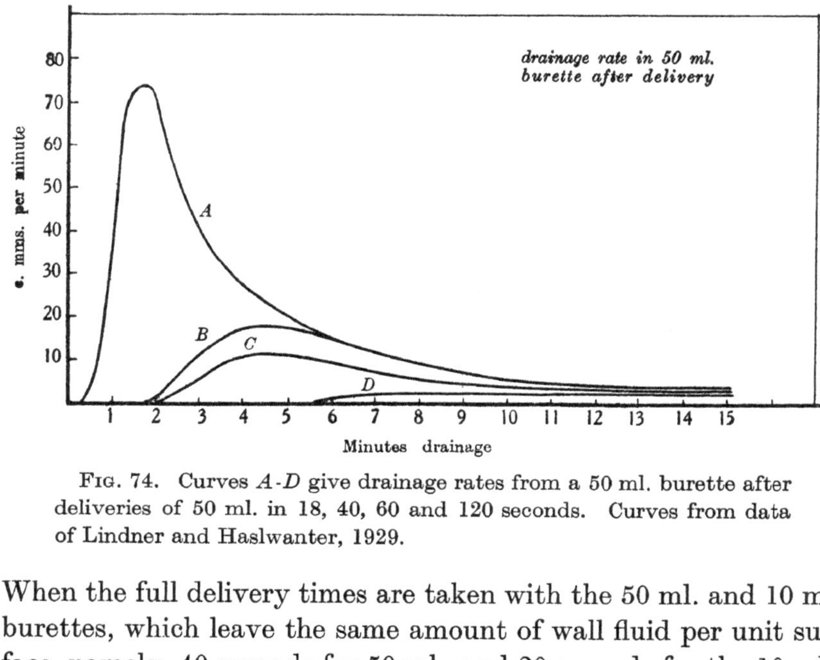

FIG. 74. Curves *A-D* give drainage rates from a 50 ml. burette after deliveries of 50 ml. in 18, 40, 60 and 120 seconds. Curves from data of Lindner and Haslwanter, 1929.

When the full delivery times are taken with the 50 ml. and 10 ml. burettes, which leave the same amount of wall fluid per unit surface, namely, 40 seconds for 50 ml., and 20 seconds for the 10 ml., the drainage after 5 minutes amounts to 0·08% and 0·387% of the total volumes respectively. Even when the same time of 40 seconds is taken for full delivery from each burette the drainage after ten minutes is 0·255% with the 10 ml. and 0·19% with the 50 ml. burette.

The manipulation error of the burette.

The end-point emergence. This for the 50 ml. burette is usually taken as a drop, which for dilute solutions and from a well-made tip should not exceed about 40 c.mm. For accurate working with micro burettes the use of the drop as the end-point emergence must be excluded. Fractions of drops are easy to manipulate, and there will be found a rough constancy of emergence for each burette on turning the tap to secure an appreciable extra quantity of fluid. With a well-made 2 ml. Bang burette about four-tenths of a final division (4 c.mm.) is the usual minimal emergence.

With the horizontal burette previously described the least emergence that is conveniently dealt with amounts to about half a division or to 0·5 c.mm., though without a suitably fine tip this may need to be increased to one division (1 c.mm.). As already considered in the introduction to this division, the end-point emergence will represent a rectangular distribution with a standard deviation given by the equation

$$\sigma^2 = \frac{l^2}{12}.$$

For the three burettes mentioned, therefore, the end-point emergence as a variable error may be expressed by 12, 1·2 and 0·15 (or 0·30) c.mm. respectively. It is possible, of course, to work with finer emergences, but these may be described as typical of careful or even fine working.

The reading error. The reading error may be said to include the initial adjustment of the meniscus to the mark and the subsequent reading after the end-point emergence. It is easier to adjust a meniscus to a definite mark than to read its position between two divisions ; it may be taken that this error is almost entirely due to the latter source. In the finest working the reading of the final position is brought with advantage to 0·2 of a division, but no practical advantage in accuracy would result from bringing it further.

The reading error could be treated as giving a rectangular distribution similar to the drop error, and treated as such, but since we have also the possibility of small parallax errors, plus errors due to lack of accurate discernment of the bottom of the meniscus, as well as errors in the initial adjustment of the meniscus to the mark, it is best treated as part of a manipulation error which can be determined experimentally, the manner of determining this being given subsequent to a consideration of the meniscus reading.

The reading of the meniscus. In reading the meniscus the avoidance of parallax errors and the securing of a uniform background are the essentials. The former can be avoided easily by the use of burettes with the lines extending completely around the barrel. The observer places his eye so that the front and back portions of the line nearest to the meniscus coincide. With regard to the background for the meniscus reading, a simple and effective procedure is shown in Fig. 75. A suitable strip of glazed white paper is taken, a horizontal line drawn in ink and the paper inked in below it. Two horizontal incisions are then made through which the burette is slipped in the manner indicated in the diagram. For the student it will be instructive to look at the meniscus firstly against a dull background, then to slip the paper into position from below, so that the white background just appears behind it, then to move it upwards until the black line is close to the meniscus. The effect is at first surprising.

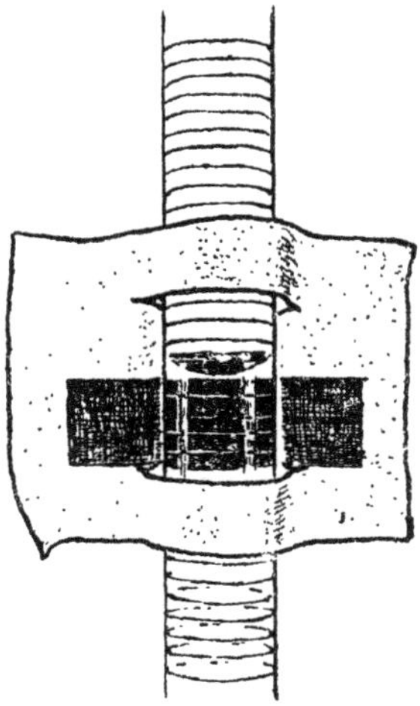

FIG. 75. Device for facilitating meniscus reading; modified from Kolthoff's '*Die Massanalyse*' (Springer, Berlin).

The determination of the total manipulation error. The total manipulation error, made up as it is from the end-point emergence error, and the reading error, is quite independent of the total amount of fluid delivered from the burette. When 5 ml. of $N/1$ alkali are used to titrate 25 ml. of $N/5$ acid delivered from a standard pipette, the manipulation error of the burette—usually about 15 c.mm. for the 50 ml. burette—amounts to 0·28% as a coefficient of variation.

The pure delivery error is only about 0·02% when the meniscus descends, as usual with the 50 ml. burette, at about 1 cm. per second ; the pipette error is 0·025%. These are quite negligible when compared with 0·28%, since they add as the square root of the sum of the squares. Consequently the variable error of titration is here practically pure manipulation error.

Similarly the manipulation errors of the other burettes may be determined by using only a small fraction of their length. The manipulation error if so determined as a coefficient of variation, may be expressed in c.mm. as follows :

$$\sigma_m = v_b \times \frac{10^3}{100} \times V = 10 V v_b, \tag{55}$$

where V = burette volume in ml.

The manipulation errors of a 50 ml., a 2 ml. and the horizontal burette were found to be 15, 1·3 and 0·15 c.mm. respectively. *The end-point emergence constitutes much the greater part of this manipulation error and would alone account for 12, 1·2 and 0·14.*

In carrying out such manipulation tests it is essential that the burette should not be filled by pouring from above. This can give rise to a comparatively large error, since the fluid poured in runs very rapidly down the sides and after the meniscus is brought exactly to the mark will continue draining from above. It may be filled from above by a pipette or from a glass tube, or from below by means of a special attachment now common with burettes.

The combined delivery and manipulation errors of the burette. These errors add according to the principles outlined in the introductory chapter. *The delivery error in actual magnitude is proportional to the volume delivered, the manipulation error being independent of the volume. Stated as percentage errors the reverse is true, the delivery error being then independent of the delivered volume and the manipulation error inversely proportional to this volume.*

The manipulation error itself (σ_m) may be always expressed as c.mm. (equation 55). The errors add then as

$$v_b^2 = v_d^2 + \left(\frac{\sigma_m}{10V}\right)^2, \tag{56}$$

where V is expressed as ml. and v_d is the coefficient of variation of the delivery. This is given by the relation already developed (equation 54) :

$$v_d = \frac{0{\cdot}022}{\sqrt{dt}},$$

where d is the diameter of the tube and t the time in seconds for meniscus descent of 1 cm.

The 50 ml. burette. With the 50 ml. burette and a meniscus descent of 1 cm. per second, v_d is 0·021 ; the manipulation errar was found to be 15 c.mm., so that it would require a delivery of 72 ml. before the delivery error as c.mm. was equal to the manipulation error. At 36 ml. delivery the delivery error is only half the manipulation error and does not much affect the total error. From 36 ml. to zero the manipulation error is dominant and the total burette error is inversely proportional to the volume delivered. At 25 ml. the total variable burette error is 0·060%, and at 10 ml. 0·15%.

The 2 ml. Bang burette. With the 2 ml. Bang burette with a bore diameter of approximately 0·24 cm. a meniscus descent of 1 cm. per second will give a delivery error of 0·045 as a coefficient of variation. The manipulation error was found to be 1·3 c.mm., requiring therefore a 3 ml. delivery for the delivery error as c.mm. to equal the manipulation error. Here also from 1·5 ml. upwards the manipulation error is dominant and the total variable error is inversely proportional to the delivered volume.

With a delivery volume of about 2·2 ml. the 2 ml. standard Bang burette is as accurate as the 50 ml. at the 25 ml. titration level with the conditions as stated. It would appear, therefore, that some advantage would be obtained by the use of a 3 ml. Bang burette instead of a 2 ml.

The horizontal burette. The horizontal burette described in the text has a manipulation error, with the conditions defined in a previous section, of 0·155 c.mm.

The internal diameter is 0·105 cm., and with a delivery rate of 1 cm. per second we obtain a delivery error of 0·068% as a coefficient of variation. It would take a delivery of 0·23 ml. before the manipulation and delivery errors became equal as c.mm. The total error then as a coefficient of variation would be 0·095%. *Using the burette in this way to deliver around 0·23 ml., it is the equal in accuracy of the 50 ml. burette titrating around 15 ml., the volume reduction being at the same time near to one-fiftieth.*

This presupposes, however, that the tube bore will show the same relative uniformity as that of the 50 ml. burette. When the finest grade thermometer tubing is used this degree of uniformity is secured. Also it may be said that the discriminating accuracy of the burette in titrating closely similar quantities is practically independent of bore uniformity.

Note on the micrometer syringe burette.

The burette of Trevan (1922; 1925) which has been further studied and somewhat modified by Hadfield (1942) has no delivery error due to varying wall fluid, but like other burettes has a manipulation error, arising from a reading error (of the micrometer) and that of the end-point emergence. In the general accurate usage of the standard 50 ml. and 2 ml. Bang burette delivering, on the one hand over the range 20–50 ml. or from 1–2 ml. with the second, the delivery error (with 1 cm. per second mean flow) is much less significant than the manipulation error. This dominates the total variable error, and the same could be said of the Conway burette delivering at the rate of 1 cm. in 2 seconds (if a very high order of accuracy were required). The absence of the pure delivery error from the micrometer syringe burette is therefore not of much relative significance.

In such comparisons of the variable error it is assumed that the tube bore and that of the piston of the syringe burette are quite uniform, or are exactly calibrated.

CHAPTER LVII

THE TOTAL VARIABLE GLASS ERROR AND ITS CONTROL

THE total variable glass error (v_g) is given by the equation

$$v_g^2 = v_p^2 + v_b^2 + \left(\frac{\sigma_m}{10V}\right)^2 \qquad (57)$$

where v_p = pipette error (coefficient of variation),
v_b = delivery error of burette (coefficient of variation).
σ_m = the manipulation error of burette in c.mm.,
V = the burette volume (ml.) delivered in titration.

In considering the control of this total glass error we may take it at different volume levels. For convenience one may speak of a 25 ml., a 1 ml., a 0·1 ml. and a 0·01 ml. technique, and consider each of these in turn.

The 25 ml. technique. We shall suppose a 50 ml. burette used with a meniscus descent of 1 cm. per second, and a manipulation error of 15 c.mm. and that 25 ml. acid are pipetted from a 25 ml. standard pipette and titrated with 25 ml. alkali from the burette. The variable error may then be computed as

$$v_g^2 = 0{\cdot}025^2 + 0{\cdot}022^2 + 0{\cdot}060^2,$$

0·025 being the pipette error, and 0·022 the pure delivery error of the burette as considered in the previous chapter. It will be seen that the error is very largely due to the manipulation error of the burette, which would alone give 0·060%. If in the titration we were to use 45 ml. from the burette (with suitable alteration of alkali strength) the manipulation error would then give a percentage error of 0·033 and the total glass error would be 0·047%, the manipulation error being here in much lesser proportion. If we sought to reduce the total glass error further by using more than 50 ml. from the burette this would require a refilling with the consequent errors of two meniscus adjustments. The practical procedure would consist in a reduction of the manipulation error by diminished end-point emergence, proceeding with some fraction of a drop. Proceeding with about half a drop or 20 c.mm. removed with a fine glass rod, this would reduce the total error with 25 ml. delivery from 0·07 to 0·05% and the 45 ml. delivery from 0·05 to 0·04%, the advantage in the latter case being small.

Discrimination error. If we were to consider a series of titrations with the same glassware and solutions, and grouped the burette readings in a succession of duplicates, then it may be shown as an elementary problem in statistics that the standard deviation of the differences between the members of the duplicates is $\sqrt{2}\sigma$, where σ is the standard deviation of the single titration from the mean of the whole series. This we may define as the discrimination error, and expressed as a percentage it may be written $\sqrt{2}v_g$, where v_g is the total glass error. If we made two successive titrations with the same glass and the same alkali, but two unknown acidic solutions, these could be regarded as significantly different if their titration figures differed by more than $3 \times \sqrt{2}v_g$. Where duplicate titrations are carried out with each acid then the range of discrimination error is $3v_g$, or the same as that of the total glass error for a single titration.

Using then the 25 ml. pipette for duplicate titrations and titrating 25 ml. from the burette, the discrimination range is $3 \times 0{\cdot}069$ or 0·21%, and for the 45 ml. titration it is $3 \times 0{\cdot}047$ or 0·14%. In other words, the two acids can be said to be different when they differ by more than 0·21 and 0·14% respectively. Such discrimination would be independent of the bore tolerance of the burette, since over the small volume range of such errors the bore diameter may be considered unvarying.

Absolute error. This involves a consideration of the constant error dealt with more fully in a later section and referred to in the chapter on the pipette. The constant glass error can be satisfactorily determined and allowed for by determining the mean weight of five deliveries and the true volume in accordance with the principles of Chapter LX. As a practical exception to this easy elimination of the constant error, we have the varying bore diameter of the burette.

Burette tolerances are given in the following table:

TABLE XXVI

Burette ml.	*Tolerances in c.mm.*		
	Nat. Phys. Lab.	*Bureau of Standards.*	*K.N.E.K.*
50	60	50	40
10	20	20	20
5	—	10	—
2	10	—	8

The tolerance with the N.P.L. extends to the whole or any part of the burette, but for the K.N.E.K. half the burette volume carries half the tolerance, but no further subdivision is allowed.

There would be no great difficulty, only an inconvenience, in standardising each section of the burette to find the exact divergence from the nominal value, and a burette certificate may be had in which this is already done as far as is practical. For such laboratory practice the total tolerance of the burette is comparatively unimportant, since like that of the pipette the exact volume is easily obtained, and it would seem therefore that with regard to the finest titrations an advance could be made here in a greater bore uniformity, such as will approach the variable error of the delivery.

If we standardise the burette for the whole delivery, it may be presumed that the error for any section is not greater than half the full tolerance value (60 c.mm.) existing without any standardisation. On this basis and collecting the previous data, Table XXVII has been made out.

TABLE XXVII. *Titrations with 25–50 ml. deliveries*

50 ml. burette. End-point emergence, 40 c.mm. Manipulation error of 15 c.mm. (standard deviation).

Pipette.	*Variable error of pipette.*	*Volume titrated from burette ml.*	*Total variable glass error.*	*Discriminating error range. Duplicate titrations.*	*Absolute error range for the single titrations.*
25 ml.	0·025 %	25	0·069 %	0·21 %	0·33 %
25 ml.	0·025	45	0·047	0·14	0·21
5 ml.	0·058	25	0·086	0·26	0·37
5 ml.	0·058	45	0·074	0·22	0·29

To the range of the discriminating error given there the 30 c.mm. (half the full tolerance) has been added, as $3 \cdot 0/V(=\%)$, the result given then as the absolute error range, or the amount which any single reading may diverge from the mean of a large number of similar titrations. It has been assumed at the same time that all constant errors have been allowed for and that the acid and alkali are of sufficient strength to render negligible the chemical error arising from an indefinite end-point.

The 1 ml. technique. The representative burette here is the 2 ml. Bang burette. We may consider firstly the magnitude of the error, using a similar procedure as with the 25 ml. technique, but at the 1 ml. level. Let us suppose a 1 ml. standard pipette is used to deliver the acid, and the meniscus descent from the burette to have a rate of 1 cm. per second. A marked difference will lie in the end-point emergence, being here a fraction of a drop or 4 c.mm., as already considered, the manipulation error being of the order of 1·3 c.mm. (st. dev.).

Considering the equation 57

$$v_g^2 = v_p^2 + v_b^2 + \left(\frac{\sigma_m}{10V}\right)^2,$$

the usual value of v_p or coefficient of variation of the 1 ml. standard pipette is here 0·13%, and from equation 54 the pure delivery error of the burette is 0·045%, so that with the manipulation error (σ_m) of 1·3 c.mm. the value of

$$v_g^2 = 0 \cdot 13^2 + 0 \cdot 045^2 + (0 \cdot 13)^2$$

or

$$v_g = 0 \cdot 19.$$

This is nearly three times the corresponding error of the 25 ml. technique.

Variable titration error in an experimental series. Before proceeding to a consideration of the manner in which this error may be reduced to that for the 25 ml. level, it is of interest to see how far equation 57 predicts the error in an actual series of titrations in which each of the three values has been experimentally determined as 0·11, 0·11 and 1·3, the rate of meniscus descent in the burette being faster than 1 cm. per second and the pipette used other than the standard type.

TABLE XXVIII

Acid used.	*Variable titration error (experimental).*	*Variable titration error (theoretical).*	*No of titrations.*
0·1*N*	0·20%	0·20%	52
0·01*N*	0·19	0·20	26
0·001*N*	0·27	0·25	20

With the 0·001*N* titrations the variable chemical error is beginning to influence the total error. For the 0·1 and 0·01*N* titrations the experimental and theoretical errors are practically identical.

Approach to the accuracy of the 25 ml. technique. Considering the end-point emergence of the burette as 1·3 c.mm., this alone will give 0·13%, whereas the total variable error for the 25 ml. technique is 0·069% as we have seen. Refining, therefore, on the pipette or burette deliveries will be useless until either the end-point emergence is reduced further or the volume titrated from the burette is increased, in which latter case there will be a departure from a purely 1 ml. technique. The end-point emergence may without much difficulty be reduced to 2 c.mm. if the burette has been provided with a tip sufficiently fine, and burettes of such type may now be had from the makers. Alternatively the use of the Shohl (1928) needle tip with Luer adapter could be considered. To the delivery end is sealed a glass Luer adapter into which fits a Luer hypodermic needle *B* of 18 to 23 gauge, cut horizontally and ground on a stone, platinum needles being obviously preferable.

With diminution of the end-point emergence to 2 c.mm. and a manipulation error then of approximately 0·7, an increased burette delivery to at least 1·5 ml. would be necessary. This would give a variable error from the burette manipulation of 0·047%. The pure delivery error of 0·045% could be allowed to remain at this level, and suitable reduction of the remaining pipette error is easily accomplished. As shown in Chapter V of Part I and in Chapter LVII of Part III, a simple straight tube pipette delivering 0·7 ml. in 40 seconds from a straight tube pipette (16 cm. length) has an error of only 0·029%, and the error may in general be deduced from equation 54, already established in previous chapter :

$$v = \frac{0{\cdot}022}{\sqrt{dt}},$$

d being the bore diameter and t the time of descent of 1 cm. Assuming then a 1 ml. pipette delivering with an error of 0·029%, a burette delivery of 0·045% and a manipulation error of 7 c.mm. with 1·5 ml. titrated, from equation 57 we get a total glass error of 0·068%, which is practically identical with the error of titrating 25 ml. acid from a standard pipette with 25 ml. alkali from the 50 ml. standard burette, using a drop as end-point emergence. Nothing exceptionally difficult is introduced in this approach, the emphasis being on delivery time, fineness of tip and of end-point emergences.

By the use of syringe pipettes and burettes the same result could also be readily obtained, but with apparatus of a higher degree of complexity.

Discrimination range and absolute error. With the technique just discussed, and which approached the 25 ml. accuracy, the discrimination range will be exactly as given for this level in Table XXVII. The absolute error range will be higher owing to a greater percentage tolerance for the smaller burette. To keep pace with the worker's possibilities here, a much reduced tolerance is required, but not in the whole burette volume, merely in the bore uniformity. The tolerance of the 2 ml. burette is 10 or 8 c.mm. On standardising the whole 2 ml. delivery it may be assumed as before that this tolerance is reduced to approximately 4 or 5 c.mm. The numerical values are assembled as before in Table XXIX.

With increasing bore uniformity in the burette the absolute error range approaches that of the discriminating range. As before, in this table the constant errors except that of bore tolerance are not considered.

The 0·1 ml. technique. With the Conway burette in present use an end-point emergence of 0·5 c.mm. is secured without much difficulty, and 1 c.mm. rather easily, but in fact almost any degree of fineness in delivery can be reached by constricting the delivery tip more and by lessening the hydrostatic pressure—shortening the distance of the delivery tip from the graduated scale. With the tip drawn out so that the rate of delivery is 3 seconds for 0·01 ml. with fully open tap, any degree of control is possible without resort to diminution in hydrostatic pressure.

TABLE XXIX. *2 ml. Bang burette. Titrations around the 1 ml. level*

Manipulation error 1·3 c.mm.—or 0·7 c.mm. in the approach to the 25 ml. accuracy—the corresponding end-point emergences being 4 and 2 c.mm. respectively. 4 c.mm. is also considered as possible error due to lack of bore uniformity after standardisation of full delivery.

Pipette used.	*Variable error of pipette.*	*Burette vol. used ml.*	*Total variable glass error.*	*Discriminating range (duplicate titrations).*	*Absolute error range of the single titration.*
1·0 ml.	0·13%	1·0	0·19%	0·57%	0·84%
0·7 ml. (40 seconds to deliver)	0·029	1·5	0·068 (2 c.mm. end-point emergence)	0·20	0·47

The transference with fine delivery tip may be made by directly touching the underneath fluid, and stirring either with fine glass rod—in the ‘ unit ’ technique or otherwise with a bubbling tube (as in Fig. 49). With the 0·5 c.mm. end-point emergence the manipulation error is 0·15 c.mm. (st. dev.) and with the 1·0 c.mm. it is 0·30 c.mm.

Before the full influence of time on the delivery error was understood, this was obtained as 0·12% in a long series by using the titration method (Chapter LVI). The delivery here would be about 2 cm. per second. It is very likely that the delivery error of the horizontal burette obeys the general relation,

$$v = \frac{0{\cdot}022}{\sqrt{dt}},$$

since the wall fluid goes in the same relation as for the vertical burettes. The particular advantages of the horizontal burette would consist then in avoidance of any drainage error, ease of securing slow and uniform flow without too fine a constriction of tip and previous contact of the emerging fluid with pyrex walls only (described in Part I).

In considering the pipette error here, we have no standard pipette to use in a similar manner to the previous volume levels. Examination of the error of a 0·1 ml. pipette of Ostwald type delivering in 2 seconds before blowing out gave a coefficient of variation of 0·39, but it may be assumed that this error would be considerably less with slow delivery.

Easily attainable accuracy with the 0·1 ml. technique. Here we shall take the error of the 0·1 ml. pipette as determined in a given instance, namely 0·39%, the pure delivery error of the burette at about 2 cm. per second as 0·12% and a manipulation error associated with an end-point emergence of 1 small division on the burette giving 0·30 c.mm. (st. dev.). With such data and delivering 0·1 ml. from the burette we obtain a total variable glass error of 0·51% (equation 57).

Where a 0·2 ml. pipette of similar type is used with 0·27% delivery error and titrating 0·2 ml. from the burette, we get a total variable glass error of 0·33%. The total variable error is for these two instances very largely made up of the pipette error. If we used a 1 ml. standard pipette and titrated with 0·2 ml. from the burette the error would be 0·23%.

Approach to the accuracy of the 25 ml. technique (using 0·1 ml.).

The 0·1 ml. pipette delivery. There is every reason to suppose that with suitable reduction in delivery time and fineness of tip, the error here can be made to approach that of the 25 ml. standard pipette, the delivery error proceeding in accordance with the relation already described.

With an internal diameter of 0·09 cm. and delivery length of 16 cm., 96 seconds should be required for the delivery to give an error of 0·03%.

The use of syringe pipettes for 0·1 deliveries.

If one is considering the delivery of standard acid of which there is a large amount available, no difficulty arises in delivering 0·1 ml. with the Krogh-Keys pipette, to a coefficient of variation of 0·15%; using with the most careful working to 0·03%. The difficulty arises when a sample for analysis is to be pipetted, and of which there is only a small amount available. Owing to the relatively large 'dead space' when the piston is pressed down to its fullest extent, the uptake of 0·1 ml. will be diluted with the fluid already present to a considerable extent. This difficulty in the use of a syringe pipette was avoided by Titus and Gray (1930) by sucking up 0·2 ml. mercury, which could then serve as a piston, with an air bubble between the mercury and the solution. Various other workers (Chambers, 1929 ; Rosebury and Heyninger, 1942 ; Kesten, Rittenberg and Schoenheimer, 1937 and Benedetti Pichler 1937) have filled the syringe with water separated from the solution by an air bubble.

The burette delivery error (0·1 ml. range). Seeing that the delivery error with the Conway burette was experimentally found to be 0·12% with a comparatively fast rate of delivery, it may be expected that this error will also go in accordance with equation 54, a delivery of 1 cm. in four seconds, giving a presumed error of 0·037%.

The manipulation error. Reduction of the hydrostatic pressure for slow delivery with a fine tip touching the fluid surface and stirring by bubbling (Fig. 49), electromagnetic ' flea ' or other such device, may be expected to reduce the manipulation error to the 0·075 c.mm. (st. dev.) region. A reading limit of 0·2 of a division or 0·2 c.mm. would itself cause an error of about 0·05% (coefficient of variation).

At the same time it will be necessary to deliver 0·15 ml. from the burette to give the required equality with the 25 ml. technique.

(With the micrometer syringe burette the required accuracy may be obtained without much difficulty, the pure delivery error, arising from varying wall fluid, being entirely excluded.)

The combination of these three errors of pipette (0·030%) and burette (0·037%) deliveries with the manipulation error (0·075 c.mm.), and adding 0·15 ml. from the burette will give a total glass error of 0·069% in accordance with equation 57. This would be practically identical with the error of titrating 25 ml. acid with 25 ml. alkali as above.

Discriminating and absolute error range with the 0·1 ml. technique. The discriminating range using duplicate titrations is taken as before at 3 times the total glass error as a coefficient of variation. The difference between this and the absolute error range will depend on the bore uniformity of the burette, assuming the other constant errors suitably allowed for. Here fortunately a high degree of uniformity can be secured by the use of the best thermometer tubing.

The above data for the 0·1 ml. technique are collected in the table on p. 387.

Note on blood pipetting. In the approach to the accuracy of the 25 ml. technique one of the main controls lay in slowing the delivery rate. In delivering whole blood, a fine tip is likely to become clogged with inconvenient frequency. Where the procedure allows of increase of fluid volume up to about five times the blood volume delivered, the easiest way out of such difficulty would consist in using wider tips and carefully washing out. Alternatively some such device as a stopcock could be used to retard flow with conveniently wide tip.

The 0·01 ml. technique.

Even at the 0·1 ml. level certain new sources of error appear. In pipetting, for example, owing to capillarity the meniscus at the end of the delivery after simple run out, stands high, and the expulsion at the end becomes an appreciable error source with the probably varying rate of the liquid movement. The effect becomes very marked as we pass to the 0·01 ml. level.

Chiefly from this cause a different technique of pipetting becomes advisable, besides the ordinary direct kind. It is true that by several washings this can be made to work very accurately, but owing to the narrowness of bore, efficient washing is not so readily carried out as would appear, and reference may be made again to the note on page 366 concerning the washing out of pipettes.

Alternative pipetting would incorporate some device for slow uniform and complete extrusion of the intended volume (apart from what may be left on the walls). The horizontal burette described in Chapter VI may be made to serve the purposes of such pipetting both where there is a large amount of fluid, as in the pipetting of acid for minute microdiffusion procedures, and also where only volumes of the order of 20 or 10 c. mm. are available for the analytic sample. The technique of such pipetting is described in Chapter VII.

Approach to the accuracy of the 25 ml. technique (using 0·01 ml.).

Proceeding in accordance with the directions in Chapter VII, not much difficulty need be anticipated in reaching a total glass error of about 0·3–0·4% as a coefficient of variation. With the Krogh syringe pipette the finest working would give a coefficient of variation of 0·3% when pipetting 10 c.mm. fluid (though it is

TABLE XXX

Pipette used.	*Variable error of pipette.* %	*Burette volume titrated.* *ml.*	*Total variable glass error.* %	*Discriminating range.* %
0·1 (Ostwald type) -	0·39	0·1	0·51	1·5
0·1 (same) - - -	0·39	0·2	0·43	1·3
0·1 (st. tube, 96 secs. to deliver through 16 cm.)	0·03 (theor.)	0·15	0·069 (theor.)	0·21 (theor.)
0·2 (Ostwald type) - -	0·27	0·2	0·33	1·0
1·0 (st. tube, 2·5 secs. to deliver 16 cm.) - -	0·13	0·2	0·23	0·7

0·25 ml. horizontal burette. Fine grade thermometer tubing. Internal bore 0·09 cm. Manipulation error 0·3 c.mm. or 0·075 c.mm. in the special procedure for the third series in the above Table.

The Ostwald pipettes used delivered with comparative rapidity (about 2 seconds), and are not therefore true to the original Ostwald type.

possible that such an apparatus on a reduced scale would give a higher accuracy). It is obvious therefore that even the most careful use of such a pipette along with the micrometer syringe burettes of Trevan, Hadfield or others would not improve on this degree of variation. To refine further on such accuracy in pipetting 10 c.mm. of an analytical sample with the Conway burette, a prior washing out of the graduated tube with alcohol and ether and drying (by air suction) could be tried so that the volume taken in is measured by the advance of the meniscus along a dry surface. Three times the volume taken in may then be pipetted out from the burette to ensure against any error source from traces of the analytic sample remaining. For ammonia estimations the pipetting error of the absorbing acid may be entirely removed when the ammonia for estimate is of the order of 0·1 g. NH_3-N using the boric acid absorbing reagent described on page 96. The delivery error is then of no significance. The indicator may be doubled in strength and the titration carried out in comparison with a companion droplet with the same strength of indicator in buffer solution at, say, 4·7 *p*H. With an extremely slow delivery rate from the burette and the tip of the delivery tube immersed in the droplet, the major error source in the whole procedure will now lie in the delivery error of the volume buretted out in the titration, and provided this amounts to about 20 large divisions, it is anticipated that it should not exceed about 0·1% error as a coefficient of variation, and it would appear possible to have it still less.

CHAPTER LVIII

THE VARIABLE CHEMICAL ERROR IN TITRATION

IF in a series of titrations we were to reduce the original strength of the acid progressively and the alkali correspondingly we would have a growing difficulty in deciding the exact end-point. An error from this source would be introduced, becoming more and more prominent as the original acid strength decreased. This error may be expressed as a standard deviation (σ_c) and conveniently in millimols, the coefficient of variation being $\frac{\sigma_c}{q} \times 10^2$, where q is the total number of millimols titrated.

The total variable error of the titration (v_s) would then be expressible as

$$v_s^2 = v_g^2 + \left(\frac{\sigma_c}{q} \times 100\right)^2, \qquad (58)$$

v_g being the total glass error as previously considered.

Experimental determination of the variable chemical error. The value of σ_c could be obtained experimentally for any type of titration where indicator is added by reducing q so far, that the glass error is negligibly small, the total error of the titration giving then the chemical error. In such procedure it would be necessary to keep the indicator strength the same throughout, and as low as conveniently possible.

In determining the variable chemical error for acidimetric work, this is largely influenced by the carbon dioxide contained in the water. The carbon dioxide of distilled water, when this is in equilibrium with the carbon dioxide of the air at room temperature, may be calculated to be approximately $1{\cdot}7 \times 10^{-5}$ molar from the known solubility of carbon dioxide (1·90 g./litre at 760 mm.), and taking the carbon dioxide of the air as 0·04%. Kolthoff (1931) has estimated it directly as $1{\cdot}5 \times 10^{-5}$. Distilled water in bulk may, however, contain much greater concentrations of carbon dioxide, arising presumably from gas burners used in the distilling plant and in bulk only very slowly attains equilibrium with the atmosphere, even with exposed surface. Water from electrically run distilling plants should be practically in equilibrium with the atmosphere, and the use of such water will be considered in the ensuing account.

Also it may be noted that for micro titrations in the central chamber of the 'unit', equilibrium with the air may be always assumed owing to the large surface and small depth. In an actual series of such micro titrations of 0·7 ml. of 0·0002N acid containing 0·0008% methyl red and 0·0016% methylene blue, with 20% alcohol and using 0·35 ml. barium hydroxide solution from the horizontal burette, a coefficient of variation of 0·78 was obtained which, allowing for the glass error (found to be 0·02% under the conditions) becomes 0·75% as the variable chemical error (equation 58). This corresponds to $1{\cdot}05 \times 10^{-6}$ millimols in the titrating alkali ($1{\cdot}4 \times 10^{-4}$ millimols being titrated).

Theoretical evaluation of the variable chemical error.

The end-point error in pH *units.* If we suppose titrations carried out with very weak solutions such as 1 ml. of $N/5000$ or of $N/10{,}000$, and that the exact pH of the end-point reached is subsequently determined, this latter will show a distribution for a series with a certain standard deviation. We should get the same or practically the same distribution if we were titrating a heavily buffered solution around the end-point and adding strong alkali. Such a distribution or standard deviation is immediately related to our visual discrimination, and hence may be expected to vary somewhat from one individual to another. For a single indicator used at or within 0·5 pH from its mid-zone of colour change, the limits of such distribution for careful working is $\pm 0{\cdot}1$—as generally observed (e.g. Bjerrum, 1914). This range corresponds also to the differences between buffered indicator tubes in the simplest pH comparators, these differing as a rule by increments of 0·2 pH.

The distribution of the end-point pH will very probably be not quite 'normal', since there must be a central region over which there is the same probability of an observation falling, but such a deviation from the normal curve will not seriously affect our treatment.

A distribution with practical limits of plus or minus 0·1 *p*H is given by a standard deviation of 0·033 to 0·050 *p*H as we take six or four times its value to cover the range. We may take 0·04 as a probable figure.

If we were to convert the *p*H standard deviation of the end-point into millimols alkali added we could do so in the following equation :

$$\sigma_c = 0{\cdot}04\beta \times V, \tag{59}$$

where β is $\dfrac{dB}{dp\text{H}}$ —the buffering power—or the quantity of alkali in millimols necessary to produce unit change of *p*H around the end-point, and may be presumed constant over the small error range in end-point *p*H. This buffering will arise from at least three factors, the hydrogen ion concentration, the carbon dioxide in equilibrium with the air, and the indicator, so that equation 59 may be expanded to

$$\sigma_c = 0{\cdot}04V(\beta_W + \beta_{CO_2} + \beta_I). \tag{60}$$

We can evaluate each of these buffering factors theoretically.

The hydrogen ion buffering. Considering the following two equations,

$$[\text{B}^{\cdot}] + [\text{H}^{\cdot}] = [\text{OH}'] + [\text{A}'] \tag{61}$$

and

$$[\text{H}^{\cdot}] \times [\text{OH}'] = K_W, \tag{62}$$

where equation 61 expresses the necessary relation for electrical neutrality when base is added to a fixed quantity of strong acid and equation 62 the water dissociation, then in differentiating 61 and 62 with respect to $[\text{H}^{\cdot}]$ and re-arranging, we get

$$\frac{d[\text{B}^{\cdot}]}{d[\text{H}^{\cdot}]} = -\frac{[\text{H}^{\cdot}] + [\text{OH}']}{[\text{H}^{\cdot}]}. \tag{63}$$

Also, since

$$\begin{aligned} p\text{H} &= -\log[\text{H}^{\cdot}] \\ &= -0{\cdot}434 \ln[\text{H}^{\cdot}], \end{aligned}$$

then, on differentiating this equation with respect to $[\text{H}^{\cdot}]$ and rearranging, we get

$$\frac{d[\text{H}^{\cdot}]}{dp\text{H}} = -2{\cdot}3[\text{H}^{\cdot}]. \tag{64}$$

From this and equation 63 we obtain

$$\frac{dB}{dp\mathrm{H}}=2{\cdot}3\{[\mathrm{H}^{\cdot}]+[\mathrm{OH}']\} \tag{65}$$

(where B is mols. base added per litre of solution) as millimols per ml. (or mols. per litre) and on the acid side of neutrality we may write

$$\frac{dB}{dp\mathrm{H}}=\beta_H=2{\cdot}3[\mathrm{H}^{\cdot}]. \tag{66}$$

At a pH of 5·5 we obtain β_H as $2{\cdot}3\times10^{-5{\cdot}5}=7{\cdot}3\times10^{-6}$.

The buffering of the carbon dioxide in equilibrium with the atmosphere. Considering the equation

$$[\mathrm{H}^{\cdot}]\times[\mathrm{HCO'}_3]=K[\mathrm{H_2CO_3}], \tag{67}$$

$[H_2CO_3]$ is here a constant when there is equilibrium with the air.

Differentiating with respect to $[\mathrm{H}^{\cdot}]$ and re-arranging, we get

$$\frac{d[\mathrm{HCO'}_3]}{d[\mathrm{H}^{\cdot}]}=-\frac{K[\mathrm{H_2CO_3}]}{[\mathrm{H}^{\cdot}]^2} \tag{68}$$

and from equation 66,

$$\frac{d[\mathrm{HCO'}_3]}{dp\mathrm{H}}=\frac{2{\cdot}3[\mathrm{H_2CO_3}]K}{[\mathrm{H}^{\cdot}]}. \tag{69}$$

Here, owing to the separate consideration of the hydrogen ion buffering above, the change of $[\mathrm{HCO'}_3]$ may be taken as being exactly equivalent to a corresponding addition of base (it being easily demonstrable that for dilute solutions these various buffer factors can be treated quite independently and additively), so that the required buffering equation is

$$\frac{dB}{dp\mathrm{H}}=\beta_{CO_2}=\frac{2{\cdot}3[\mathrm{H_2CO_3}]K}{[\mathrm{H}^{\cdot}]} \tag{70}$$

in millimols per ml.

Where the pH is 5·5 and $[H_2CO_3]$ is $1{\cdot}7\times10^{-5}$ as given above, we obtain the carbon dioxide buffering for methyl red titrations at 5·5 end-point as

$$\frac{2{\cdot}3\times1{\cdot}7\times10^{-5}\times10^{-6{\cdot}4}}{10^{-5{\cdot}5}}=4{\cdot}9\times10^{-6}\text{ millimols per ml.}$$

The indicator buffering. Considering the equation

$$[\mathrm{H}^{\cdot}] \times [\mathrm{I}'] = [\mathrm{HI}] \times K_I$$
$$= (C_I - [\mathrm{I}'])K_I, \tag{71}$$

where C_I is the total indicator concentration, and $[\mathrm{I}']$ the concentration of the ionised substance.

Differentiating with respect to $[\mathrm{H}^{\cdot}]$, where C_I is a constant and equating $d[\mathrm{I}']$ with $d[\mathrm{B}^{\cdot}]$ as in previous paragraph :

$$\frac{d[\mathrm{B}^{\cdot}]}{d[\mathrm{H}^{\cdot}]} = \frac{-C_I K_I}{([\mathrm{H}^{\cdot}] + K_I)^2}$$

and
$$\frac{dB}{dp\mathrm{H}} = \frac{2{\cdot}3 C_I K_I [\mathrm{H}^{\cdot}]}{([\mathrm{H}^{\cdot}] + K_I)^2}. \tag{72}$$

The K_I value of methyl red is $10^{-5{\cdot}1}$ (Clarke, 1928). For methyl red with a molecular weight of 269, an indicator strength of 0·0008% is equivalent to a molar concentration of $2{\cdot}98 \times 10^{-5}$. Inserting these values in equation 72, we get at a pH of 5·5

$$\frac{dB}{dp\mathrm{H}} = \frac{2{\cdot}3 \times 2{\cdot}98 \times 10^{-5{\cdot}0} \times 10^{-5{\cdot}1} \times 10^{-5{\cdot}5}}{(10^{-5{\cdot}5} + 10^{-5{\cdot}1})^2}$$
$$= 14{\cdot}0 \times 10^{-6}.$$

In the example given above for the experimental determination of the variable chemical error, the deliveries of 0·7 ml. of 0·0002N acid containing 0·0008% methyl red, etc., were diluted to 1·05 ml. at the end-point, so that the indicator buffering for this dilution would be

$$14{\cdot}0 \times 0{\cdot}67 \times 10^{-6} = 9{\cdot}4 \times 10^{-6} \text{ millimols per ml.}$$

Assembling the values of

$$\beta_H = 7{\cdot}3 \times 10^{-6}\ ;\ \beta_{CO_2} = 4{\cdot}9 \times 10^{-6} \text{ and } \beta_I = 9{\cdot}4 \times 10^{-6}$$

in the equation

$$\sigma_c = 0{\cdot}04 \times V \times (\beta_H + \beta_{CO_2} + \beta_I),$$

we obtain $0{\cdot}9 \times 10^{-6}$ millimols as the theoretical value of the variable chemical error. The experimental value was $1{\cdot}05 \times 10^{-6}$. The agreement here is reasonably good, and an exact correspondence would in fact be only fortuitous.

Removal or diminution of the variable chemical error. In the foregoing account the titrations were taken as conducted in the presence of atmospheric carbon dioxide. By titrating in a stream of air free from carbon dioxide the effect of this buffer is removed and we can approach nearer to the neutral point. We are still left, however, with some buffering both by hydrogen and hydroxyl ions plus that of the indicator. The latter remains unchanged provided the colour change with pH can be appreciated with equal delicacy. Under such circumstances we may consider what is the variable chemical error at a pH of 7·0. We have

$$\sigma_c = 0{\cdot}04(\beta_{(H^{\cdot}+OH')} + \beta_I) \times V.$$

In general we could titrate micro volumes, using indicator concentrations of about $1{\cdot}5 \times 10^{-5}M$ or even less at the end-point, that is, with about half the indicator buffering previously considered, so that

$$\sigma_c = 0{\cdot}04(2 \times 10^{-7} + 7 \times 10^{-6}) = 0{\cdot}3 \times 10^{-6} \text{ millimols/ml.},$$

which represents the minimum variable chemical error or a figure reasonably close thereto.

In such titrations it is assumed that the alkali used is perfectly free from carbon dioxide, the merest trace of which will introduce a comparatively serious error. Such an ideal alkali is approached in practice only when we use barium hydroxide.

The effect of absorbed ammonia or other weak base on the variable chemical error in acidic titrations. Where the acid has been used to absorb ammonia, on approaching a pH of 7·0 we introduce a new error due to the buffering produced by the absorbed ammonia. It is of some interest to inquire into the magnitude of such an error.

The ammonium ion may be conveniently regarded as a weak acid dissociating into ammonia gas and hydrogen ion, with a dissociation constant of approximately $10^{-9{\cdot}4}$. Writing then the general buffering equation for a weak acid as developed above for the indicator buffering (equation 72), we have

$$\frac{dB}{dp\text{H}} = \frac{2{\cdot}3 \times 10^{-9{\cdot}4} C_{am} \times [\text{H}^{\cdot}]}{(10^{-9{\cdot}4} + [\text{H}^{\cdot}])^2}$$

and (from 59),

$$\sigma_c = 0{\cdot}04\frac{dB}{dp\text{H}} \times V = \frac{0{\cdot}092 \times 10^{-9{\cdot}4} C_{am} \times [\text{H}^{\cdot}]}{(10^{-9{\cdot}4} + [\text{H}^{\cdot}])^2} \times V.$$

At pH values at or below 7·0, $10^{-9\cdot4}$ in the denominator is negligibly small when compared with [H˙], so that

$$\sigma_c = \frac{0\cdot092 \times C_{am} \times 10^{-9\cdot4}}{[\mathrm{H}^{\cdot}]} \times V. \tag{73}$$

The variable error at the end-point is therefore proportional to the amount of absorbed ammonia present. If, however, it is the ammonia and not the acid that is being determined, then the *relative* percentage error will be

$$\frac{0\cdot092 \times 10^{-9\cdot4}}{[\mathrm{H}^{\cdot}]} \times 100 \tag{74}$$

or a figure independent of the ammonia absorbed and dependent only on the pH and the dissociation constant. Putting in the end-point pH, we get $0\cdot11 \times 10^{-2\cdot0}$ or a quite negligible figure at 5·5 pH. At a pH of 7·0 it becomes 0·04% or something just appreciable.

A similar treatment may be extended to other weak bases.

CHAPTER LIX

THE RATIONALE OF MICRO TITRATION

THE general variable error of titration is given by the equation already considered (58),

$$v_s^2 = v_g^2 + \left(\frac{\sigma_c}{q} \times 10^2\right)^2$$

σ_c, apart from the indicator, being proportional to the volume at the end-point. For special conditions the form of the equation will alter somewhat, as considered below.

We have seen that for exposed titrations with methyl red σ_c is of the order 1×10^{-6} millimol for each ml. of end-point volume with a possible reduction to $0{\cdot}3 \times 10^{-6}$ in a stream of carbon dioxide free air when titrating at a pH of 7·0, though for volumes over a few ml. the titration in a current of carbon dioxide free air, if it is to be conducted efficiently, will be found an inconvenient procedure. These figures apply when the only buffering present at the end-point is the hydrogen and hydroxyl ions of the water, the carbon dioxide of the atmosphere and the indicator. It has also been shown that when ammonia is being determined it begins to introduce a just appreciable buffering of its own at a pH of 7·0, the variable error being then 0·04% and independent of the ammonia concentration.

A certain glass error becomes unavoidable in titration, but where there is no restriction on quantity of material the chemical error above can be reduced to any level considered negligible by increasing q or the quantity examined. The introduction of an appreciable variable chemical error of the type considered would therefore be merely incorrect technique when we have a large amount of substance. When, on the contrary, there is only a small amount of material for analysis, we may be unable to increase q sufficiently to avoid an appreciable chemical error. In this case we must proceed by diminishing the end-point volume. *This is the chief reason for the employment of a micro volumetric procedure.*

Were it not for this question of variable chemical error, after an initial micro measurement either of volume or quantity we could conduct the remaining analysis by macro volumetric procedure subsequent to dilution.

It will be seen that at the same time any constant deviation of the final *p*H from that of the ideal end-point becomes proportionally less as we increase the concentration for titration, or decrease the volume with fixed quantity. This applies to the titration of weak acids and bases as well as to strong, and may be derived formally from the equations in the chapter on the constant chemical error.

The value of a micro technique in this restricted sense has been already pointed out by Rehberg (1925), but since the titration error referred to is a constant error and may be theoretically determined and so allowed for, it is not so important a reason as that previously given.

With many precipitates, when these are subsequently worked up volumetrically or otherwise, it is necessary also to keep the fluid volume low to avoid appreciable solution.

Besides such reasons of accuracy, we have others also which favour volume reduction with reduced quantities. Many standard solutions, for example, maintain their effective concentrations better in concentrated than in weak solutions, and this applies not only to reagents like permanganate and thiosulphate, but also to ordinary standard acids and alkalis.

Finally, we have reasons of economy, not only of material but often of time, in favour of a micro procedure. Summarising, we may say that the following reasons may determine the choice of a micro volumetric procedure throughout :

(1) The diminution of the variable chemical error, often the compelling reason for accurate work.

(2) Diminution of the effect of constant chemical error of the end-point when this deviates from the ideal.

(3) Diminution of loss by solution with small precipitates intended for quantitative measurement.

(4) The better keeping value of concentrated solutions.

(5) Reasons of economy in material, and often in time.

In the reasons 1 and 2, acidimetric titration is implied, though analogous considerations hold for other volumetric titrations in which there is a change of one spectral colour to another at the end-point. Here as before we will have a variable error of colour discrimination which will be related to concentration of the titrating reagent in excess or defect. An apparent exception is provided in the change from no colour to a coloured end-point. This will be virtually a change from white to some spectral colour, and we can increase the depth of the latter and hence our discrimination by merely increasing the depth of fluid looked through without altering the concentration. The choice of a micro volumetric technique will here depend rather on the other reasons given and the inconvenience of titrating long columns of fluid.

Experimental exemplification of the interplay of the variable glass and chemical errors. A large series of titrations—167 in all—were carried out with the same glassware. At the time of these titrations the fuller implications of the error of titration as developed here were not known, but they serve as an illustration of the main reason on which the use of the micro volumetric procedure turns, and of the interplay of the variable glass and chemical errors.

In the series 1 ml. of standard acid was delivered from a Bang burette into a ' unit ', 0·1 ml. Tashiro's reagent pipetted into the acid and the mixture titrated with standard carbon dioxide free NaOH of the same strength as the acid. Successive series of such titrations were carried out varying the acid and alkali strengths in each series, but keeping the conditions otherwise the same.

These titration series may be examined in relation to the equation (already considered in Chapter LVIII),

$$v_s^2 = v_g^2 + \left(\frac{\sigma_c}{q} \times 10^2\right)^2,$$

where v_s is the variable error of the full titration, v_g that of the glass, σ_c the variable chemical error, and q the quantity titrated. The total glass error can be determined by titrating a series of acid deliveries so strong that the chemical error is negligible, but it may also be determined from its individual components of pipette error (here the delivery error from a Bang burette), burette delivery error and manipulation error. These are related, as already discussed in Chapter XXXIX, by the equation :

$$v_g^2 = v_p^2 + v_b^2 + \left(\frac{\sigma_m}{10V}\right)^2.$$

The following values were found experimentally

$$v_p = 0{\cdot}11\%,$$
$$v_b = 0{\cdot}11\%,$$
$$\sigma_m = 1{\cdot}3 \text{ c.mm.},$$
$$V_b = 1{\cdot}0 \text{ ml.}$$

The total variable glass error therefore becomes

$$v_g = \sqrt{0{\cdot}11^2 + 0{\cdot}11^2 + 0{\cdot}13^2}$$
$$= 0{\cdot}203\%.$$

In a similar way we could build up the variable chemical error from the water, carbon dioxide and indicator buffering and the variable pH error of the end-point as a standard deviation. The total variable chemical error was, however, determined from a titration series of acid so weak that the glass error did not appreciably enter, and was found to be $1{\cdot}92 \times 10^{-6}$ millimols.

The result of the series of titrations are then seen in the following table (and in Fig. 76), in which the experimental and theoretical results (from equation 58) are given.

TABLE XXXI

Normality of acid and alkali (N).	*No. of titrations.*	*Total variable error of titration (v_s).*		*Titration error expressed as millimols of alkali.*	
		Exp.	*Theor.*	*Exp.*	*Theor.*
0·1	52	0·20	0·20	$2{\cdot}00 \times 10^{-4}$	$2{\cdot}00 \times 10^{-4}$
0·01	26	0·19	0·20	$1{\cdot}90 \times 10^{-4}$	$2{\cdot}00 \times 10^{-5}$
0·001	20	0·30	0·27	$3{\cdot}00 \times 10^{-6}$	$2{\cdot}70 \times 10^{-6}$
0·0005	18	0·45	0·42	$2{\cdot}25 \times 10^{-6}$	$2{\cdot}10 \times 10^{-6}$
0·0002	51	0·95	0·96	$1{\cdot}90 \times 10^{-6}$	$1{\cdot}92 \times 10^{-6}$
0·0001	—	—	1·88	—	$1{\cdot}88 \times 10^{-6}$

The question of ideal volumes. When the need of a micro technique occurs in securing a desired accuracy of determination, the question arises as to how far we are to reduce the volume. Unless we have some guiding principles here to go by, the extent of this reduction can only be arbitrary, or determined mostly by the micro apparatus at our disposal.

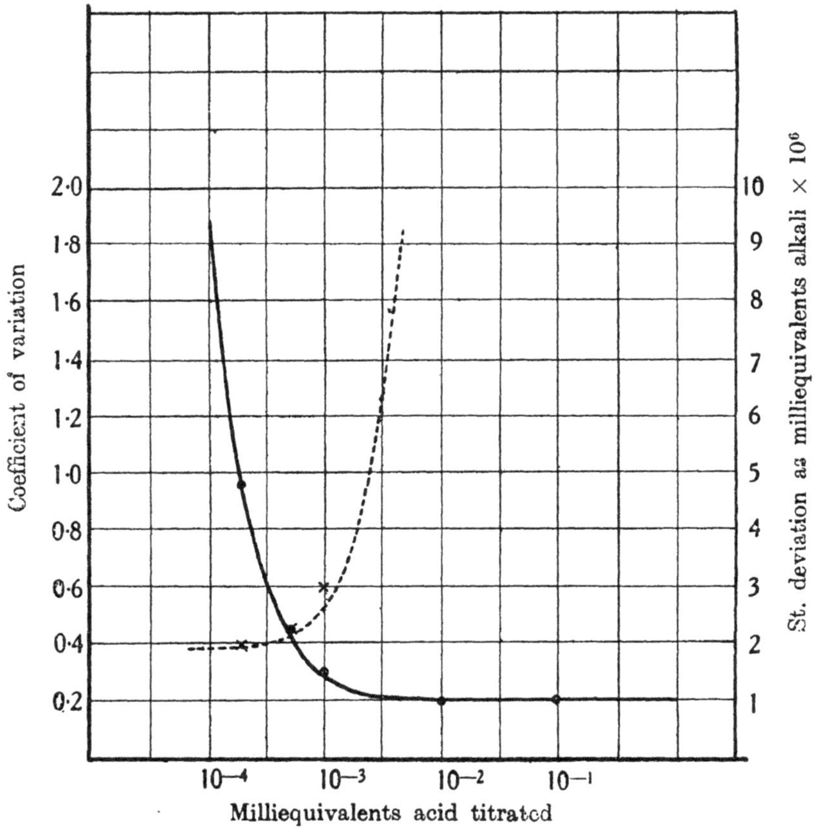

FIG. 76. The curves show the variable error in titrating 1 ml. hydrochloric acid with 1 ml. carbon dioxide free alkali from a 2 ml. Bang burette, using the Tashiro indicator (0·1 ml.). (*Vide* text.)

Dots and crosses give the experimental points, the curves being theoretical.

The dotted curve gives the variable error in millimols alkali (*KOH*).

If we consider the pipettes and burettes in actual use, we shall find that as the glassware is reduced in volume to deliver smaller and smaller amounts, the percentage error of the delivery increases.

As previously shown, this need not be so, but it is a fact of the conventional usage.

It has been shown in Chapter LVIII that the variable chemical error decreases as the volume decreases, so that at some point we must have an ideal volume when the error is least. We could solve for such a volume if a relation were found connecting the conventional increase of error with decrease of *glass volume.* This, however, would be a doubtful procedure, for within limits we can have almost any degree of glass accuracy, and it is preferable to consider firstly what degree of accuracy we want, and then to relate our volumes accordingly, and this we may do approximately as follows.

A general principle of micro volumetric limitation. In general it may be said that the end-point volume in micro volumetric procedure should be such that the variable chemical error is at most half the variable glass error that we require. From our glassware we can then secure the accuracy we want in accordance with the principles already outlined. If, for example, an accuracy at the 0·1% (coefficient of variation) level is desired, the end-point volume should give a variable chemical error not greater than 0·05%, when the total error of titration with a glass error of 0·1% will not exceed 0·11%. Further volume reduction will secure practically no advantage and only serve to increase the difficulty with which is obtained the necessary glass accuracy.

The question then remains as to how we are to predict the volume that gives us a variable chemical error of, say, 0·05%. Here we may assume as a working relation that the variable chemical error goes proportionally to the end-point volume, though this strictly necessitates the indicator concentration being always the same therein. We may write then

$$\frac{V\sigma_c}{q} \times 10^2 = 0{\cdot}5v_g, \tag{75}$$

where V is the end-point volume, σ_c the variable chemical error in millimols per ml. end-point volume, q the millimols titrated and v_g the glass error at the level required.

For acidimetric titration with methyl red σ_c may be taken as approximately 1×10^{-6}, so that in the above example if q were 10^{-3} and $v_g = 0{\cdot}1$, V would require to be 0·5 ml. If our figures were exact it would not be an advantage to have the end-point volume less than 0·5 ml. Since, however, the variable chemical error cannot be accurately predicted for a given set of conditions, it would here be safer to take somewhat less than the volume so derived.

A special principle of micro volumetric limitation for use with a burette, and fixed pipetted volumes. When we are restricted to the use of a special burette and a given pipetted volume it may perhaps be of some interest to note that there is an ideal volume of alkaline solution to add from the burette corresponding to the quantity of acid titrated. The consideration of such ideal volumes is useful only as a working guide and within a certain range of q. For the methyl red with methylene blue titrations already considered the figure may be arrived at as follows.

We may suppose, for example, that the pipetted acid always contains methyl red in the concentration of 0·0008% with the usual relation of methylene blue, etc. The total error (v_t) of the titration may be written

$$v_t^2 = v_p^2 + v_b^2 + \left(\frac{\sigma_m}{10 V_b}\right)^2 + \frac{10^4}{q^2}\{(V_b + V_p) \times w + V_p \times i\}^2. \quad (76)$$

In this equation v_p and v_b are the pipetted and buretted delivery errors respectively as coefficients of variation, σ_m is the manipulation error in c.mm., V_b and V_p the buretted and pipetted volumes respectively, w is the water and carbon dioxide buffering and i is the indicator buffering, both being multiplied by the standard deviation of the colour change at the end-point expressed as pH units.

On differentiating, equating to zero and re-arranging, we get

$$V_b \left\{1 + \frac{V_p}{V_b}\left(1 + \frac{i}{w}\right)\right\}^{0\cdot 25} = 0\cdot 032\sqrt{\frac{q\sigma_m}{w}}. \quad (77)$$

When V_b is of the same order or greater than V_p and the indicator strength is maintained as low as conveniently possible, the fourth root of the bracketed expression on the left will approximate to unity, and we get as a working rule

$$V_b = 0\cdot 03\sqrt{\frac{q\sigma_m}{w}}. \quad (78)$$

Let us suppose, for example, we wished to know the ideal buretted volume in titrating 1 ml. of $10^{-3}\,N$ acid, using the methyl red (or methyl red/methylene blue) end-point. At this end-point the water and carbon dioxide (in equilibrium with air) buffering is 12×10^{-6} millimols. If a standard deviation of the colour change at the end-point of 0·04 pH is assumed, then w becomes $0{\cdot}5 \times 10^{-6}$.

We may then write

$$V_b = 42\sqrt{\sigma_m q} \tag{79}$$
$$= 1{\cdot}4\sqrt{\sigma_m},$$

whence $$q = 10^{-3}.$$

With the Bang 2 ml. burette when σ_m is 1·3 (c.mm.) the value of V is 1·5 ml. With the horizontal burette when σ_m is 0·30 (minimal emergence of 1 c.mm.) and q is $0{\cdot}2 \times 10^{-3}$ we get 0·24. The value of σ_m or the manipulation error is very largely due to the end-point emergence which gives an error, as already mentioned, expressed by $\sqrt{\frac{l^2}{12}}$, where l is the emergence in c.mm.

It is scarcely necessary to say that these figures are only approximate, but afford at least a rational basis for the order of limitation in micro volumetric titration.

CHAPTER LX

THE CONSTANT GLASS ERROR

THE constant glass error as here defined is the difference between the mean value of a large number of single deliveries (or fillings) and the nominal value under fixed conditions of temperature, rate of delivery, drainage, etc.

The constant error may always be exactly allowed for in contrast to the variable error. We can know the probable magnitude of this latter, but under the fixed conditions of our procedure we cannot alter or remove it.

Determination of the constant glass error. This determination amounts to the standardisation of the glassware in the usual sense, and may be carried out by determining the mean weight of five deliveries or fillings under the fixed working conditions of the particular glassware.

The standard deviation of the mean value taken from a series of observations is given by

$$\sigma_m = \frac{\sigma}{\sqrt{5-1}} = 0{\cdot}5\sigma, \qquad (80)$$

where five deliveries are taken and σ is the standard deviation of the single delivery from the true mean. σ_m will add to the standard deviation of the single delivery as if it were itself an uncorrelated variable error, so that we have the sum as

$$\sigma_s = \sqrt{\sigma^2 + (0{\cdot}5\sigma)^2} \qquad (81)$$
$$= 1{\cdot}12\sigma.$$

It will be seen that by taking five deliveries the sum is very little different from the true standard deviation, so that the five delivery mean functions here for all practical purposes as the true mean. Where a volumetric flask is being standardised only one filling need be weighed since the variable error is negligible.

The five deliveries are made into a tared weighing bottle or into a beaker covered with a watch glass, and from the mean weight the volume is determined. In the ensuing section we shall consider how the volume is obtained from the weight at a given temperature and how it may be corrected for a standard temperature.

Determination and standardisation of the volume from the mean weight at a given temperature. When the mean weight has been found, the volume at the given temperature or the volume at some standard temperature is then determined from suitable tables (Tables XXXII, XXXIII).

We may firstly consider a useful rule for working with factors close to unity. The multiplication or division of such factors can be treated as additions and subtractions.

Thus $$(1+x)(1+y)=1+x+y+xy, \tag{82}$$

but if x and y are less than one part in 100 or less than 1%, then xy may be neglected and we have

$$(1+x)(1+y)=1+x+y. \tag{83}$$

Similarly $$\frac{1+x}{1+y}=(1+x)(1+y)^{-1}$$
$$=(1+x)(1-y+y^2-y^3, \text{ etc.})$$
$$=(1+x)(1-y),$$

(since y^2, etc., may be neglected)

$$=1+x-y, \tag{84}$$

as in the first example.

If we consider the 1 in these factors as understood we can then always work as if they subtracted or added irrespective of their being in fact divisions or multiplications. With the various small corrections to be considered we need only write these down at once with due consideration of their signs, and instead of being parts in 1, we shall consider them to be parts in 100 to avoid undue writing of zeros. Not only can the various temperature and density corrections be treated in this fashion, but also the final allowances for the constant errors in titrations.

If, for example, we have a number of such to deal with and write them as

$$a+b-c+d, \text{ etc.},$$

and require to correct a figure X, then

$$X \times \left(1 + \frac{a+b-c+d, \text{ etc.}}{100}\right)$$

gives the required correction.

In the correction of an observed weight of a delivery or a filling of distilled water, we have four possible corrections to consider.

(1) **The density correction.** If the weight of a given volume of water *in vacuo* is W, then

$$V = \frac{W}{\rho}, \tag{85}$$

where V is the volume and ρ is the specific gravity. ρ may be written as $\left(1 - \frac{s}{100}\right)$ and, according to the above treatment, we may write

$$V = W\left(1 + \frac{s}{100}\right), \tag{86}$$

where s is 100 times the specific gravity subtracted from unity. The value of s is given in Table XXXII.

(2) **The air displacement corrections.** The weight in air will be smaller than the weight *in vacuo* by the weight of the air displaced by the water over that of the brass weights. This correction, though small, is larger than may seem likely on a first approach, being approximately 0·1% of the total weight.

If V is the volume of water, then $V\left(1 - \frac{\rho_W}{\rho_b}\right)$ is the volume of air displaced where ρ_W is the density of the water and ρ_b the density of the brass weights. This volume of air multiplied by the density of air $(=\rho_a)$ at the given temperature and barometric pressure will give the necessary correction.

The full expression for the weight of air displaced in excess of the weights by 100 ml. water is given by

$$a = 100\left(1 - \frac{\rho_W}{\rho_b}\right) \times \rho_a \times \frac{273}{T} \times \frac{P}{760}, \tag{87}$$

where ρ_a is the density of the air at zero centigrade and T and P are the temperature (absolute) and barometric pressure.

Here ρ_b is 8·4 and ρ_a, 0·00129.

The correction for differences of barometric pressure may be neglected since it will introduce at ordinary levels only a negligible variation.

TABLE XXXII

Temp.	*s*	*a*	*g*	*f*
10°	0·027	0·109	0·026	0·064
11	0·037	0·109	0·023	0·058
12	0·048	0·109	0·021	0·051
13	0·060	0·108	0·018	0·045
14	0·073	0·108	0·016	0·038
15	0·087	0·107	0·013	0·032
16	0·103	0·107	0·010	0·026
17	0·120	0·107	0·008	0·019
18	0·138	0·106	0·005	0·013
19	0·157	0·106	0·003	0·006
20	0·177	0·105	0·000	0·000
21	0·198	0·105	– 0·003	– 0·006
22	0·220	0·105	– 0·005	– 0·013
23	0·244	0·104	– 0·008	– 0·019
24	0·268	0·104	– 0·010	– 0·026
25	0·293	0·103	– 0·013	– 0·032
26	0·319	0·103	– 0·016	– 0·038
27	0·346	0·103	– 0·018	– 0·045
28	0·374	0·102	– 0·021	– 0·051
29	0·403	0·102	– 0·023	– 0·058
30	0·433	0·101	– 0·026	– 0·064

For the purpose of this small correction also the volume of water may be taken as equivalent to the weight of the water.

The required correction *a* is listed in Table XXXII, and is there 100 times the correction for 1 g. water.

Combining this allowance with the previous we may write

$$V = W_a\left(1 + \frac{s+a}{100}\right), \tag{88}$$

where W_a is now the weight in air and V the volume.

(3) **The correction for glass expansion.** If we have determined the volume of the delivery or filling at a given temperature, when we use the same glassware at another temperature the volume of the glass will have altered, requiring a correction, therefore, for the glass expansion. This is obviously a correction for use when we are permanently standardising the glass and is quite unnecessary where we want the actual volume after weighing of the immediate delivery or filling. It is convenient to make the

correction to some standard temperature and to correct subsequently from this by a table for varying room temperatures.

It is clear that the internal volume of the glassware will alter just to the same extent as if the whole of the internal air space of the empty vessel were solid glass. We must consider therefore the cubical expansion of V the volume of glass.

The cubical expansion of glass is given as 0·000026 per ml. per degree in the International Critical Tables, so that for V ml. at $t°$, we have the excess volume at 20° :

$$(20 - t°) \times 0{\cdot}000026\,V,$$

and for 100 ml., or practically 100 g. of the weighed water, it will be

$$(20 - t°) \times 0{\cdot}0026.$$

This percentage correction is listed in Table XXXII as g, so that

$$V = W_a \left(1 + \frac{s + a + g}{100}\right), \tag{89}$$

where V is the volume of a volumetric flask to the mark at 20° when the weight in air of the filling at $t°$ is W_a.

The magnitude of this g correction for a divergence of plus or minus 10° from 20° is 0·026, and could therefore be neglected in practice, but is usually included in a formal treatment.

(4) **The correction for changes in wall fluid with temperature.** This correction does not apply to the volumetric flask in which the fluid volume is not delivered but made up to the mark. The volume of fluid left behind on the walls changes with temperature as determined from the measurement of this volume with the 25 ml. standard pipette. This effect has already been given in Table XXIII.

The wall fluid increases as the temperature decreases, and like this will diminish the volume delivered. For every 100 ml. delivered the wall fluid decreases 0·0064 ml. for every degree rise of temperature.* The correction is listed in Table XXXII as f and for a range of plus or minus five degrees this correction amounts to 0·036%. We have then for the combination of the four corrections to standardise a pipette or burette

$$V = W_a \left(1 + \frac{s + a + g + f}{100}\right). \tag{90}$$

Using these four basic corrections, we may consider the various

* See note at end of chapter.

instances in which they arise. These will be considered formally, with subsequent consideration of the corrections we may omit or consider unnecessary to apply in general.

From the principle of treating small errors given at the beginning of the chapter, the required combinations of the corrections can be written down at once as additions or subtractions after a consideration of sign.

I. Actual volume of a weighed quantity of water at any room temperature.

Here we have s, the density correction, and a, the air displacement correction, to consider. Both will obviously be positive, so that

$$V = W_a\left(1 + \frac{\alpha}{100}\right), \tag{91}$$

where $\alpha = s + a$. α is given in Table XXXIII.

II. Standardisation to 20° of a volumetric flask, the weight of the filling being taken at t°.

Here we have to consider s, a and g, the last being the glass expansion correction. We have to consider only the sign of g. If it is less than 20° the effect of g must be positive since the volume will expand on heating to 20°. Making g always positive in the equation will mean positive values below 20° in table and negative values above, as already considered. We have then

$$V = W_a\left(1 + \frac{\beta}{100}\right), \tag{92}$$

where β is $(s + a + g)$ and is listed in Table XXXIII.

III. Standardisation of pipette or burette mean delivery to a standard temperature of 20° ; the mean weight of five deliveries being taken at t°.

Here we have s, a, g and f to consider. We have then

$$V = W_a\left(1 + \frac{\gamma}{100}\right), \tag{93}$$

where γ is $(s + a + g + f)$.

The value of γ is listed in Table XXXIII.

IV. A volumetric flask standardised for 20° is used at some other temperature, t°, the volume correction being required.

The sum of the corrections for IV and II must give that for

TABLE XXXIII

Temp.	α	β	γ	δ	ϵ	ζ	η	θ
10°	0·136	0·162	0·226	− 0·026	− 0·090	− 0·124	− 0·060	0·150
11	0·146	0·169	0·227	− 0·023	− 0·081	− 0·117	− 0·059	0·140
12	0·157	0·178	0·229	− 0·021	− 0·072	− 0·108	− 0·057	0·129
13	0·168	0·186	0·231	− 0·018	− 0·063	− 0·099	− 0·054	0·117
14	0·181	0·197	0·235	− 0·016	− 0·054	− 0·088	− 0·050	0·104
15	0·194	0·207	0·239	− 0·013	− 0·045	− 0·077	− 0·045	0·090
16	0·210	0·220	0·246	− 0·010	− 0·036	− 0·066	− 0·038	0·074
17	0·227	0·235	0·254	− 0·008	− 0·027	− 0·049	− 0·030	0·057
18	0·244	0·249	0·262	− 0·005	− 0·018	− 0·034	− 0·021	0·039
19	0·263	0·265	0·271	− 0·003	− 0·009	− 0·017	− 0·011	0·020
20	0·282	0·282	0·282	0·000	0·000	0·000	0·000	0·000
21	0·303	0·300	0·294	0·003	0·009	0·018	0·012	− 0·021
22	0·325	0·320	0·307	0·005	0·018	0·038	0·025	− 0·043
23	0·348	0·340	0·321	0·008	0·027	0·059	0·040	− 0·067
24	0·372	0·362	0·336	0·010	0·036	0·081	0·055	− 0·091
25	0·396	0·383	0·351	0·013	0·045	0·103	0·071	− 0·116
26	0·422	0·406	0·368	0·016	0·054	0·126	0·088	− 0·142
27	0·449	0·431	0·386	0·018	0·063	0·151	0·106	− 0·169
28	0·476	0·455	0·404	0·021	0·072	0·176	0·125	− 0·197
29	0·505	0·482	0·424	0·023	0·081	0·203	0·145	− 0·226
30	0·534	0·508	0·444	0·026	0·090	0·230	0·166	− 0·256

I, since if a pipette is standardised to 20° at a room temperature of $t°$ and then wanted for immediate use, obviously on correcting back to $t°$, we would obtain the actual volume of the delivery examined, or the value considered in I, consequently the required correction is

$$V_t = V_{20}\left(1 + \frac{\delta}{100}\right), \tag{94}$$

where δ is $(s+a)-(s+a+g) = -g$.

δ is listed in Table XXXIII and is merely minus the g value of the previous table.

V. A pipette or burette standardised for 20° is used at some other temperature, t°, the extra volume correction for delivery being required.

Here as with IV the sum of the corrections in II and V must give that in I, so that we may write

$$V_t = V_{20}\left(1 + \frac{\epsilon}{100}\right), \tag{95}$$

where ϵ is $(s+a)-(s+a+g+f)$ or $-(f+g)$.

This correction is listed in Table XXXIII.

Reduction in the number of the volume corrections. For work short of the finest all the above corrections may be reduced to one, namely, α, i.e. for β and γ we may write α, and consider δ and ϵ as negligible. α is the percentage factor for converting a weighed quantity of water to a volume.

Example of the corrections and factors in use.

Mean weight of five deliveries from a 25 ml. pipette 24·9275 g.

Temperature of water 23·5°.

It is required to standardise the pipette for 20°.

The value of $\frac{(24{\cdot}9275 - 25{\cdot}0000)}{25} \times 100 = -0{\cdot}290$,

so that
$$W = 25\left(1 + \frac{(-0{\cdot}290)}{100}\right).$$

From I above the γ correction is required, so that

$$V = 25\left(1 + \frac{\gamma - 0{\cdot}290}{100}\right)$$
$$= 25\left(1 + \frac{0{\cdot}038}{100}\right),$$

γ from Table XXXIII being 0·328 at 23·5°.

The factor of the pipette for 20° is therefore most conveniently expressed as $P_{20} = +0{\cdot}038$.

This factor may then be kept as a constant standardising factor of the 25 ml. pipette at 20°.

The actual volume of the delivery at 20° from the above is

$$25\left(1 + \frac{0{\cdot}038}{100}\right) \quad \text{or} \quad 25{\cdot}0097 \text{ ml.}$$

2nd example. A 25 ml. pipette has a factor at 20° (P_{20} as in above example) of $+0{\cdot}084$, and is used at a room and water temperature of 25°. What is the factor for this temperature?

We have from IV above

$$V = V_{20}\left(1 + \frac{\delta}{100}\right)$$
$$= 25\left(1 + \frac{P_{20}}{100}\right)\left(1 + \frac{\delta}{100}\right)$$
$$= 25\left(1 + \frac{P_{20} + \delta}{100}\right).$$

P_{20} is $+0{\cdot}084$ and δ from Table XXXIII is 0·013, so that the factor for 25 degrees is

$$0{\cdot}084+0{\cdot}013=0{\cdot}097.$$

Thus the final factor is got at once by reading δ from Table XXXIII and adding to P_{20}.

From the above two examples it will be seen that all the glassware should be standardised for 20° with their respective percentage factors noted. From these any other correction is very easily obtained simply by adding or subtracting the figure obtained from the table.

The other factors of Table XXXIII are explained in the ensuing chapter concerning the constant chemical correction.

The corrections with analytical solutions. The corrections considered above may be taken to apply when using all volumetric solutions up to $0{\cdot}1N$ concentrations, and even with solutions of normal concentration, with some few exceptions.

With regard to these exceptions, the volumes delivered from a 100 ml. pipette of a number of ordinary reagents up to normal strength have been investigated by Schloesser (1903), and with the exception of normal sodium carbonate, Fehling's solution II, ferric chloride (1 ml. $=0{\cdot}012$ g. Fe) and $0{\cdot}1N$ iodine, the volume was inappreciably different from a similar delivery of water. Even these exceptions differed only by the following amounts: $-0{\cdot}035\%$, $-0{\cdot}198\%$, $-0{\cdot}035\%$ and $+0{\cdot}025\%$ respectively.

Concerning the density changes of such solutions with temperature, these practically agree with water up to $0{\cdot}1N$ strength, and again even at $1{\cdot}0N$ strength differ but little. An exception to this is $1{\cdot}0N$ sodium hydroxide, the density change of which is about 50% greater than that of water around 20°.

Note on the temperature correction for wall fluid (f, *Table XXXII*).

Since the temperature effect is probably proportional to the volume of wall fluid, this varying in accordance with the factors described in Chapter LV, it follows that the f correction listed is valid only for standard glassware within a certain volume range. For 20–50 ml. deliveries, however, it may be taken as giving an accuracy to 0·01% over ± 5 degrees C. Further adjustment of the correction to varying wall fluid may be made from the relation this bears to the variable error. It may be assumed in short that the factor f will vary as $x/0{\cdot}025$ where x is the variable error and 0·025 that of the standard 25 ml. pipette. Factors γ, ϵ and η in Table XXXIII will also be affected since they contain the correction f.

CHAPTER LXI

THE CONSTANT CHEMICAL ERROR

THE constant chemical error in titration may be considered under three headings :

(1) Constant error of impurity in the standardising substance.

(2) Constant error in the standard solutions under the given conditions, the actual strength differing from the nominal.

(3) Constant error resulting from deviation of the mean end-point from the ideal end-point.

(1) **Constant impurity in the standardising substance.** The standardising substances such as sodium oxalate, sodium carbonate, potassium biphthalate, benzoic acid, etc., are now prepared commercially to a degree of purity of about 99·9%, referring to the dried substance.

For acidimetric standardisation of standard acidic solutions sodium oxalate (equivalent weight 66·98) is probably the best, and has been strongly recommended by Sørensen (1897), by Kolthoff (1931) and others.

It has the advantage over sodium bicarbonate (which is converted to carbonate on heating and subsequently weighed as such) in that an excess loss of carbon dioxide with conversion to hydroxide on heating is of no consequence, the substance being accurately weighed for use before it is decomposed.

Sodium oxalate can be prepared to an accuracy of at least 99·96% purity without much difficulty.

It may also be mentioned that the preparation of standard hydrochloric acid to a possible accuracy of 99·99% is claimed for the Hulett and Bonner (1909) constant boiling mixture methods.

Expression of the constant error in the standardising substance. This may be expressed in a similar way to the constant errors of the glassware, thus :

$$M = M_0 \left(1 + \frac{i + a_m}{100}\right). \qquad (96)$$

Here M is the actual weight of the perfectly pure substance, M_0 is the nominal weight in air, i and a_m are the impurity and air displacement errors as percentages. In the particular instance of sodium oxalate as prepared above, these two are opposite in sign and practically identical in amount, so that the whole correction for impurity and air displacement could be neglected.

(2) **Constant error in the standard solutions.**

(*a*) *An equivalent weight of a solid or some fraction thereof is weighed and dissolved in water to the mark in a volumetric flask at t° ; it is required to standardise the solution for 20°.*

The flask may be at first supposed to have exactly the nominal volume at $t°$. After making up the solution to the mark, if the flask and contents are changed from $t°$ to 20° the meniscus will rise or fall above or below the mark, the glass expansion itself having no effect on the alteration of strength produced. This is referable only to the density change.

If the temperature is below 20°, then on changing to 20° the volume increases and the solution strength decreases, so that the ultimate sign of the correction must then be minus, and we may write $(s_t - s_{20})$ as the factor, since this from Table XXXII gives a minus sign.

In the above we proceeded as if the flask had no factor at $t°$. Assuming its factor at 20° to be F_{20}, then its factor at $t°$ (F_t) is given by $F_{20} - g$, where g is the glass expansion factor in Table XXXII. If the sign of this factor is plus its effect on the strength of the solution must be minus, so that the whole factor is

$$\begin{aligned} S_{20} &= (s_t - s_{20}) - (F_{20} - g) \\ &= s_t - s_{20} + g - F_{20} \\ &= \zeta - F_{20}, \end{aligned} \tag{97}$$

where ζ is $(s_t - s_{20} + g)$ and is listed in Table XXXIII.

(*b*) *An exactly weighed standardising solid is dissolved and titrated from a burette at t°. It is required to standardise the burette solution for 20°.*

The solution is standardised firstly as follows :

If X is the expected number of ml. and $(X + x)$ the actual, then we may write

$$\frac{(X + x) - X}{X} \times 100 = D. \tag{98}$$

D will be positive if x is positive, but the strength of the solution must then decrease from the nominal, so that D will require a minus sign.

For standardisation we must now combine the factor for the density change and the factor of the burette.

The density and glass change will enter in the same way as in the previous example, but the effect of temperature on the wall fluid must be also considered, so that

$$S_{20} = \zeta + f - B_{20} - D,$$

(where B_{20} is the burette percentage factor)

$$= \eta - B_{20} - D. \quad (99)$$

η is listed in Table XXXIII and is given by $(s_t - s_{20} + g + f)$ from Table XXXII.*

(c) *A pipette delivery of a fluid standardised for 20° is titrated from a burette at t°. It is required to standardise the burette fluid for 20°.*

Here the effect of the density change and of wall fluid will be the same for both fluids and may be omitted. We can regard the whole titration as taking place at 20°.

The required standardisation factor is here

$$S_{20} = P_{20} - B_{20} - D, \quad (100)$$

where P_{20} and B_{20} are the pipette and burette factors, and D the value from equation 98, being the immediate titration factor.

It is obvious that the pipette and burette factors must go in opposite ways. If the pipetted fluid has already a factor at 20 (S'_{20}), this is included as

$$S_{20} = P_{20} - B_{20} + S'_{20} - D. \quad (101)$$

The above three are standardisation corrections for solutions.

If the solutions are standardised for 20°, then on titrating at other temperatures no further temperature corrections need be considered, since glass and solutions in the different glassware will

* See note at end of previous chapter.

be equally affected. It may be noted, however, that wherever we require to titrate a measured volume against a weighed quantity we then need to correct for altered temperature.

(*d*) *A solution in bottle has a factor for 20°* ($=S_{20}$) ; *its factor at t° is required.*

Here as special factor we require to consider only the density change. The colder the solution the more it increases in strength, and since $s_{20}-s_t$ is positive for t values below 20, the required factor is obtained from

$$\begin{aligned} S_t &= S + (s_{20} - s_t) \\ &= S + \theta. \end{aligned} \tag{102}$$

The value of $(s_{20}-s_t)=\theta$ is listed in Table XXXIII.

(*e*) *A burette or pipette at t° delivers a certain volume of a standard solution ; it is required to find the exact volume of the nominal strength (N/10, N/20 or other such) delivered.*

The special factor entering here will be necessarily made up of $s_{20}-s_t$ and $g+f$; we have also the burette and solution factors for 20° to consider in the full equation.

In a similar way to the foregoing sections, if V_t is the required volume correction, then

$$\begin{aligned} V_t &= S_{20} + B_{20} + (s_{20} - s_t) - (g+f) \\ &= S_{20} + B_{20} - \eta. \end{aligned} \tag{103}$$

(η is already listed in Table XXXIII.)

Some examples of the use of foregoing constant chemical corrections.

EXAMPLE 1. 25 ml. of a $0{\cdot}1N$ hydrochloric acid are pipetted into a flask and titrated with $0{\cdot}1N$ alkali from a burette. It is required to standardise the alkali against the acid. Room temperature 16°.

P_{20} percentage factor of pipette at 20° $= +0{\cdot}140$,
B_{20} percentage factor of burette at 20° $= -0{\cdot}067$,
S_{20} percentage factor for solution strength of acid at 20° $= +0{\cdot}348$.

From (c) above (equation 101) the percentage correction is given by

$$S_{20} = P_{20} - B_{20} + S_{20}' - D.$$

D is given (see equation 98 above) by

$$\frac{25{\cdot}15 - 25{\cdot}0}{25} \times 100 = +0{\cdot}600.$$

We have then
$$S_{20} = +0{\cdot}140 + 0{\cdot}167 + 0{\cdot}348 - 0{\cdot}600$$
$$= +0{\cdot}055,$$

which is the percentage factor for the alkali at 20°.

The strength of the $0{\cdot}1N$ alkali at 20 is therefore

$$\frac{N}{10} \times \left(1 + \frac{0{\cdot}055}{100}\right) = \frac{N}{10}(1{\cdot}00055).$$

Example 2. A 10 ml. pipette is used to deliver $0{\cdot}1N$ sulphuric acid at 26·0°. What is the exact equivalent of the acid delivered?

P_{20} is $+0{\cdot}134$.

S_{20} is $-0{\cdot}467$.

From section (*e*) (equation 103), $V_t = P_{20} + S_{20} - \eta$ and from Table XXXIII η is 0·088, so that the percentage factor is

$$+0{\cdot}134 - 0{\cdot}467 + 0{\cdot}088 = -0{\cdot}245.$$

$10 \times 0{\cdot}1 \times \left(1 - \frac{0{\cdot}245}{100}\right)$ milliequivalents are therefore delivered

$= 0{\cdot}9975.$

Résumé of constant corrections for glassware and standard solutions. The various special factors, etc., are conveniently grouped in the following table :

Table XXXIV

Requirement.	*Special factor* (%).	*Full factor* (%).
1. Actual volume of weighed water.	$\alpha = (s + a)$.	α
Glass standardisation to 20°.		
2. Volumetric flask at t°.	$\beta = (s + a + g)$.	$F_{20} = \beta + D$, where $D = \frac{W - V_N}{W} \times 100$. V_N is the nominal volume ; W the weight in air.
3. Pipette or burette at t°.	$\gamma = (s + a + g + f^*)$.	P_{20} (or B_{20}) $= \gamma + D$. (D as in previous.)
Volume of glass filling or delivery at t° from standard value at 20°.		
4. Volumetric flask (standardised at 20°).	$\delta = (-g)$.	$F_t = F_{20} + \delta$.
5. Pipette (or burette).	$\epsilon = (-f - g)$.	$P_t = P_{20} + \epsilon$.

* See note at end of previous chapter.

(*Continued overleaf*)

TABLE XXXIV—*Continued*

Requirement.	*Special factor* (%).	*Full factor* (%).
Solution standardisation to 20°.		
6. Solid dissolved to mark in flask at $t°$.	$\zeta = (-s_{20} + s_t + g)$.	$S_{20} = \zeta - F_{20}$.
7. Burette solution *v* solid at $t°$.	$\eta = (-s_{20} + s_t + g + f)$.	$S_{20} = \eta - B_{20} - D$. $D = \frac{T - T_N}{T} \times 100$, where T is the actual titration figure and T_N the nominally expected figure.
8. Burette solution *v* pipetted solution.	None.	$S_{20} = P_{20} - B_{20} + S_{20}' - D$ (symbols as in previous). S_{20}' = strength of pipetted solution.
Correction for molarity or equivalents at t° from value at 20°.		
9. Solution strength at $t°$.	$\theta = (s_{20} - s_t)$.	$s_t = S_{20} + \theta$.
10. Equivalents delivered at $t°$ by pipette or burette.	$-\eta = (s_{20} - s_t - g - f)$.	$V_T = -\eta + P_{20} + S_{20}$.

Reduction of number of factors. If g and f be considered negligibly small, all the special factors above reduce to two, α and ζ. For α, β, γ we have simply α, δ and ϵ disappearing while ζ, η, θ become ζ, ζ and $-\zeta$. A study of Table XXXIII will help to decide what factors may be neglected in special circumstances.

(3) **Constant error arising from deviation of end-point pH from the ideal figure.** In titrating strong acid with strong base in pure dilute solutions from

$$[B^{\cdot}] + [H^{\cdot}] = [OH'] + [A'], \tag{104}$$

which is a necessary relationship for electric neutrality, we have $[H^{\cdot}]$ equal to $[OH']$ when $[B^{\cdot}]$ is equal to $[A']$, or where there is present an exact equivalence in the titration, the ideal end-point being therefore the theoretically neutral point, which we may take here as having a pH of exactly 7·0.

Where there is a divergence from this equivalence, we have

$$\frac{[B^{\cdot}] - [A']}{[A']} \times 100 = \frac{[OH'] - [H^{\cdot}]}{[A']} \times 100. \tag{105}$$

The expression on the right may be written

$$\frac{-[\mathrm{H}^{\cdot}]}{[\mathrm{A}']} \times 100, \tag{106}$$

when the end-point is one or more pH units from 7·0, and gives the percentage error. Here

$$[\mathrm{A}'] = C_A \times \frac{V_0}{V};$$

C_A being the original acid concentration and V_0 and V the original and end-point volumes. It is of some interest to assess this error for the methyl red indicator with varying strengths of acid for titration. We shall suppose that the acid is titrated with an equal volume of alkali, so that the error becomes

$$\frac{-10^{-5\cdot5}}{0\cdot5C_A} \times 100 = \frac{-2 \times 10^{-3\cdot5}}{C_A}, \tag{107}$$

where C_A is the original concentration of the acid.

The following short table gives the values of the error for different acidic concentrations (the error being independent of actual amounts).

TABLE XXXV

Acid concentration.	*% error with equal volume of alkali.*	*% error titrating with alkali volume one-fifth the acid volume.*
0·1*N*	0·0063	0·0038
0·01*N*	0·063	0·038
0·001*N*	0·63	0·38
0·0005*N*	1·26	0·76
0·0002*N*	3·15	1·90
0·0001*N*	6·30	3·80

When titrating 0·0002*N* acid in the ‘ unit ’ therefore (as for the blood ammonia) we have a constant error in the alkali used of 1·9%. This does not enter into the ammonia determination, since it is the difference between the two titrations brought to the same end-point that is used to measure the ammonia. It does enter, however, into the standardisation of the alkali with the acid, 1·9 per cent. more alkali being required for exact equivalence.

When titrating with methyl red using equal alkaline volume, the constant error reaches 0·01% when the original acid strength is $0{\cdot}063N$ or in the region of $N/20$. The variable chemical error when the end-point mixture is in equilibrium with the atmospheric carbon dioxide is (from Chapter LVIII) approximately

$$\frac{1\times 10^{-6}\times 2V}{V\times C_A}\times 100=\frac{10^{-4}\times 2}{C_A}=0{\cdot}003\%,$$

with a range of about 0·01% or the same order as the constant error. This variable error maintains the same relation to the constant error throughout, but whereas the constant error may be always allowed for, the variable error cannot.

Constant error in the titration of a weak acid or alkali. The pH of exact equivalence when titrating a weak acid is not 7·0, but approaches this as the concentration for titration diminishes. For dilute solutions the theory of this equivalence and associated errors may be treated as follows.

With a weak acid, the equation

$$[\mathrm{HA}]+[\mathrm{A}']=[\mathrm{B}^{\cdot}] \tag{108}$$

must hold at the equivalent point—the amount of strong base added being equal to the total weak acid, dissociated and undissociated.

Also the total negative charges must be equal to the total positive,

$$[\mathrm{A}']+[\mathrm{OH}']=[\mathrm{B}^{\cdot}]+[\mathrm{H}^{\cdot}]. \tag{109}$$

From these two equations, we have

$$\begin{aligned}[\mathrm{H}^{\cdot}]&=[\mathrm{OH}']-[\mathrm{HA}]\\ &=\frac{K_W}{[\mathrm{H}^{\cdot}]}-[\mathrm{HA}],\end{aligned} \tag{110}$$

where $[\mathrm{H}^{\cdot}]$ is the hydrogen ion concentration of the ideal end-point.

[HA] may now be expressed in terms of $[\mathrm{H}^{\cdot}]$ and the total weak acid concentration $C_A\{=[\mathrm{A}']+[\mathrm{HA}]\}$,

for

$$[\mathrm{H}^{\cdot}]\times[\mathrm{A}']=K_a\times[\mathrm{HA}],$$

so that

$$[\mathrm{H}^{\cdot}](C_A-[\mathrm{HA}])=K_a[\mathrm{HA}]. \tag{111}$$

From equations 111 and 110 we get

$$[\mathrm{H}^{\cdot}]^3+[\mathrm{H}]^2(K_a+C_a)-K_W[\mathrm{H}^{\cdot}]-K_WK_a=0, \tag{112}$$

where K_a is greater than 10^{-7} this reduces to

$$[\mathrm{H}^{\cdot}]=\sqrt{\frac{K_W K_a}{C_a+K_a}}, \tag{113}$$

$[\mathrm{H}^{\cdot}]$ being the hydrogen ion of the ideal end-point. The pH value is given by

$$p\mathrm{H}=0{\cdot}5 \log \left(1+\frac{C_a}{K_a}\right)+7{\cdot}0. \tag{114}$$

Similarly for the titration of a weak alkali, where K_b is greater than $10^{-7\cdot0}$,

$$[\mathrm{OH}']=\sqrt{\frac{K_W K_b}{C_b+K_b}}$$

or

$$[\mathrm{H}^{\cdot}]=\sqrt{\frac{K_W(C_b+K_b)}{K_b}}, \tag{115}$$

the ideal pH is given for the weak alkali by

$$p\mathrm{H}=-0{\cdot}5 \log \left(1+\frac{C_b}{K_b}\right)+7{\cdot}0. \tag{116}$$

Constant error resulting from deviations from the ideal end-point in titrating weak acid or weak base. For acids or alkalis with ionization constant of the order of 10^{-7} or greater, we can devise useful approximate expressions. If we take the total buffering of a weak acid in water it may be written (according to Chapter LVIII) as

$$\frac{dB}{dp\mathrm{H}}=\frac{2{\cdot}3C_aK_a[\mathrm{H}^{\cdot}]}{\{[\mathrm{H}^{\cdot}]+K_a\}^2}+2{\cdot}3\{[\mathrm{H}^{\cdot}]+[\mathrm{OH}']\}, \tag{117}$$

whence considering $\quad dp\mathrm{H}=\dfrac{-d[\mathrm{H}^{\cdot}]}{2{\cdot}3[\mathrm{H}^{\cdot}]}$,

we obtain

$$\frac{dB}{d[\mathrm{H}^{\cdot}]}=\frac{-C_aK_a}{\{[\mathrm{H}^{\cdot}]+K_a\}^2}-\left\{\frac{[\mathrm{H}^{\cdot}]+[\mathrm{OH}']}{[\mathrm{H}^{\cdot}]}\right\}. \tag{118}$$

Since the ideal end-point for a weak acid $[\mathrm{OH}']$ will be much greater than $[\mathrm{H}^{\cdot}]$, the latter may be neglected in the second expression on right. Then on integrating, we get

$$B_1-B_2=\frac{C_aK_a}{[\mathrm{H}^{\cdot}]_1+K_a}-\frac{C_aK_a}{[\mathrm{H}^{\cdot}]_2+K_a}+[\mathrm{OH}']_1-[\mathrm{OH}']_2$$

$$=C_aK_a\left\{\frac{[\mathrm{H}_2^{\cdot}]-[\mathrm{H}^{\cdot}]}{([\mathrm{H}^{\cdot}]_1+K_a)([\mathrm{H}^{\cdot}]_2+K_a)}\right\}+[\mathrm{OH}']_1-[\mathrm{OH}']_2. \tag{119}$$

In this the value of $[H^{\cdot}]_1$ and $[H^{\cdot}]_2$ may be taken as negligibly small compared with K, so that

$$B_1 - B_2 = \frac{C_a}{K_a}\{[H^{\cdot}]_2 - [H^{\cdot}]_1\} + [OH']_1 - [OH']_2. \tag{120}$$

Here C_a must be read as the total weak acid concentration in the end-point volume.

The base required per unit volume to produce a change in the actual $[H^{\cdot}]$ of the end-point to the ideal value given above may then be calculated from this equation, and expressed as a percentage of the total base added or the total acidic equivalent.

In this treatment the carbon dioxide of the air has been ignored, which can only be done when for weak concentrations we titrate in a current of carbon dioxide free air. Otherwise we should have extreme errors in titrating a weak acid to an alkaline end-point.

CHAPTER LXII

VOLUMETRIC ERROR IN KJELDAHL NITROGEN ANALYSES

An analysis of the volumetric error in the classical macro-Kjeldahl determination and a comparison with that using the microdiffusion principle. This is taken as providing a practical example for the foregoing system of error analysis. It will be considered firstly in relation to the error between two successive determinations carried out in duplicate, using the same glassware and the same solutions and quantities of material. This will give a measure of the discriminating delicacy of the procedure. Secondly, the absolute error will be considered, involving the introduction or allowance for the constant glass and chemical errors.

Since it will make it somewhat more general to consider the estimation carried out on a fluid which is delivered from a pipette, this procedure will be considered. The following example then may be cited.

3 ml. urine pipetted into incinerating flask.
20 ml. 0·1N HCl pipetted into receiving flask.
40 ml. 0·05N NaOH required to titrate a trial delivery of acid.
10 ml. 0·05N NaOH required to titrate back after distilling.
Standardisation of alkali. 45·0 ml. used in standardisation against potassium biphthalate or benzoic acid (or alternatively of acid against incinerated oxalate, with subsequent standardisation of alkali against acid).

1. **The discriminating range.** By discriminating range is meant here the percentage range of error that could occur between two determinations carried out on the same solution and with the same glass and standard solutions. In Chapter LVII this range was considered for duplicate titrations being independent of the constant errors, including the bore variation of the burette used.

The discriminating range when dealing with two duplicates is the same as that arising for the variable error of a single determination from the mean of a large number (Chapter LVII). Here there enters practically no variable chemical error owing to the strength of the solutions used, so that it may be considered entirely as a glass and glass manipulation error. We have then the following constituent errors to consider :

	Variable error, coeff. of variation.	*Variable error as millimols (NaOH).*
3 ml. pipetted	0·072	$30 \times 0{\cdot}05 \times 7{\cdot}2 \times 10^{-4} = 10{\cdot}8 \times 10^{-3}$
20 ml. acid delivered	0·028	$40 \times 0{\cdot}05 \times 2{\cdot}8 \times 10^{-4} = 5{\cdot}6 \times {}^{-}10^{4}$
10 ml. back titration	0·152	$10 \times 0{\cdot}05 \times 15{\cdot}2 \times 10^{-4} = 7{\cdot}6 \times {}^{-}10^{4}$

Summation of the above variable errors gives as millimols

$$\sigma^2 = 10^{-8}(10{\cdot}8^2 + 5{\cdot}6^2 + 7{\cdot}6^2)$$
$$= 10^{-8}(205)$$

and $$\sigma = 14{\cdot}3 \times 10^{-4}.$$

The total millimols of N (as NH_3) are $(40 - 10) \times 0{\cdot}05 = 1{\cdot}5$. The variable error of the determination as a coefficient of variation is therefore

$$\frac{14{\cdot}3 \times 10^{-4}}{1{\cdot}5} \times 100 = 0{\cdot}095\%.$$

The standard deviation from the mean of a large number of similar determinations is hence 0·095% with a range of ±0·28%. If a 5 ml. standard pipette were used instead of the 3 ml. it would be ±0·24. This range is also the discriminating range if duplicates are carried out with each solution.

Absolute range of error. In the discriminating error range, the variable error of titration of a delivery of unchanged acid by the alkali did not enter, since the same titration would suffice for the two analyses to be discriminated.

Concerning the absolute error of the single analysis, however, we require to know the exact value in equivalents of the acid delivered and its corresponding alkali value, and in determining this, further variable errors are necessarily introduced.

We could suppose the acid standardised by titration of a weighed solid and the alkali in turn by titration of the preliminary acid delivery. It will mean then the summation of the variable error of standardising a weighed solid, and the variable error of titrating a pipetted volume, a sum slightly (but scarcely significantly) larger than where the alkali is also standardised against a solid.

We have then

	Variable error (from considerations in Chapter LVII). Coeff. of variation	*Variable error in millimols (NaOH).*
Preliminary titration of 20 ml. of 0·1*N* acid, using 40 ml. alkali - -	0·052	$10{\cdot}4 \times 10^{-4}$

This added to the previous variable discriminating error of $14{\cdot}3 \times 10^{-4}$, gives

$$\sigma = 10^{-4}\sqrt{10{\cdot}4^2 + 14{\cdot}3^2}$$
$$= 17{\cdot}7 \times 10^{-4}.$$

This gives a coefficient of variation of

$$\frac{17{\cdot}7 \times 10^{-4}}{(40-10) \times 0{\cdot}05} = 0{\cdot}118\%.$$

To this now adds the coefficient of variation due to the acid standardisation, and we may assume this is carried out with a 40 ml. delivery from a 50 ml. burette in the usual way, and giving a coefficient of variation of 0·044, so that the total becomes

$$v_t = \sqrt{0{\cdot}118^2 + 0{\cdot}044^2}$$
$$= 0{\cdot}126,$$

giving a range of 0·38%.

(The variable error of acid standardisation adds in this way, because it will affect each member of the previous error—expressed as millimols—in a proportionate manner, but will at the same time be uncorrelated with or independent of the total previous error, i.e. of the 0·118%.)

Bore tolerance of burette. In the above it has been taken that the bore variation of the burette came under the heading of constant error, but the bore variation may be inconvenient to determine, though given in the certificate of standardisation supplied with the best class burettes. If we go instead by the full 'tolerance' of the burette, this can be included in a consideration of the limits of the variable error. In the above example, the difference between the 40 ml. of the preliminary titration and the 10 ml. of the back titration may be up to this full tolerance limit or 40–60 c.mm. with different kinds of first-class burette; 40 c.mm. tolerance will be taken here, since such burettes may be readily procured.

At the same time an allowance for bore or (as here) full tolerance comes into the standardisation, but will act in the opposite direction. If we take it that the 40 ml. on the burette is really $40 + 0{\cdot}040$ and the 10 ml. exact, then the alkali standardisation correction is

$$\frac{-0{\cdot}040}{40} \times 100 = -0{\cdot}100\%,$$

whereas the correction for the ammonia absorbed would amount to

$$\frac{+0{\cdot}040}{(40{\cdot}04 - 10)} \times 100 = +0{\cdot}133\%.$$

Here there is a total effect of $+0{\cdot}033\%$. If, however, the 40 ml. is exactly correct and the 10 ml. is $10 + 0{\cdot}040$, there is no standardisation correction, but the correction for ammonia absorbed is

$$\frac{-0{\cdot}040}{(40 - 10{\cdot}04)} \times 100 = -0{\cdot}133\%.$$

The maximum error here is therefore expressed by

$$\frac{\text{Tolerance}}{\text{acid neutralised}} \times 100,$$

the tolerance and acid neutralised being expressed as burette volumes of alkali. Adding this tolerance error of 0·13 to 0·38, we get 0·51% as the total range.

In the above example it was supposed that 10 ml. or 25% of the acid-equivalent were used in back titration. In the same way we may consider the total variable error for different percentage back titrations, and these are illustrated in Fig. 77.

It will be seen that a very marked effect on the error is produced as the back titration increases, and it is always advisable to keep this below the 25% level wherever possible.

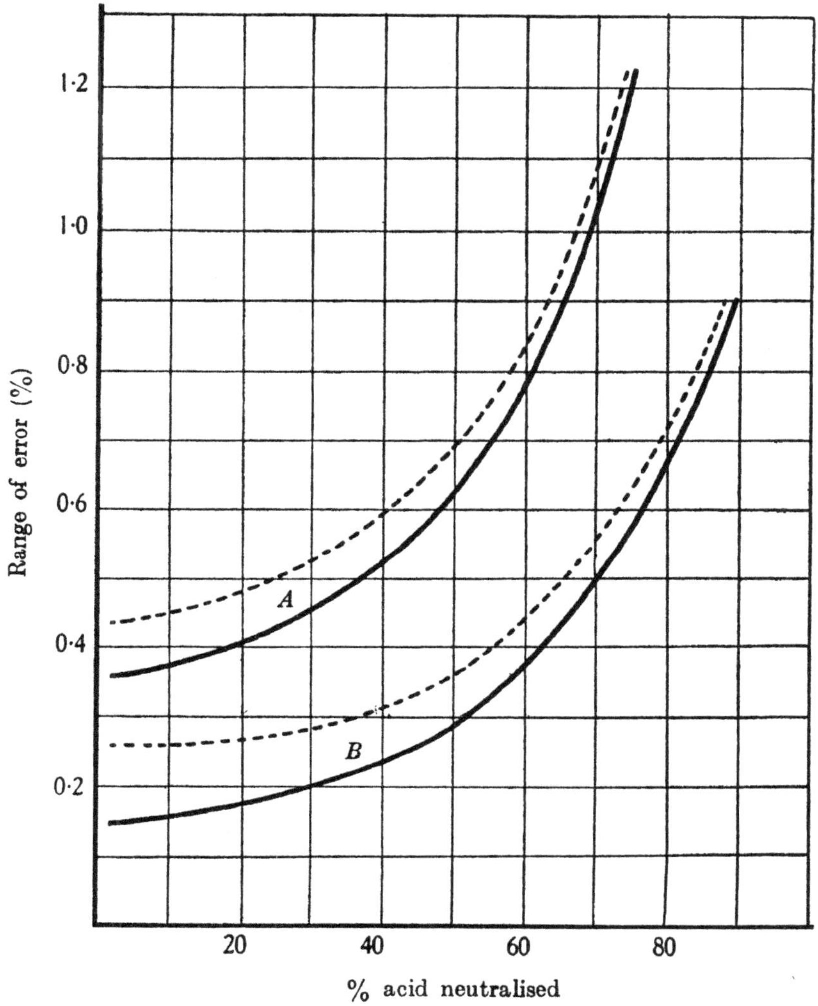

FIG. 77. Curves of volumetric error in the classical Kjeldahl determination.

The dotted lines are for pipetted fluid (3 ml.) and the smooth curves for weighed solid.

The upper two curves *A* give the *total* error range, including standardisation errors and bore tolerance of the 50 ml. burette. The lower two curves *B* give the full error range of the discrimination error as described in text.

Fig. 77 shows also (thick line) the error involved where a weighed solid is used for nitrogen analysis instead of a pipetted volume of 3 ml.

The curves give the *full range* of error, the upper curves including that for standardisation and bore tolerance (dotted lines) errors, the lower curves giving the discrimination error.

From the above treatment the impurity error in the standardising substance has been omitted. In the case of sodium oxalate as prepared above, and responding to the tests given, the maximum error is 0·04%, which may be treated as the limit of a variable error and added to the above range.

The constant error allowances. The method to be described reduces all the constant errors or allowances to a simple and easily workable form.

If X be the ml. of alkali solution from the burette corresponding to the ammonia analysed, then

$$X \times 0{\cdot}05 \times 14 = \text{mg. ammonia } N \text{ absorbed}$$

$$(\text{using } 0{\cdot}05N \text{ alkali})$$

and

$$\frac{0{\cdot}7X}{V} = Y,$$

where Y is the mg. N/ml. of fluid analysed, of which V ml. were pipetted for analysis. To allow for the constant errors we multiply above and below by the factors required to convert pipetted and buretted fluid from nominal to real equivalents at $t°$ (Table XXXIV), so that, in full

$$0{\cdot}7X\left(1+\frac{B_{20}+S_{20}-\eta}{100}\right) = VY\left(1+\frac{P_{20}+S'_{20}-\eta}{100}\right),$$

where B_{20}, P_{20}, S_{20} and S'_{20} are the % factors for the burette, pipette, alkali, and solution for analysis, at 20°, η being the special temperature factor as in Table XXXIII.

From the above equation (using the principle for small errors in Chapter LX), we have

$$\frac{0{\cdot}7X}{V}\left(1+\frac{B_{20}-P_{20}+S_{20}}{100}\right) = Y\left(1+\frac{S'_{20}}{100}\right).$$

The required final factor S'_{20} is therefore given by

$$S'_{20} = B_{20} - P_{20} + S_{20}.$$

This may be expressed in a final form by expanding the standardising factor for the alkali S_{20}. This depends on whether it has been standardised against solid or using the preliminary acid titration.

From Table XXXIV, and inserting the expansion of S_{20} for each case, we get

$S'_{20} = -D - P_{20} + \eta$ (solid standardisation)

or $S'_{20} = -D + S_{20}^{ac} + P_{20}^{ac} - P_{20}$, (standardisation from preliminary acid titration).

In the second equation S_{20}^{ac} and P_{20}^{ac} are the % constant errors of the acid strength and the 25 ml. pipette for delivering the acid.

If we neglected a standardisation of the two pipettes and assumed that they could vary in plus or minus value up to the maximum of their tolerance, then the two will add as 0·1% for the 25 ml. and 0·3% for the 3 ml. standard pipettes or a total of 0·4%. By the omission therefore of this easy standardisation from the five delivery mean, we double our variable error range, the 0·4% being added on to the limits of the variable error considered above.

Constant error allowances using weighed solid for analysis. The form of the allowance may be obtained as above. Here the value of the burette reading corresponding to the neutralised ammonia is multiplied by the factor for converting such burette deliveries from nominal to actual equivalent value. This factor is given in Table XXXIV as

$$(-\eta + B_{20} + S_{20}).$$

We may write therefore

$$F(\text{solid}) = -\eta + B_{20} + S_{20}.$$

On expanding S_{20} the alkali factor according as it has been standardised against solid or by the preliminary titration (from Table XXXIV), we get

$F(\text{solid}) = -D$ (alkali standardised by solid)

or $= -D + S_{20}^{ac} + P_{20}^{ac} - \eta$ (alkali standardised by the preliminary acid titration).

In Table XXXVI the constant error allowances and the variable error ranges for given conditions are summarised.

Table XXXVI

Analysed substance.	*Conditions of standardisation.*	*Discriminating range as back titration approaches zero. (Full duplicates for each sample.)* (%)	*Absolute error range of determination allowing for standardisation and burette tolerance.* (%)	*Factor for constant error* (%).
Weighed solid.	Alkali standardised by solid.	0·14	0·35	$-D$
Weighed solid.	Alkali standardised by preliminary titration.	0·14	0·36	$-D+S^{ac}_{20}+P^{ac}_{20}-\eta$
Pipetted fluid (3 ml.).	Alkali standardised by solid.	0·26	0·42	$-D-P_{20}+\eta$
Pipetted fluid (3 ml.).	Alkali standardised by preliminary titration.	0·27	0·43	$-D+S^{ac}_{20}+P^{ac}_{20}$ P_{20}

In the table it is considered that 20 ml. of acid are pipetted into the receiving flask by a 20 ml. pipette and that the alkali used to titrate is half the acid strength. It is also taken that the burette is read to 0·2 of a final division.

Reading of this accuracy is very unusual, but if read to 0·5 of a small division (and there is no point in further inaccuracy), this will cause a change in the figure 0·26 above to 0·29% and the 0·42% to 0·45%. The symbols in the final column are described throughout the account in the text.

P_{20} pipette factor for analysed fluid at 20°,

P^{ac}_{20} ,, ,, ,, acid ,, 20°,

S^{ac}_{20} acid factor for 20°,

$D=\dfrac{\text{titration fig.}-40}{40}\times 100$, where 40 ml. is the nominal expected value before standardising.

η is a temperature correction given in Table XXXIII.

Analysis of the volumetric error of a micro Kjeldahl procedure using the microdiffusion technique. Here we may consider the pipetting of 0·2 ml. fluid, such as urine, and its incineration, using 1 ml. of concentrated sulphuric acid, making up the incinerated mixture to mark with water in a 50 ml. volumetric flask and the removal of 0·5 ml. of this for analysis. The use of 1 ml. acid for the absorption, and alkali from the Conway burette in accordance with the solution strengths of Range I, Table V, p. 92, is assumed.

In a consideration of the error involved only a procedure will be considered which is convenient and easily attained, and which at the same time may be expected to give the high accuracies stated. We shall suppose 3 simple tube pipettes constructed to deliver 0·2, 0·5 and 1·0 ml. respectively, and with fine tips as commented upon in Chapters V and LV. The length of each pipette is considered to be 20 cm. up to the mark and to take 20 seconds to deliver. As already shown (Chapter LV), the coefficients of variation are then 0·065, 0·041 and 0·029 respectively.

With regard to the burette, on titrating we shall assume that an end-point emergence of one small division on the burette can be made easily (though much less can in fact be controlled) and transferred by a fine pyrex rod or by direct contact, and that the burette reading is taken to half of one small division. This gives a manipulation error of 0·3 c.mm. (standard deviation).

In delivering from the burette it is supposed that the meniscus travels at 1 cm. per second, giving a delivery error of 0·065%. We may consider then the following :

Variable error of pipetting 0·2 ml. fluid	$=0{\cdot}065 \times 50\gamma N \times 10^{-2}$.
Variable error of taking 0·5 ml. of the 50 ml. mixture	$=0{\cdot}041 \times 50\gamma N \times 10^{-2}$.
Variable error of pipetting 1·0 ml. acid	$=0{\cdot}029 \times 70\gamma N \times 10^{-2}$.
Variable error of titrating back 0·2 ml. (including delivery and manipulation errors)	$=0{\cdot}163 \times 20\gamma N \times 10^{-2}$.

$$\text{whence } \sigma = 10^{-2}\sqrt{3{\cdot}25^2 + 2{\cdot}05^2 + 2{\cdot}03^2 + 3{\cdot}26^2}\ \gamma \text{ am. } N.$$

$$= 5{\cdot}4 \times 10^{-2}.$$

Amount analysed is $50\gamma\ NH_3-N$, therefore the coefficient of variation is 0·108%.

Hence the total range is 0·324 and is the discriminating range using a duplicate determination, being therefore practically the same as in the full macro Kjeldahl.

If a duplicate analysis with each sample is carried out on the diluted fluid after incineration (and the ' unit ' procedure is particularly suitable for this), we would get

$$\sigma = 10^{-2}\sqrt{3{\cdot}25^2 + 0{\cdot}5 \times 2{\cdot}05^2 + 0{\cdot}5 \times 2{\cdot}03^2 + 0{\cdot}5 \times 3{\cdot}26^2}\gamma$$
$$= 4{\cdot}5 \times 10^{-2}\gamma.$$

This gives with 50γ ammonia N analysed a coefficient of variation of 0·090 and a range of 0·270 for the full duplicate of such a procedure.

This error range (using double ' unit ' determinations of the diluted incinerated fluid) is therefore on the purely volumetric side, as accurate as the full macro Kjeldahl procedure, since from Fig. 53 it will be seen that for the analysis of 3·0 ml. pipetted fluid by the classical Kjeldahl procedure a discriminating error of 0·27% is obtained with a 20% back titration.

If we analyse, instead of pipetted fluid, a weighed sample by a micro balance so that the variable error of introducing sample is practically eliminated, we get for a back titration of 20% :

$$v_t = \sqrt{0{\cdot}090^2 - 0{\cdot}065^2}$$
$$= 0{\cdot}062,$$

or a range of $\pm 0{\cdot}19$ which is closely similar to that with the macro Kjeldahl procedure, which gives $\pm 0{\cdot}17$.

This comparison refers only to the volumetric procedure, but it is practically certain that the simple passive diffusion absorptions in the ' unit ' will introduce less further error (in fact practically negligible) than the large scale high temperature distillations of the classical Kjeldahl procedure.

Absolute error range. This will include the variable error of standardisation as before, and the bore tolerance of the burette if this is considered inconvenient to determine, though here it is much easier than with larger burettes and can be done by measuring a length of mercury at various positions. At the same time it may be noted that the use of high quality thermometer tubing practically eliminates the need of this allowance or investigation, and gives therefore one further advantage. The standardisation correction follows the previous account. There is a small further variable error included since the standardised acid of strength about $0{\cdot}1N$ will need to be diluted to $N/200$, but when this is done using a 10 ml. standard pipette, the extra error has no appreciable effect on the whole.

The constant error allowances may be also used as in Table XXXIV. The effect of the dilution on the S_{20}^{ac} factor of the $\frac{M}{10}$ acid may be obtained from $S'^{ac}_{20} = S^{ac}_{20} - F_{20} + P_{20}$, where F_{20}, P_{20} are the factors for pipette and flask and S^{ac}_{20} of the comparatively strong solution.

APPENDIX

CHAPTER LXIII

UREA EXCRETION AS RENAL FUNCTION TEST

THE excretion of urea as the most important nitrogenous end product in man has been much used as a test of renal function. It acts as a very useful measure of that part of the total renal functioning which is concerned with the excretion of nitrogenous waste products. Obviously it does not in itself give any index of the water or electrolyte regulation.

The urea excretion by the kidney may be easily shown to be influenced by the blood urea concentration, and the urine rate at constant body weight. Clearly therefore we must consider any renal output of urea with regard to the blood urea concentration and the urine volume.

What we are seeking here is a comparison of the given urea excretion with the average normal figure. This we can do in two ways, apparently different, but which are in fact identical. We could calculate what the excretion of the given subject would be at standard blood urea concentration and standard urine volume, allowing for the effect of the difference from the standard in these variables; and then compare such a calculated excretion with the average normal figure at such standard values.

On the other hand we could compare directly with the average normal excretion at the observed blood concentration and urine rate. The mean normal excretion so obtained may be written VC_u', where V is the subject's urine volume in unit time and C_u' the average normal urine concentration at this volume and given blood concentration. The subject's actual excretion may be written VC_u. In comparing these we may obviously neglect V, which is the same in both, so that our requirement is the figure for the average normal urine concentration under the given conditions of blood concentration and urine rate. *Such a figure could be obtained most readily from tables compiled from a study of the available figures for excretion in the normal human subject.*

TABLE XXXVII. *Average normal urea concentration ratios for a body weight of* 70 *kilos* (*urine/blood*)

Volume of urine ml. per hour.	0	1	2	3	4	5	6	7	8	9	*Volume of urine ml. per hour.*
20	96·2	93·6	91·2	88·9	86·8	85·0	83·3	81·8	80·3	78·9	20
30	77·6	76·3	75·1	73·9	72·7	71·6	70·6	69·5	68·5	67·5	30
40	66·5	65·6	64·8	64·0	63·1	62·3	61·6	60·9	60·2	59·4	40
50	58·8	58·1	57·4	56·8	56·2	55·7	55·1	54·5	53·9	53·3	50
60	52·8	52·3	51·8	51·3	50·8	50·4	50·0	49·6	49·1	48·7	60
70	48·3	47·9	47·5	47·1	46·7	46·3	46·0	45·6	45·2	44·8	70
80	44·4	44·1	43·8	43·5	43·2	42·9	42·6	42·3	42·0	41·7	80
90	41·4	41·1	40·8	40·6	40·3	40·0	39·8	39·5	39·3	39·0	90
100	38·8	38·5	38·2	38·0	37·8	37·6	37·3	37·1	36·9	36·7	100
110	36·5	36·3	36·1	35·9	35·7	35·5	35·3	35·1	35·0	34·8	110
120	34·6	34·4	34·2	34·0	33·8	33·6	33·4	33·3	33·1	32·9	120
130	32·7	32·5	32·3	32·2	32·0	31·8	31·6	31·5	31·3	31·2	130
140	31·1	30·9	30·7	30·6	30·4	30·3	30·2	30·0	29·8	29·7	140
150	29·6	29·5	29·3	29·2	29·1	28·9	28·8	28·7	28·6	28·4	150
160	28·4	28·3	28·2	28·0	27·9	27·8	27·6	27·5	27·4	27·3	160
170	27·1	27·0	26·8	26·7	26·6	26·5	26·4	26·3	26·1	26·0	170
180	25·9	25·8	25·7	25·6	25·5	25·4	25·3	25·2	25·1	25·0	180
190	25·0	24·9	24·8	24·7	24·5	24·4	24·3	24·2	24·1	24·0	190
	00	10	20	30	40	50	60	70	80	90	
200	23·9	23·0	22·2	21·4	20·7	20·0	19·3	18·7	18·1	17·6	200
300	17·1	16·6	16·1	15·6	15·2	14·8	14·4	14·0	13·7	13·3	300
400	13·0	12·6	12·3	12·0	11·7	11·4	11·1	10·8	10·6	10·4	400
500	10·2	9·98	9·76	9·54	9·32	9·10	8·90	8·72	8·53	8·35	500
600	8·19	8·04	7·90	7·75	7·60	7·45	7·30	7·15	7·00	6·85	600
700	6·72	6·60	6·49	6·38	6·26	6·14	6·03	5·92	5·81	5·71	700

Such tables though convenient would be lengthy. However, a much shorter table, and nearly as convenient (Table XXXVII), may be used, giving the average number of times the blood urea is concentrated in the urine of the normal subject. This multiplied by the blood concentration gives the required urea concentration in the urine, and dividing this figure into the actual concentration for the subject and multiplying by 100 gives the urea excretory function expressed as a mean '**normality ratio**'. The use of Table XXXVII in this way depends on the fact—upon which all are agreed—that over the clinical range the urea excretion at constant urine rate is proportional to the blood concentration.

The urea normality ratio for a given subject may be formally defined then as his urea excretion (or urine concentration) expressed as a percentage of the average normal value at the same blood urea and urine rate.

Equations or formulae need have no connection with the determination of such a figure, which we could suppose obtained with the aid of normal excretion tables. It will be shown subsequently that for the purpose of determining renal function all the equation does, even when ideally correct, is to provide us with a means whereby the average normal figure may be obtained without tables, or alternatively to assist in their construction.

TABLE XXXVIII

Range of V ml./hr.	*Mean V ml./hr.*	*N (No. of sets).*	$\frac{VC_u}{C_b}$.	*A general formula (mean values).*	*Clearance formulae (mean values).* II.	I.
0–20	15·9	8	13·6	78·8	81·5	
20–40	31·2	67	23·6	99·4	101·1	
40–60	48·5	102	29·4	101·1	100·9	
60–80	68·1	77	33·7	100·7	97·6	
80–100	87·5	40	36·7	99·1	93·8	
100–120	108	19	39·9	99·8	91·6	
120–160	139	36	45·9	105·0		101·8
160–200	180	15	46·7	100·2		103·4
200–300	235	22	47·5	96·1		105·1
300–400	375	16	51·7	99·3		113·5
400–500	448	34	49·8	96·1		110·4
500–600	551	30	50·5	100·7		112·0
600–700	640	26	46·5	95·5		103·1
$>$ 700	753	15	48·5	105·6		107·1

Sources of the data :

	Sets.	*Subjects.*
Walker and Rowe (1927) - - - - - -	52	15
Addis and Watanabe (1916) - - - -	106	33
McLean (1915) - - - - - - - -	107	34
Austin *et al.* (1921) - - - - - - - -	32	2
Conway and Dillon (1933) - - - - - -	77	77
Addis and Drury (1923) - - - - - -	72	4
Conway and O'Donovan (unpublished data) -	22	2
Polland (1928) - - - - - - -	39	1
	507	168

A general formula may be written as follows

$$\frac{\sqrt{V}(C_u - C_b)}{441 C_b e^{-\alpha v}} = 100 \text{ (mean)}$$

where $\alpha = 0{\cdot}00152$.

'Clearance' equations, below and above a V of 120, are:—

$$\frac{\sqrt{V} C_u}{4{\cdot}18 C_b} = 100 \text{ (mean)}; \quad \frac{V C_u}{45{\cdot}1 C_b} = 100 \text{ (mean)}.$$

The comparison of the general equation and the 'clearance' equations in predicting the output of urea with change of urine volume at constant blood concentration is given in Table XXXVIII and Fig. 78. The use of the equations alone—without tables—will be considered in a subsequent section, though it may be noted again that tables are ideally the most exact, when constructed from the actual data of the normal excretion, and involve no theoretical interpretation in the ordinary sense.

The determination of the urea normality ratio.

The collection of the blood and urine samples. The test may be carried out at any time of the day and without intake of urea or other substance, though a glass of water prior to the test is advisable, but not essential; also since meals have the effect of raising the excretory power of the kidney within the first few hours after a meal, it would seem advisable to carry out these tests in the hour preceding a meal.

The bladder is firstly emptied, the exact time being noted and the urine discarded. After half an hour a sample of blood is taken from the finger in the manner described for the blood urea determination, and after one hour the bladder is emptied again and the volume in ml. or c.c. measured in a graduated cylinder. The weight of the subject is also noted.

In the dilution of the urine for analysis a procedure may be adopted which is very convenient both for this purpose and the subsequent calculation of renal function. The volume of urine per hour (corrected if necessary for change of body weight from a standard 70 kilos, according to the very simple formula below, though from 60–80 kilos body weight very little is gained by such corrections) is noted and the average normal concentration ratio read off from Table XXXVII, the urine being diluted this number of times. Thus if the average normal concentration ratio is found

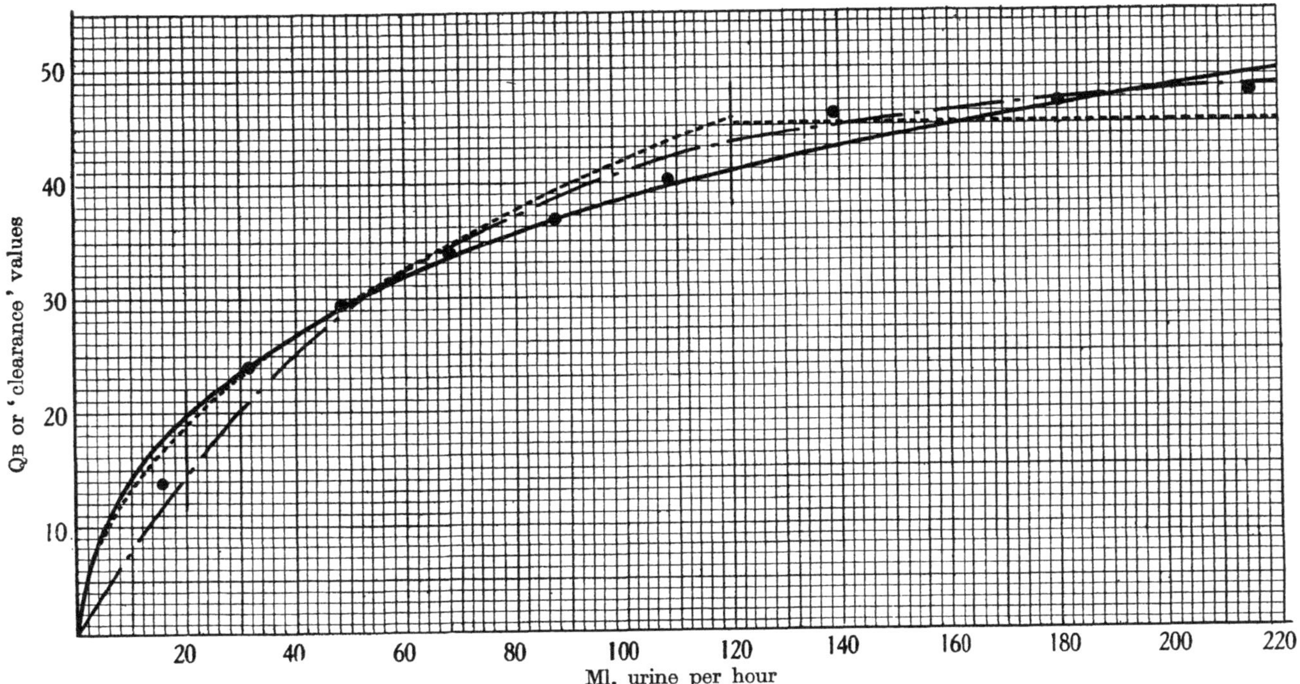

FIG. 78. The dots give the means of the excretion per unit blood concentration at varying urine rates for the normal human subject.

Line ———— curves from the general equation.
" ·········· from 'clearance' equations.
" —·—·— from the equation of Dominguez.

The best numerical values—statistically—were used for the 'clearance' curves and for that of Dominguez (1935).

Between the two upright lines from 20 ml./hr. to 120, there are 305 sets of observations from the general literature summarised. The means are almost exactly described by the general equation. The comparative success of the Dominguez exponential equation (as considered by Dominguez) is very largely due to the fortuitous correspondence of the few observations below 20 ml./hr. His statistical treatment artificially exaggerates this correspondence. Clearly his curve —at its best (as here drawn)—is not the true or best descriptive curve, and it is obvious also from this and Table XXXVIII that the general curve is superior to the two clearance curves combined.

to be 76·3 (urine volume of 31 ml./hr.), 1 ml. urine is pipetted into a 200 ml. Erlenmeyer flask and 75·3 ml. water added from a burette. The water and urine are well and carefully mixed. This urine should then have exactly the same concentration as the blood if the subject's excretion corresponded to the average normal value. *The figure for the urea excretory function or the normality ratio is given at once on dividing the concentration of the blood urea into that of this diluted urine and multiplying by 100,* the procedure being always the same, excepting the different figures for the necessary dilution required. This calculation involves as a rule only one division (pipette and alkali factors cancel out) and is in principle the most accurate form in which the urea excretory test can be applied.

If tables are not available, the calculation from equations is given below.

Correction of the urine volume for body weight, or surface. This correction should be made on a surface relation since the urine volume goes proportionally thereto. The following simple formula gives a suitable correction factor.

$$\frac{W+30}{100}=F. \qquad (1a)$$

The value of F, which can be instantly written down, is divided into the urine volume obtained. The result gives a $W^{\frac{2}{3}}$ correction to a standard body weight of 70 kilos over the range of 40 to 90 kilos, to within 2% of the exactly calculated value at the extremes of this range (amounting usually to less than one per cent. in the final ratio or deviation). A more exact correction for surface is unnecessary with the adult subject since the correlation between the mean urine volume and body weight is very low. In fact there is no appreciable increase in accuracy introduced by such corrections when the body weight is between 60–80 kilos.

A corresponding formula for the range of 12 to 35 kilos may be given as

$$\frac{W+10}{70}=F. \qquad (2a)$$

This has the same accuracy as the first, showing at the extremes the same divergence from an exact $W^{\frac{2}{3}}$ correction.

Examples of the determination of urea normality ratios (R_u).

Subject 1. Body weight - - - 70 kilos.
Urine volume - - - 180 ml./hour.
Blood collected midway.

From Table XXXVII at 180 ml./hr. the average normal concentration ratio is 25·9, so that to 1 ml. urine 24·9 ml. water were added.

Diluted urine concentration - - 32·3 mg./100 ml.
Blood concentration - - - 28·5 mg./100 ml.

$$\text{Normality ratio } (R_u) = \frac{32{\cdot}3}{28{\cdot}5} \times 100 = 113.$$

The tables are made out for a body weight of 70 kilos, so that no correction for weight occurs here.

If the urine were not diluted as above, and the concentration returned finally as 0·837%, the normality ratio is obtained by multiplying the blood concentration by 25·9 from Table XXXVII (corresponding to 180 ml./hr.) and dividing into 0·837 × 100, thus obtaining as before 113.

Subject 2. Body weight - - - 86 kilos.
Urine volume - - - 62·5 ml. per hour.
Blood collected midway.

From the weight formula above, we may write down immediately 1·16 as the divisor into 62·5 ml. and obtain 54 ml. as the corrected volume. From Table XXXVII this corresponds to a concentration ratio of 56·2, 55·2 ml. water being added to 1 ml. urine.

Diluted urine concentration - - 26·3 mg./100 ml.
Blood concentration - - - 46·7 mg./100 ml.

$$\text{Normality ratio } (R_u) = \frac{26{\cdot}3}{46{\cdot}7} \times 100 = 56.$$

Subject 3. Body weight - - - 64 kilos.
Urine volume - - - 14·0 ml./hr.
Blood collected midway.

Here no correction need be made for body weight as it lies in the region 60–80 kilos, though some may prefer to make it as a routine.

Making the correction will alter the volume to 14·9 ml./hr.

Table XXXVII begins at 20 ml./hr. since there is no apparent increase in urine concentration below this level, though more data are here desirable. Consequently the urine is regarded as being 20 ml./hr. and a dilution of 96·2 made ; 95·2 ml. water being added to 1 ml. urine.

Diluted urine concentration - - 21·4 mg./100 ml.
Blood concentration - - - 30·0 mg./100 ml.

$$\text{Normality ratio } (R_u) = \frac{21{\cdot}4}{30{\cdot}0} \times 100 = 71.$$

Normal limits and distribution of the normality ratio (R_u). Fig. 66I gives the general distribution of the normal ratios or R_u values, and also the distribution for a single individual (histogram III, in which the dotted line gives the effect of a meal). The chief interest for the pathologist or clinician in such assessment of function lies in the consideration of a lower normal limit (already considered).

With regard to the normal limits, 5% of the R_u values fall below 70. These include no doubt a certain percentage of experimental errors. From 60 to 70 may be taken as a region of suspected subnormal functioning, to be certified by further determinations.

Concerning the variability of R_u it may be said that even for the single individual there occur very appreciable variations in the renal functional power. When the observations are scattered throughout the day, and up to 120 ml. per hour, the standard deviation is 12·3 (probable error = 8·3). Nearly all the observations will then lie between ±25% of the mean. When the observations are carried out each day at the same time the standard deviation is 9·5 (probable error 6·4). The difference is largely due to the effect of meals—but differences even in the normal range of body temperature appear also to have appreciable influence on the excretion. Distributions of R_u values are given in Fig. 79.

The use of equations instead of tables for renal function tests. *Demonstration that the value of the equation for this purpose lies in providing a figure for the average normal urine concentration under the given conditions.* Any urea excretion formula which may be considered as at all significant, and the mean normal value of which is represented as 100, can be written in the form

$$VC_u = 100 \times f(C_b,\ V,\ W), \tag{3a}$$

where VC_u is the excretion in unit time, and $f(C_b,\ V,\ W)$ some function of the urine volume (V), the blood concentration (C_b) and the body weight (W).

If on the right we insert the values for a given individual during an observation, we shall get nearly always values different from

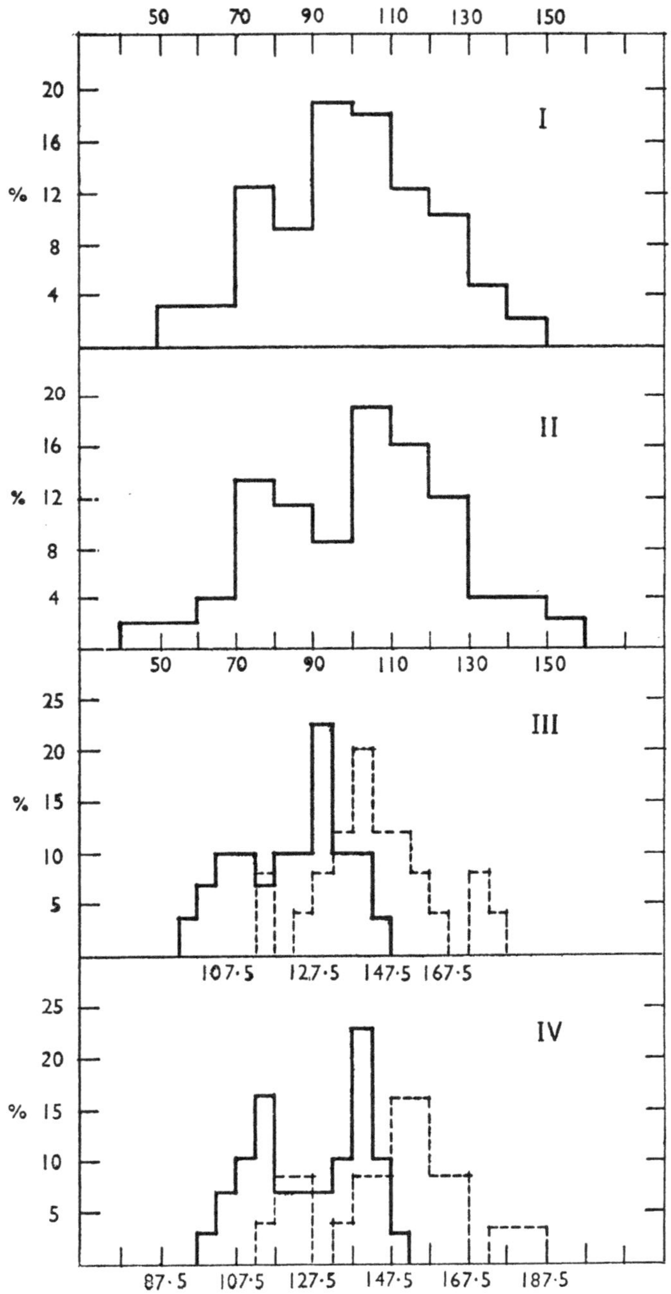

Urea normality ratios

FIG. 79. Distribution of normality ratios for normal human subject.

I. 169 obs.—data from general literature. Urine rate limited to 120 ml./hr. to show peaked distribution best.

II. The same obs. in which the ratios were calculated from equation 7*a* in text. This corresponds to a 'clearance' distribution.

III. 56 obs. on a subject before and after a light lunch (dotted line—after lunch with urine collection up to 90 mins.). Urine rate limited to 120 ml./hr.

IV. The same obs. as in III, using equation 7*a*, corresponding to a 'clearance' distribution.

VC_u the actual excretion rate. However, the value of the expression $\Sigma(VC_u' - VC_u)^2$ is (or should be) a minimum. Here C_u is the actual urine concentration and C_u' the calculated value; VC_u' being the value of the expression on the right of the previous equation. The value of the above expression can only be a minimum when VC_u' is the average normal figure under the given conditions (of blood concentration, etc.). Equation 3*a* may now be written

$$\frac{VC_u}{f(V, C_b, W)} = R_u (= 100 \text{ as the mean value}),$$

or, as we have seen

$$100 \times \frac{VC_u}{VC_u'} = R_u (= 100, \text{ mean value}), \quad (4a)$$

where VC_u' is the average normal excretion at given values of C_b, V and W.

It is obvious that the equation—for the purpose in question—only provides a figure for the average normal output.

The theoretical significance of the equation whereby VC_u' is obtained is a second stage in the interpretation.

REFERENCES

Abegg and Riesenfeld (1902), *Z. physikal. Chem.*, **40**, 100.
Abelin (1938), *Biochem. Z.*, **297**, 2035.
Adams and Foley (1953), *Res. Publ. Assn. New. Ment. Dis.*, **32**, 198.
Addis, Barnett and Shevky (1918), *Amer. J. Physiol.*, **46**, 1.
Addis and Drury (1923), *J. Biol. Chem.*, **45**, 639.
Addis and Watanabe (1916), *J. Biol. Chem.*, **24**, 203.
Alexander, Landwehr and Seligman (1945), *J. Biol. Chem.*, **160**, 51.
Ambard (1910), *Compt. rend. Soc. de Biol.*, **69**, 411.
Ambard and Weil (1912), *J. Physiol. et Path. gen.*, **14**, 753.
Amberg and Jones (1911), *Z. für Physiol. Chem.*, **73**, 407.
Archer, Chapman, Rhoden and Warren (1948), *Biochem. J.*, **42**, 58.
Archibald (1944), *J. Biol. Chem.*
—— (1945), *J. Biol. Chem.*, **159**, 693.
Atkinson, Barnett, Sherlock and Steiner (1955), *Quart. J. Med.* in press.
Audoynaud (1875), *Compt. rend.*, **81**, 619.
Austin, Stillman and Van Slyke (1921), *J. Biol. Chem.*, **46**, 91.

Bacharach, A. L. (1947), *Analyst*, p. 244.
Bandemer and Schaible (1936), *Ind. and Eng. Chem. Analytical Ed.*, **8**, 201.
Bang (1920), 'Micromethoden zur Blutuntersuchung', München, 192.
Bartley (1953), *Biochem. J.*, **53**, 306.
Bass (1914), *Arch. exp. Path. u. Pharmakol.*, **76**, 40.
Bassil and Hain (1950), *Nature*, 525.
Baubigny (1897), *Compt. rend. Acad. Sc.*, **125**, 654.
Beatty and Gibbons (1937), *Jour. of the Biol. Board of Canada*, Vol. 3.
Benedict, Osterburg and Neuwirth (1918), *J. Biol. Chem.*, **34**, 217.
Bentley and Kirk (1936), *Mikrochemie*, **21**, 260.
Berka (1955), *Acta Medica Scandinavica*, CLII, fasc. vi.
—— —— (1959), *Pravcovni lekarstvi*, **1**.
Bernhardt and Ucko (1925), *Biochem. Z.*, **155**, 174.
—— —— (1926), *Biochem. Z.*, **170**, 459.
Bertram (1933), *Biochem. Z.*, **261**, 202.
Bjerrum (1914), 'Die Theorie der alkalimetrischen und azidimereischen Titrierung', Stuttgart.
Black (1949), *Archives of Biochemistry*, Vol. **23**, 347.
—— *Archiv. of Biochemistry*, **23**, 3.
Blaschko, Tichter and Schlossman (1937), *J. Physiol.*, **90**, 1.

Bohr and Bock, cited in the *Chemiker Kalendar*, **1930**, Table 242, Part II. Berlin.

Boos (1948), *Anal. Chem.*, **20**, 964.

Borsook (1935), *J. Biol. Chem.*, **110**, 481.

Borsook and Dubnoff (1939), *J. Biol. Chem.*, **131**, 163.

Boussingault (1853), *Ann. Chim. Phys.*, **39**, 257.

—— (1854), *Ann. Chim. Phys.*, **40**, 129.

Boxer and Richards (1951), *Archiv. Biochem.*, **30**, 372.

—— —— (1952), *Archiv. Biochem.*, **39**, 292.

Boyland and Rhoden (1949), *Biochem. J.*, **44**, 528.

Bradstreet (1940), *Chem. Revs.*, **27**, 331.

Brady and O'Callaghan (1955), unpublished obs.

Brinkman (1933), *J. Physiol.*, **80**, 171.

Brinkman, Margaria and Roughton (1933), *Philos. Trans. Roy. Soc.*, A232, 65.

Brocklesby and Riddell (1937), Progress Report of Pacific Fisheries Expt. Station, Prince Rupert, B.C., 17.

Brooks and Norymberski (1952), *Chem. and Ind.*, p. 804.

—— —— (1953), *Biochem. J.*, **55**, 371.

Brown *et al.* (1957), *Arch. Biochem. Biophys.*, **66**, 301.

Brüel, Holter, Linderstrøm-Lang and Rositz (1946), *Compt. rend. trav. lab. Carlsberg, Ser. chim.*, **25**, 289.

Buell (1935), *J. Biol. Chem.*, **108**, 273.

Burbridge, Hine and Schick (1950), *Jour. Lab. and Clin. Med.*, **35**, 983.

Burgen (1948), *Brit. Med. Jour.* (June Number), 1238.

Calkins (1956), *J. Lab. Clin. Med.*, **47**, 343.

Cavett (1938), *J. Lab. and Clin. Med.*, **23**, 543.

Chain (1952), *Biochem. J.*, **50**, 716.

Chambers (1929), as referred to in McLung's *Handbook of Microscopical Technique*, New York.

Charnley (1945), *Analyst*, **70**, 14.

Chibnall, Rees and Williams (1943), *Biochem. J.*, **37**, 354.

Christensen, Riggs and Ray (1951), *Anal. Chem.*, **23**, 1521.

Christman and Randall (1933), *J. Biol. Chem.*, **102**, 595.

Clark and Collip (1925), *J. Biol. Chem.*, **63**, 461.

Clarke (1928), 'The determination of hydrogen ions', p. 102, London.

Claude, Harvey Lectures (1947–8).

Cleghorn and Jendrassik (1934), *Biochem. Z.*, **274**, 189.

Common (1936), *J. Agri. Sc.*, **26**, 85.

Conway (1925), *J. Physiol.*, **60**, 30.

—— (1929), *Amer. J. Physiol.*, **88**, 1.

—— (1930), *Z. für. Inn. Med.*, **13**, 225.

—— (1931), *Proc. Roy. Irish Acad.*, B.**28**, 574.

—— (1933), *Biochem. J.*, **27**, 430.

—— (1934), *Biochem. J.*, **28**, 283.
—— (1935), *Biochem. J.*, **29**, 2221.
—— (1935), *Biochem. J.*, **29**, 275.
—— (1938), *Nature*, **141**, 471.
—— (1947), *Microdiffusion Analysis and Volumetric Error*, 2nd edit., Crosby Lockwood, London.
Conway, Brady and Hingerty (1945), unpublished results.
Conway and Cooke (1937), *Nature*, **139**, 627.
—— —— (1938), *Nature*, **142**, 720.
—— —— (1939), *Biochem. J.*, **33**, 457.
Conway and Cruess-Callaghan (1937), *Biochem. J.*, **31**, 828.
Conway and Dolan (1953), *Thesis to National University.*
Conway and Downey (1950), *Biochem. Soc. Proceedings*, in press.
Conway and Dillon (1933), *Irish J. Med. Sc.*, **76**, 33.
Conway and Fitzgerald (1938), unpublished obs.
—— —— (1942), *J. Physiol.*, **101**, 86.
Conway, Fitzgerald and MacDougall (1946), *J. of Gen. Physiol.*, **29**, 305.
Conway and Hingerty (1943), unpublished results.
Conway and Kane (1926), *J. Physiol.*, **61**, 595.
—— —— (1935), *Biochem. J.*, **29**, 1446.
Conway and MacCarvill (1949), 1*st International Congress of Biochemistry* (Abstracts).
Conway and McVerry (1937), unpublished obs.
Conway, Moriarty and O'Connor (1932), *Pflügers Archiv.*, **231**, 280.
Conway and Nolan (1948), *Biochem. J.*, *Proc. of Biochem. Soc.*, **42**.
Conway, O'Connor and O'Donovan (1937), *Proc. Roy. Irish Acad.*, **44**B, 1.
Conway and O'Malley (1942), *Biochem. J.*, **36**, 655.
—— —— (1943), unpublished obs.
Conway and O'Rourke (1945), unpublished obs.
Conway and Ryan (1948), *Biochem. J.*, *Proc. of Biochem. Soc.*, **42**.
Corcoran and Page (1948), *J. Lab. Clin. Med.*, **33**, 1327.
Cotzias and Dole (1951), *J. Biol. Chem.*, **190**, 2.
—— —— (1952), *J. Biol. Chem.*, **196**, 240.
Crismer (1937), *Bull de la Soc. de Chimie Biologique*, XIX, 100.

Dale (1949), *Biochem. J.*, **44**, xxvii.
Daroga and Pollard (1941), *J. Soc. Chim. Ind.*, **60**, 218.
Daughaday, Jaffe and Williams (1948), *J. Clin. Endocrin*, **8**, 166.
Davidson (1952), *New Eng. J. Med.*, **247**, 239.
Debye and Hückel (1923), *Physik.*, **24**, 185.
DeGroote and Vandenbroucke (1958), *Am. J. Digest. Diseases*, **3**, 502.
Deniges (1930), ' Précis de Chimie Analytique', 213.
Dieulafait (1878), *Ann. Chim. Phys.*, **13**, 374.
Dill, Daly and Forbes (1937), *J. Biol. Chem.*, **117**, 569.
Dintzis and Baird Hastings, *Proc. Nat. Acad. Sciences* (1953), No. 7.

Dixon (1935), *Biochem. J.*, **29**, 86.
Dominguez (1935), *Amer. J. Physiol.*, **112**, 529.
Doolin (1955), unpublished obs.
Dubowski and Witherow (1952).

Eegriwe, E. (1937), *Z. Anal. Chem.*, **110**, 22.
Edlbacher and Wiss (1944), *Helv. Chim. Acta*, **27**, 1824.
Edwards and Kellie (1954), *Biochem. J.*, **56**, 2.
Elsden and Gibson (1954), *Biochem. J.*, **58**, 1.
Embden (1926), ' Vortrag auf der Naturfirschen Versammlung in Dusseldorf.'
Embden, Cartensen and Schumacher (1928), *Z. für Physiol. Chem.*, **179**, 200.
Embden and Zimmerman (1927), *Z. Physiol. Chem.*, **167**, 137.
Emich (1932), *Microchemical Laboratory Manual* (trans. by F. Schneider, New York).

Fearon and Mitchell (1932), *The Analyst*, **57**, 372.
Fearon and Webb (1936), *The Analyst*, **61**, 539.
Feigl (1937), ' Quantitative Analysis by Spot Tests ', trans. by J. W. Matthews, Amsterdam.
Feigl and Krumholz (1929), *Mikrochemie, Pregl. Fest.*, **83**.
Feldstein and Klendshoj (1954), *Anal. Chem.*, **26**, 932.
—— —— (1954), *Jour. Lab. and Clin. Med.*, **44**, No. I, 166.
—— —— (1954), Presented at 6th Annual Meeting of the American Academy of Forensic Sciences, Chicago, Illinois, Feb. 25–27th.
Fleischaker and Scheiderer (1932), *Klin. Wochenschr.*, **11**, 1550.
Florkin and Bosson (1939), *Bull. de la Société de Chem. Biol.*, XIX, 665.
Florkin and Frappez (1940), *Archiv. Internat. de Physiologie*, **50**, 197.
Florkin and Houet (1938), *Archiv. Internat. de Physiologie*, **47**, 125.
Florkin and Renwart (1939), *Archiv. Internat. de Physiologie*, **49**, 127.
Folin (1932), *J. Biol. Chem.*, **97**, 141.
Folin and Berglund (1922), *J. Biol. Chem.*, **51**, 213.
Folin and Ciocalteu (1929), *J. Biol.*, **73**, 627.
Folin and Denis (1912), *J. Biol. Chem.*, **11**, 161, 527.
Folin and Wu (1919), *J. Biol. Chem.*, **38**, 81.
Friedlander (1907), *M. H. Chemie*, **28**, 991.
—— (1922), *Ber. Chem. Ges.*, **55**, 1655.
Friedemann, Cotonio and Shaffer (1927), *J. Biol. Chem.*, **73**, 335.
Fu Chuan (1953), *Proc. Soc. Exp. Biol. Med.*, **84**, 406.
Fujita and Iwatake (1931), *Biochem. Z.*, **242**, 43.
Fujiwara (1914), *S.B. naturf. Ges. Rostock*, **6**, 33.
Fuld (1933), *Klin. Wochschr.*, **11**, 1364.

Gale (1945), *Biochem. J.*, **39**, 46.
—— (1947), *J. Gen. Microbiol.*, **1**, 53.

Gattermann and Wieland, *Lab. Methods of Organic Chemistry*, 19th Edit., London, Macmillan.
Gaus (1900), *Zeit, anorg. chem.*, **25**, 236.
Gibbs and Kirk (1934), *Mikrochemie*, **16**, 25.
Glick (1949), *Techniques of Histo- and Cytochemistry*, Interscience Inc., New York.
Gordon (1951), *Anal. Chem.*, **23**, 1853.
—— (1952), *Anal. Chem.*, **24**, 857.
Gordon and Quastal (1939), *Biochem. J.*, **33**, 1332.
Gray and Sandiford (1946), *The Analyst*, **71**, 107.
Greenwald, Gross and Samet (1924), *J. Biol. Chem.*, **62**, 401.
Guareschi (1913), *Fresenius' Z. f. Analyt. Chem.*, **52**, 454, 538, 609.

Habgood and Powell (1945), *Brit. J. Indust. Med.*, **2**, 39.
Hadfield (1942), *J. Soc. Chem. Ind.*, **61**, 45.
Hahnel (1935), *Svensk. Kem. Tids.*, **47**, 4.
Hallam (1955), *Analyst*, **80**, 552.
Hallett (1942), *Ind. and Eng. Chem. Anal., Ed.*, **14**, 956.
Harding and Selby (1931), *J. Biol. Chem.*, **25**, 1915.
Harman and Webster (1948), *Amer. J. Clin. Path.*
Hasselbalch (1916), *Biochem. Z.*, **74**, 18.
Hawes and Skavinski (1942), *Ind. and Eng. Chem. Anal. Ed.*, **14**, 917.
Heatley (1935), *Biochem. J.*, **29**, 626.
—— (1939), *Mikrochemie*, **26**, 147.
Heilmayer (1929), Abderhalden's ' Handbuch der biologischen Arbeitsmethoden ', Abt. 11, Teil 2–11, 2337.
Holden (1937), *J. Biol. Chem.*, **119**, 347.
Hollander, DiMauro and Pearsons (1951), *Endocrin.*, **49**, 617.
Holt and Callow (1942), *J. Soc. Chem. Ind.*, **61**, 99.
Holter (1943), *Compt. rend. Lab. Carlsberg, Sér. Chim.*
Horvath and Roughton (1942), *J. Biol. Chem.*, **144**, 747.
Hulett and Bonner (1909), *J. Amer. Chem. Soc.*, **31**, 309.
Hulme and Cooper (1956), *Jour. Med. Lab. Tech.*, **13**, No. 8, p. 543.
Hutchins and Kaas (1949), *Jour. Biol. Chem.*, **177**, 575.

Indovina (1935), *Biochem. Z.*, **275**, 286.
Isaccson (1936), Thesis to the National University, Ireland.

Jacquez *et al.* (1959), *Science*, **129**, 269.
Jetter (1941), *Quart. J. Alcohol.*, **2**, 572.
Jørgensen (1956), *Scand. Jour. Clin. and Lab. Investig.*, Vol. 8, No. 2, p. 168.

Kalk and Bonis (1933), *Z. klin. med.*, **123**, 731.
Kane and O'Connor (1937), *J. Physiol.*, **91**, 48.
Kapeller-Adler (1949), *Biochem. J.*, **44**, 70.

Kawerau (1941), *Sci. Proc. Roy. Dub. Soc.*, **22**, 405.
Kay and Reid (1934), *Biochem. J.*, **28**, 1798.
Kay and Sheehan (1934), *Biochem. J.*, **28**, 1784.
Keilin and Mann (1940), *Biochem. J.*, **34**, 1163.
Keston, Rittenberg and Schoenheimer (1937), *J. Biol. Chem.*, **122**, 227.
Kirk (1933), *Mikrochemie*, **14**, 13.
—— (1947), *Advances in Protein Chemistry*, Vol. 3, p. 139, Interscience Inc., New York.
—— (1950), *Quantitative Ultramicroanalysis*, Wiley, New York.
—— (1950), *Anal. Chem.*, **22**, 611.
Kinsey and Robison (1946).
Klein (1940), *J. Biol. Chem.*, **135**, 143.
Klein and Steiner (1928), *Jb. wiss. Bot.*, **68**, 631.
Knop (1868), ' Kreislauf des Stoffes ', Leipzig.
Kolthoff (1931), ' Die Massanalyse ', *Zweiter Teil*, Berlin.
Kolthoff and Furman (1936), ' Potentiometric Titrations ', New York.
Koprowski and Uninski (1939), *Biochem. J.*, **33**, 747.
Krebs (1948), *Biochem. J.*, **43**, 51.
Krogh and Keys (1931), *J. Chem. Soc.*, 2436.

Labbé, Nepveux and Hejda (1929), *C. r. Acad. Sc. Paris*, **183**, 738.
Lee and Widdowson (1937), *Biochem. J.*, **31**, 2035.
Lehmann (1944), *Svenska Lakartidningen*, **41**, 230.
—— (1944), *Nordisk Medicin*, **21**, 317.
Leipert (1935), *Biochem. Z.*, **280**, 416.
Leipert and Watzlawek (1934), *Z. physiol. Chem.*, **226**, 108.
—— —— (1935), *Biochem. Z.*, **280**, 434.
Levine and Bodansky (1939), *Amer. J. Clin. Path. Tech. Section*, **3**, 159.
Levvy (1943), *Biochem. J.*, **37**, 598.
—— (1945), *Chem. and Ind.*, No. 1, p. 4.
—— (1946), *Biochem. J.*, **40**.
—— (1948), *Biochem. J.*, **42**, 396.
Linderstrøm-Lang and Holter, (1931), *Compt. rend. Lab. Carlsberg*, IV.
—— —— (1933), *Compt. rend. Lab. Carlsberg*, XIX.
—— —— (1931), *Zeitschr. für Physiol. Chem.*, **201**, 18.
—— —— (1933), *Zeitsch. für Physiol. Chem.*, **220**, 15.
Lindner and Haslwanter (1929), *Ztchrr. für Physiol. Chem.*, **220**, 15.
Lohmann (1929), *Biochem. Z.*, **202**, 466.
Long (1946), *Biochem. J.*, **40**, 27.
Loschmidt (1870), *Sitzber. Akad. Wien.*, **61**, 367.
Lowenstein, Corcoran and Page (1946), *Endocrin.*, **39**, 82.

MacFadyen (1945), *J. Biol. Chem.*, **158**, 107.
MacLeod (1949), *J. Biol. Chem.*, 181.

Mann, Bollmann, Huizenga, Farrar and Grindley, *Gastroenterology*, **27**, 4, 399.

Marshall and Davis (1914), *J. Biol. Chem.*, 18, 53.

Martin and Synge (1941), *Biochem. J.*, **35**, 294.

Mayrhofer (1923–8), ' Mikrochemie der Arzneimittel und Gifte ', Vienna and Berlin.

McCauley, Conway and Dillon (1944), unpublished obs.

McCready, Mitchell and Kirk (1939), *Mikrochemie*, **28**, 23.

McDermott and Adams (1954), *J. Clin. Invest.*, **33**, 1.

McDermott, Adams and Riddell (1954), *Annals of Surgery*, 140, No. 4, 539.

McFarlane (1936), *Ind. and Eng. Chem. Anal. Ed.*, Vol. 8, p. 125.

McLean (1915), *J. Exp. Med.*, **22**, 212.

Meldrum and Roughton (1933), *J. Physiol.*, **80**, 113.

Mellor (1928), *Inorg. and Theor. Chem.*, Vol. VIII.

Milton and Waters (1949), *Methods of Quant. Micro-Analysis*, Arnold and Co., London.

Mitchell (1896), *The Analyst*, **21**, 98.

Molisch (1921), ' Mikrochemie der Pflanze ', 65.

Möller, McIntosh and Van Slyke (1928), *J. Clin. Invest.*, **6**, 467.

Mörner (1913), *Z. f. physiol. Chem.*, **88**, 158.

Murnaghan (1949), *Analyst*, **94**, No. 883.

Nash and Benedict (1921), *J. Biol. Chem.*, **48**, 463.

Needham and Boell (1939), *Biochem. J.*, **33**, 149.

Nelson and Schigson (1953), *Surgery*, **34**, 1.

Neuberg (1911), *Der Harn*, Berlin.

Newman, Lehman and Cutting (1937), *J. Pharmacol.*, **61**, 158.

Obermayer (1880), *Sitzber. akad. Wien*, **81**, 162.

O'Brien (1941), Thesis to the National University of Ireland.

Öbrink (1955), *Biochem. J.*, **59**, 1, 134.

Öbrink and Maj Ulfendahl (1959), *Acta Societatis Medicorum Upsaliensis*, LXIV, No. 3–4, pp. 384–91.

O'Connor (1935), *Proc. R.I.A.*, **42**N, 351.

—— (1936), *Proc. R.I.A.*, **43**B, 23.

—— (1936) *Proc. R.I.A.*, **43**B, 34.

O'Malley, Conway and Fitzgerald (1943), *Biochem. J.*, **37**, 278.

Oppenheimer 1921, *Arch. f. exper. Path. u. Pharmm.*, **89**, 17.

Orloff and Berliner (1956), *J. Clin. Invest.*, XXXV, **2**, 223.

O'Rourke (1944), Thesis to the National University of Ireland.

Osborn and Wilkie (1933), *J. of the Ass. of Off. Agr. Chemists*, **16**, 110.

Ostwald and Luther (1902), ' Physiko-Chemische Messungen ', p. 113, Leipzig.

—— —— (1912), ' Physiko-Chemische Messungen ', p. 135, Leipzig.

Parnas (1925), *Biochem. Z.*, **155**, 247.
Parnas and Heller (1924), *Biochem. Z.*, **152**, 1.
Parnas and Klisiecki (1926), *Biochem. Z.*, **173**, 229.
Parnas, Lewinski, Jaworski and Umschweif (1930), *Biochem. Z.*, **228**, 336.
Parnas and Lutwak-Mann (1935), *Biochem. Z.*, **278**, 11.
Parnas and Mozolowski (1927), *Biochem. Z.*, **184**, 407.
Parnas, Mozolowski and Lewinski (1927), *Klin. Wochschr.*, **6**, 1710.
Parnas, Ostern and Mann (1934), *Biochem. Z.*, **272**, 64.
—— —— —— (1935), *Biochem. Z.*, **275**, 74.
Parnas and Wagner (1921), *Biochem. Z.*, **173**, 224.
—— —— (1921), *Biochem. Z.*, **125**, 235.
Pepper and Austin (1915), *J. Biol. Chem.*, **28**, 251.
Peters and Van Slyke (1932), *Quant. clin. chem.*, London.
Phillys, Schwartz, Sabbatini and Davidson (1952), *New Eng. J. Med.*, **247**, 239.
Philpot and Philpot (1936), *Biochem. J.*, **30**, 2191.
Pichler, Benedetti (1937), *Ind. Eng. Chem. Anal. Ed.*, **9**, 483.
Pincussen (1930), ' Mikromethodik ', p. 18, Leipzig.
Pincussen and Roman (1929), *Biochem. Z.*, **216**, 336.
Pirt and Chain (1952), *Biochem. J.*, **50**, 716.
Polland (1928), *Amer. J. Physiol.*, **112**, 529.
Powell (1945), *Brit. J. Ind. Med.*, **2**, 142.
Prater, Cowles and Straka (1942), *Ind. and Eng. Chem. Anal. Ed.*, **14.**

Quastal and Yates (1934), *Biochem. J.*, **28**, 1530.

Rafflin (1926), *Bull. Soc. de chim. biol.*, **8**, 352.
Ramsay (1944), *Biochem. J.*, **38**, 467.
Rappaport and Glaser (1932), *Klin. Wochschr.*, 814.
Rappaport and Gutman (1935), *Klin. Wochschr.*, **14**, 1325.
Rehberg (1925), *Biochem. J.*, **19**, 270.
Reif (1960), *Anal. Biochem.*, I, No. 4–5.
Richter (1937), *Biochem. J.*, **31**, 2022.
Riddell and McDermott (1954), *The Lancet*, June (1263).
Rosebury and Heyninger (1942), *Ind. Eng. Chem. Anal. Ed.*, **14**, 363.
Rossler (1887), *Ber.*, **20**, 2629.
Roughton (1935), *Physiol. Rev.*, **15**, 241.
Russell (1945), *J. Biol. Chem.*, **156**, 457.
Ryan (1956), unpublished obs.
Ryan (1958), *Analyst*, **83**, 990.

Samogyi (1927), *J. Biol. Chem.*, **75**, 33.
Schloesser (1903), *Z. angew. chem.*
—— (1904), *Z. f. angew. Ch.*, **16**, 953, 997, 1004.
—— (1906), *Chemiker Zeitung*, **30**, 1074.

Schmidt (1928), *Z. Physiol. Chem.*, **179**, 243.

—— (1932), *Z. Physiol. Chem.*, **208**, 185.

—— (1928), *J. Biol. Chem.*, **78**, 54.

Schneller (1928), ' Ergebnisse der Physiologie ', 37.

Scholander *et al* (1943), *J. Biol. Chem.*, **148**, 495.

Schwartz, Robertson and Holmes (1955), *J. Lab. Clin. Med.*, **46**, 657.

Seligson and Seligson (1951), *J. Lab. Clin. Med.*, **38**, 324.

Seligson and Hirahara (1957), *J. Lab. Clin. Med.*, **49**, 962.

Serlin and Cotzias (1955), *J. Biol. Chem.*, **215**, 263.

Sherlock, Summerskill, White and Phear (1954), *Lancet*, Sept. 4th.

Shinn and Nicolet (1941), *J. Biol. Chem.*, **35**, 294.

Shohl (1928), *J. Am. Chem. Soc.*, **50**, 417.

Singh, Punjab, Barclay and Cooke (1954), *Lancet*, May 15th, p. 1004.

Sisco, Cunningham and Kirk (1941), *J. Biol. Chem.*, **139**, 1–10.

Skrabal (1904), *Ztschr. f. anorg. u. allg. ch.*, **42**, 1.

Smith (1872), ' Air and Rain ', 266, London.

Sorensen (1897), *Z. f. anal. Ch.*, **36**, 639.

Stanoyevitch (1938), ' Der Physiologie und klinische Bedeutung des Blutammoniaks ', Dresden and Leipzig.

Stavsky, Harrison, Dasson and Sater (1944), *Ind. and Eng. Chem., Anal. Ed.*, **16**, 593.

Stefan (1872), *Sitzber. Akad. Wien*, **63**, 63 ; **65**, 323.

Stoll and Brenken (1934), *Biochem. Z.*, **268**, 229.

Stolz (1943), *J. Biol. Chem.*, **148**, 585.

Stone (1938), *Biochem. J.*, **32**, 1908.

Stott (1927), ' Volumetric Glassware ', London.

Strehler, Haas and Rupp (1942), *Biochem. Ztchr.*, **313**, 170.

Sunshine and Nenad (1953), *Anal. Chem.*, **25**, 653.

Tashiro (1922), *Amer. J, Physiol.*, **60**, 519.

Thomas (1912), *Bull. Soc. Chim.*, **11**, 797.

—— (1913), *Bull. Soc. Chim.*, **13**, 308.

Thorpe and Morton (1870), *Proc. Manchester Phil. Soc.*, **9**, 16.

Titus and Gray (1930), *Ind. Eng. Chem. Anal. Ed.*, **2**, 368.

Tompkins and Kirk (1942), *J. Biol. Chem.*, **142**, 477.

Traeger, Gabuzda, Ballou and Davidson (1954), *Metabolism*, III, 2.

Trevan (1922), *Lancet*, i, 786.

—— (1925), *Biochem. J.*, **19**, 1111.

Ucko (1934), *Compt. rend. soc. biol.*, **116**, 48.

—— (1934), *Biochem. J.*, **30**, 992.

Van Slyke and Cullen (1914), *J. Biol. Chem.*, **19**, 211.

—— —— (1917), *J. Biol. Chem.*, **30**, 289.

Van Slyke and Hawkins (1929), *J. Biol. Chem.*, **83**, 51.

Van Slyke and Hiller (1933), *J. Biol. Chem.*, **102**, 49.
Van Slyke, Lindeer, Hiller, Leiter and McIntosh (1925), *J. of Clin. Invest.*, **2**, 255.
Van Slyke, Page, Hiller and Kirk (1935), *J. Clin. Invest.*, **14**, 901.
Van Slyke, Sendroy, Hastings and Neill (1928), **78**, 265.
Van Slyke and Zacharias (1914), *J. Biol. Chem.*, **19**, 181.
Vierthaler (1867), *Sitzber. Akad. Wien*, **56**, 479.

Wagner (1898), ' Massanakytische Studien ', *Habilitationsschrift*, p. 17, Leipzig.
Walker and Hudson (1937), *Amer. J. Physiol.*, **118**, 153.
Walker and Rowe (1927), *Amer. J. Physiol.*, **81**, 738.
Webster and Davidson (19—), *J. Clin. Invest.*, XXXV, **2**, 191.
Weigert (1927), ' Optische Methoden der Chem ', Leipzig.
Weil-Malherbe and Green (1955), *Biochem. J.*, **61**, 210.
Wennesland (1940), *Acta Physiol. Scand.*, **1**, 49.
Werch (1940), *J. Lab. and Clin. Med.*, **25**, 414.
—— (1941), *J. Lab. and Clin. Med.*, **26**, 278.
Werner (1942), *J. Biol. Chem.*, **142**, 705 ; **142**, 725, 741.
White, Phear, Summerskill and Sherlock (1955), *J. Clin. Invest.*, **34**, 2.
Widmark (1922), *Biochem. Z.*, **131**, 473.
Widmark and Orskov (1928), *Biochem. Z.*, **201**, 15.
Wilson and Hildegarde (1953), *J. Clin. Endocrin.*, **13**, 1465.
Winkler (1913), *Angew. Chem.*, **26**, 231.
—— (1914), **27**, 630.
—— (1915), **28**, 48.
Winnick (1941), *J. Biol. Chem.*, **141**, 115.
—— (1942*a*), *J. Biol. Chem.*, **142**, 451.
—— (1942*b*), *J. Biol. Chem.*, **142**, 461.
—— (1942), *Ind. and Eng. Chem. Anal. Ed.*, **14**, 523.
Wyatt (1944), *Analyst*, **69**, 819.
—— (1944), ' A Review of Micro-volumetric Apparatus ', *The Analyst*, **69**, 816.
—— (1944), *Metallurgia*, 329.

Zeller, Stern and Wenk (1940), *Helv. chim. acta.*, **23**, 3.
Zondek and Bier (1931), *Biochem. Z.*, **241**, 491.

INDEX OF SUBJECTS

www.ingramcontent.com/pod-product-compliance
Lightning Source LLC
LaVergne TN
LVHW011532200926
848484LV00001B/240